Lecture Notes in Applied and Computational Mechanics

Volume 36

Series Editors

Prof. Dr.-Ing. Friedrich Pfeiffer
Prof. Dr.-Ing. Peter Wriggers

Lecture Notes in Applied and Computational Mechanics

Edited by F. Pfeiffer and P. Wriggers

Further volumes of this series found on our homepage: springer.com

Vol. 36: Leine, R.I.; van de Wouw, N.
Stability and Convergence of Mechanical Systems
with Unilateral Constraints
250 p. 2008 [978-3-540-76974-3]

Vol. 35: Acary, V.; Brogliato, B.
Numerical Methods for Nonsmooth Dynamical Systems:
Applications in Mechanics and Electronics
xxx p. 2008 [978-3-540-75391-9]

Vol. 34: Flores, P.; Ambrósio, J.; Pimenta Claro, J.C.;
Lankarani Hamid M.
Kinematics and Dynamics of Multibody Systems
with Imperfect Joints: Models and Case Studies
186 p. 2008 [978-3-540-74359-0]

Vol. 33: Niesłony, A.; Macha, E.
Spectral Method in Multiaxial Random Fatigue
146 p. 2007 [978-3-540-73822-0]

Vol. 32: Bardzokas, D.I.; Filshtinsky, M.L.; Filshtinsky, L.A. (Eds.)
Mathematical Methods
in Electro-Magneto-Elasticity
530 p. 2007 [978-3-540-71030-1]

Vol. 31: Lehmann, L. (Ed.)
Wave Propagation in Infinite Domains
186 p. 2007 [978-3-540-71108-7]

Vol. 30: Stupkiewicz, S. (Ed.)
Micromechanics of Contact and Interphase Layers
206 p. 2006 [978-3-540-49716-5]

Vol. 29: Schanz, M.; Steinbach, O. (Eds.)
Boundary Element Analysis
571 p. 2006 [978-3-540-47465-4]

Vol. 28: Helmig, R.; Mielke, A.; Wohlmuth, B.I. (Eds.)
Multifield Problems in Solid and Fluid Mechanics
571 p. 2006 [978-3-540-34959-4]

Vol. 27: Wriggers P., Nackenhorst U. (Eds.)
Analysis and Simulation of Contact Problems
395 p. 2006 [978-3-540-31760-9]

Vol. 26: Nowacki, J.P.
Static and Dynamic Coupled Fields in Bodies
with Piezoeffects or Polarization Gradient
209 p. 2006 [978-3-540-31668-8]

Vol. 25: Chen C.-N.
Discrete Element Analysis Methods
of Generic Differential Quadratures
282 p. 2006 [978-3-540-28947-0]

Vol. 24: Schenk, C.A., Schuëller. G
Uncertainty Assessment of Large
Finite Element Systems
165 p. 2006 [978-3-540-25343-3]

Vol. 23: Frémond M., Maceri F. (Eds.)
Mechanical Modelling and Computational Issues
in Civil Engineering
400 p. 2005 [978-3-540-25567-3]

Vol. 22: Chang C.H.
Mechanics of Elastic Structures with Inclined Members:
Analysis of Vibration, Buckling and Bending of X-Braced
Frames and Conical Shells
190 p. 2004 [978-3-540-24384-7]

Vol. 21: Hinkelmann R.
Efficient Numerical Methods and Information-Processing
Techniques for Modeling Hydro- and Environmental
Systems
305 p. 2005 [978-3-540-24146-1]

Vol. 20: Zohdi T.I., Wriggers P.
Introduction to Computational Micromechanics
196 p. 2005 [978-3-540-22820-2]

Vol. 19: McCallen R., Browand F., Ross J. (Eds.)
The Aerodynamics of Heavy Vehicles:
Trucks, Buses, and Trains
567 p. 2004 [978-3-540-22088-6]

Vol. 18: Leine, R.I., Nijmeijer, H.
Dynamics and Bifurcations
of Non-Smooth Mechanical Systems
236 p. 2004 [978-3-540-21987-3]

Vol. 17: Hurtado, J.E.
Structural Reliability: Statistical Learning Perspectives
257 p. 2004 [978-3-540-21963-7]

Vol. 16: Kienzler R., Altenbach H., Ott I. (Eds.)
Theories of Plates and Shells:
Critical Review and New Applications
238 p. 2004 [978-3-540-20997-3]

Vol. 15: Dyszlewicz, J.
Micropolar Theory of Elasticity
356 p. 2004 [978-3-540-41835-1]

Vol. 14: Frémond M., Maceri F. (Eds.)
Novel Approaches in Civil Engineering
400 p. 2003 [978-3-540-41836-8]

Vol. 13: Kolymbas D. (Eds.)
Advanced Mathematical and Computational
Geomechanics
315 p. 2003 [978-3-540-40547-4]

Vol. 12: Wendland W., Efendiev M. (Eds.)
Analysis and Simulation of Multifield Problems
381 p. 2003 [978-3-540-00696-1]

Vol. 11: Hutter K., Kirchner N. (Eds.)
Dynamic Response of Granular and Porous Materials
under Large and Catastrophic Deformations
426 p. 2003 [978-3-540-00849-1]

Vol. 10: Hutter K., Baaser H. (Eds.)
Deformation and Failure in Metallic Materials
409 p. 2003 [978-3-540-00848-4]

Stability and Convergence of Mechanical Systems with Unilateral Constraints

Remco I. Leine · Nathan van de Wouw

With 62 Figures

 Springer

Dr.ir. habil. Remco I. Leine
Institute of Mechanical Systems
Department of Mechanical
and Process Engineering
ETH Zurich
CH-8092 Zurich
Switzerland
remco.leine@imes.mavt.ethz.ch

Dr.ir. Nathan van de Wouw
Dynamics and Control Group
Department of Mechanical
Engineering
Eindhoven University of Technology
PO Box 513
5600 MB Eindhoven
The Netherlands
N.v.d.Wouw@tue.nl

ISBN   978-3-540-76974-3   e-ISBN   978-3-540-76975-0

ISSN   1613-7736

Library of Congress Control Number: 2007941933

Cover design: WMX Design GmbH, Heidelberg

Printed on acid-free paper

9 8 7 8 6 5 4 3 2 1 0

springer.com

Preface

During the last two decades a new research field has emerged: the field of non-smooth dynamical systems and in particular non-smooth mechanics. Mechanics, being one of the oldest natural sciences, has played a forerunner role in formalising a theory to describe evolution problems with some degree of non-smoothness. The classical modelling approach used in engineering sciences and physics is to express all relations within a system by equalities, a dogma which has been reinforced by the emergence of computers. A crucial point in dealing with non-smoothness is to leave the equality dogma and to allow ourselves to think and work with (variational) inequalities and inclusions. The key merit of modern non-smooth mechanics is the development of a mathematical framework, based on convex analysis, which effectively describes non-smooth or set-valued relations within a system. This mathematical framework has primarily been used to gain a proper understanding of unilateral behaviour, to reveal the inherent structure of non-smooth mechanical systems, as well as to set up numerical integration methods for the simulation of non-smooth (mechanical) systems. These numerical methods fill the need to obtain quantitative information about the motion starting from a particular initial state. Instead of the endeavour to obtain one or more approximate solutions, one may strive to obtain qualitative information about all solutions. The latter is referred to as a qualitative theory for dynamical systems and has been initiated by H. Poincaré around the end of the 19th century. Of central importance in the qualitative theory is the problem of the stability of motion. The work of A. M. Lyapunov has been seminal in the formulation of a stability theory. The aim of this monograph is to free the existing Lyapunov stability theory from the omnipresent equality dogma by exploiting the structure of non-smooth (mechanical) systems, opening the way to a qualitative analysis of non-smooth dynamical systems. Furthermore, this monograph deals with the concept of convergence as has been developed in the Russian literature in the 1960's. The convergence property reflects a stability property on system level and has recently been shown to be highly instrumental in solving many control problems, such as tracking, synchronisation and observer design. Here, we investigate the convergence property for a class of non-smooth dynamical systems, namely monotone measure differential inclusions.

This text is written to address primarily researchers interested in stability properties of non-smooth mechanical systems. However, the text has been arranged such that it is also of use for those interested in *general* non-smooth dynamical systems. To this end, the exposition of the results for mechanical systems is kept in separate chapters. First, the theory is explained for general non-smooth dynamical systems. Subsequently, these results are specialised for mechanical systems. Mechanical systems are taken as examples throughout the work.

This monograph is the result of about four years active research of the authors in the field of stability theory and non-smooth dynamical systems. Part of the monograph is based on the habilitation thesis of the first author and on joint-research papers of the authors.

Zürich, September 2007   *Remco I. Leine*

Eindhoven, September 2007   *Nathan van de Wouw*

Contents

Notation

$\mathbb{R}$	set of all real numbers
$\overline{\mathbb{R}}$	$\mathbb{R} \cup \{\infty\}$, extended set of all real numbers
$\mathbb{N}$	set of all natural numbers
$:=$	definition
inf	infinum
sup	supremum
$\forall$	for all
$\exists$	there exists
x	scalar
$\boldsymbol{x}$	column-vector in $\mathbb{R}^n$
$\boldsymbol{X}$	matrix in $\mathbb{R}^{n \times m}$
$\boldsymbol{I}$	identity matrix
$\det(\boldsymbol{A})$	determinant of $\boldsymbol{A}$
$\mathrm{cond}(\boldsymbol{A})$	condition of $\boldsymbol{A}$
$\boldsymbol{x}^{\mathrm{T}}$	transpose of $\boldsymbol{x}$
$\boldsymbol{X} = \mathrm{diag}\,\boldsymbol{x}$	diagonal matrix with $X_{ii} = x_i$, $X_{ij} = 0$, $j \neq i$
x_i	i-th element of $\boldsymbol{x}$
$|x|$	absolute value of x
$\|\boldsymbol{x}\|$	Euclidian norm of $\boldsymbol{x}$
$\|\boldsymbol{x}\|_M$	norm of $\boldsymbol{x}$ with respect to $\boldsymbol{M}$
$\dot{x}(t)$	differentiation w.r.t. time of an abs. cont. function $x(t)$ or density of $\mathrm{d}x$ with respect to the differential measure $\mathrm{d}t$
$\boldsymbol{a} \perp \boldsymbol{b}$	$\boldsymbol{a}$ is orthogonal to $\boldsymbol{b}$
B_r	the n-dimensional open ball with radius r centred at the origin
$\overline{B}_r$	the n-dimensional closed ball with radius r centred at the origin
$[a, b]$	the closed interval $\{x \in \mathbb{R} \mid a \leq x \leq b\}$
$\{a, b\}$	the set comprising the elements $\{a\}$ and $\{b\}$

C^0	class of continuous functions
C^k	class of functions differentiable up to order k
$\emptyset$	empty set
$\overline{C}$ or $\mathrm{cl}\, C$	closure of C
$\overline{\mathrm{co}}\, C$	closed convex hull of C
$\mathrm{int}\, C$	interior of C
$\mathrm{bdry}\, C$	boundary of C
$x \in C$	x is an element of C (inclusion)
$f(\boldsymbol{x})$	single-valued function $\mathbb{R}^n \to \mathbb{R}$
$\boldsymbol{f}(\boldsymbol{x})$	single-valued function $\mathbb{R}^n \to \mathbb{R}^n$
$\mathcal{F}(\boldsymbol{x})$	set-valued function $\mathbb{R}^n \to \mathbb{R}^n$
$f'(x)$	classical derivative of f with respect to x
$\nabla \boldsymbol{f}(\boldsymbol{x})$	classical gradient of $\boldsymbol{f}$ with respect to $\boldsymbol{x}$
$\partial f(\boldsymbol{x})$	generalized differential of f with respect to $\boldsymbol{x}$ (or subdifferential if f is a convex function)
$df(\boldsymbol{x})(\boldsymbol{v})$	subderivative of f at $\boldsymbol{x}$ in the direction $\boldsymbol{v}$
$\mathrm{graph}(f)$	graph of f (2.2)
$\mathrm{graph}(\mathcal{F})$	graph of a set-valued function $\mathcal{F}$ (2.20)
$\mathrm{epi}(f)$	epigraph of f (2.3)
$f^*(\boldsymbol{x}^*)$	conjugate function
$\Psi_C(\boldsymbol{x})$	indicator function of C at $\boldsymbol{x}$
$\Psi_C^*(\boldsymbol{x})$	support function of C at $\boldsymbol{x}$
$N_C(\boldsymbol{x})$	normal cone of C at $\boldsymbol{x}$
K°	polar cone to the cone K
$\mathrm{prox}_C(\boldsymbol{x})$	proximal point from $\boldsymbol{x}$ to the convex set C
$\mathrm{dist}_C(\boldsymbol{x})$	distance from $\boldsymbol{x}$ to the convex set C
$\mathrm{sign}(x)$	sign function with $\mathrm{sign}(0) = 0$
$\mathrm{Sign}(x)$	set-valued sign function
$\mathrm{var}(\boldsymbol{f}, I)$	variation of $\boldsymbol{f}$ over the interval I (Definition 2.11)
$\boldsymbol{f} \in \mathrm{lbv}(I, X)$	$\boldsymbol{f} : I \to X$ is a function of locally bounded variation on I
$\mathfrak{A}$	a σ-algebra (Definition 3.1)
μ	friction coefficient
$\mu(A)$	measure of the set A (Definition 3.2)
$\mu_f(I)$	Lebesgue-Stieltjes measure of f on an interval I (3.2)
$\chi_A(x)$	characteristic function of the set A (3.4)
$\nu(A)$	signed measure of the set A (Definition 3.3)
$\nu_f(I)$	signed Lebesgue-Stieltjes measure of f on an interval I (3.14)
$\nu = f \odot \mu$	the signed measure ν is has a density f with respect to μ
$\nu \ll \mu$	the signed measure ν is μ-continuous (Definition 3.4)
$\nu \perp \rho$	the signed measures ν and ρ are orthogonal (Definition 3.5)

υ	real measure
$\mathrm{d}f$	differential measure of f
$\mathrm{d}t$	(differential) Lebesgue measure
$\mathrm{d}\eta$	atomic measure
$\boldsymbol{\varphi}(t, t_0, \boldsymbol{x}_0)$	a solution curve $\boldsymbol{x}(t)$ with initial condition $\boldsymbol{x}(t_0) = \boldsymbol{x}_0$
$\mathcal{S}(\mathrm{d}\boldsymbol{\Gamma}, t_0, \boldsymbol{x}_0)$	set of solution curves of $\mathrm{d}\boldsymbol{x} \in \mathrm{d}\boldsymbol{\Gamma}(t, \boldsymbol{x})$
	with initial condition $\boldsymbol{x}(t_0) = \boldsymbol{x}_0$
Ω_c	level set (6.11)
$\mathcal{L}_c$	level surface (6.12)
$F \circ G$	the composition of the two maps F and G
$\boldsymbol{x}^*$	equilibrium point of a dynamical system
$\boldsymbol{q}^*$	equilibrium position of a mechanical system
$\mathcal{E}$	equilibrium set (Definition 6.8)
$\mathcal{E}_{\boldsymbol{q}}$	the set of equilibrium positions
$\mathcal{A}$	admissible set of a measure differential inclusion
$\mathcal{A}^c$	the complement of the set $\mathcal{A}$
$\mathcal{K}$	the set of admissible generalised coordinates
$U_\varepsilon(\mathcal{M})$	the open ε-neighbourhood of the set $\mathcal{M}$ (6.23)
$\overline{U}_\varepsilon(\mathcal{M})$	the closed ε-neighbourhood of the set $\mathcal{M}$

1

Introduction

When looking at the title of this work, one may wonder why we need another text on the stability of dynamical systems. Why do we need one more, when the number of existing textbooks on the topic of stability is already very large? This monograph is not just about stability, it deals in particular with non-smooth systems. In this introductory chapter a motivation will be given for a monograph on the stability of non-smooth (mechanical) systems (see Section 1.1). Historical notes on the theory of stability in Section 1.2 might shed some light on the origin of terminology in stability theory. In Section 1.3, we provide a brief, and unavoidably incomplete, discussion on different mathematical formalisms used to model non-smooth dynamical systems. After introducing the reader to some basic stability notions and the concept of convergent systems in Section 1.4, a concise literature survey is presented in Section 1.5 on the topics of non-smooth analysis, (measure) differential inclusions, non-smooth mechanics, stability of non-smooth systems and convergent systems. The objective of this monograph will be put forth in Section 1.6. Finally, a closing section containing the outline of the succeeding chapters is included.

1.1 Motivation

Is the solar system stable? Under which load will a beam buckle? Is the figure of equilibrium of a steady rotating fluid stable? These fundamental questions were some of the major problems that motivated scientists such as Euler, Lagrange, Poincaré and Lyapunov to think about the concept of stability of motion and how to prove it. The origin of stability theory must clearly be sought in mechanics. The interest in the stability of motion is now greater than ever and is no longer confined to mechanics. Stability issues play a role in economical models, numerical algorithms, quantum mechanics, nuclear physics, and control theory as fruitfully applied in for example the fields of mechanical and electrical engineering. The theory of stability is now studied by many branches of science and is well established as can be inferred from the vast amount of textbooks, conferences and journal articles on the subject.

A much younger field of interest is the study of the dynamics of non-smooth systems. A look at the programs of international conferences from the past few years reveals a rapidly increasing interest in non-smooth systems. It is perhaps not exaggerated to state that it is nowadays fashion to study non-smooth systems. Non-smooth models appear in many different disciplines. Mechanical engineers study stick-slip oscillations in systems with dry friction and the dynamics of impact phenomena and unilateral constraints. Electrical circuits contain diodes and transistors, which ideally behave in a non-smooth way. Control theorists also have to deal with other sources of non-smoothness such as switching control laws and actuator saturation affecting the resulting closed-loop dynamics. Switching systems also arise in other fields such as air traffic management, economic models of markets and scheduling of automated railway systems. Although these examples come from very different fields, the mathematical structure and the related questions of interest are very similar. In particular, the time evolution is often described by non-smooth differential equations or (measure) differential inclusions.

This research monograph puts itself squarely between stability theory and the field of non-smooth dynamics. As opposed to stability theory for smooth dynamical systems, the research on stability properties of non-smooth dynamical systems is still in its infancy. Here, we will focus primarily on the development of stability results for measure differential inclusions, which naturally arise when considering (mechanical) systems with inequality (unilateral) constraints. The impulsive nature of the vectorfield and the resulting state jumps (discontinuities in the state evolution) challenge the extension of the existing stability results for smooth systems to measure differential inclusions. First of all, it is not even trivial how to define stability, attractivity and other properties of an equilibrium if the dynamics of the system has unilateral (inequality) constraints, which restrict the state of the system to an admissible domain and may induce state jumps. Moreover, the loss of uniqueness and existence of solutions (of the initial value problem) greatly complicates the unambiguous definition of these stability properties. Typically, the terminology for stability theory is designed for smooth dynamical systems and its direct application to non-smooth systems does sometimes not make sense or leads to confusion. The classical results on stability properties have been formulated for smooth dynamical systems and intrinsically make use of the smoothness assumption. Clearly, when trying to study stability properties in non-smooth dynamical systems, one should proceed very carefully and can not rely on standard textbooks. The motivation of this monograph has now become apparent: there exists a need for the development of stability and attractivity results for non-smooth systems formulated as measure differential inclusions.

As one of the oldest natural sciences, mechanics occupies a certain pioneering role in determining the development of exact sciences through its interaction with mathematics (quote from [63]). The historical notes given below show that mechanics has also played a dominant role in the development of the theory of stability. It is therefore only natural to reserve a special place for non-smooth mechanical systems within this monograph and to adopt it as the prime application field in the current work.

1.2 Historical Notes on the Theory of Stability

Many different stability concepts exist in literature and are often named after an illustrious scientist of a long forgotten time. Some historical notes on how these stability concepts came into being therefore seem appropriate. The following is based on original texts as well as numerous textbooks on the history of mechanics, see for instance [45, 54, 69, 161, 164].

The problem of stability in dynamical systems has attracted much interest of physicists, mathematicians, engineers and astronomers during the last centuries. In 1644, E. Torricelli (1608-1647) postulated his axiom [163]:

> **Duo grauia fimul coniuncta ex fe mo-
> ueri non poffe, nifi centrum commune
> grauitatis ipforum defcendat.**

E. Torricelli, Opera Geometrica, De Motu Gravium, p. 99.

which we might translate as

> *Two interconnected weights cannot start moving by themselves simultaneously if their common centre of gravity does not descend.*

Torricelli simply uses the words 'start moving by themselves' without speaking of an initial disturbance from an equilibrium. Although Torricelli did not use the word 'stability', his axiom certainly preludes a stability concept based on the (gravitational) potential energy.

Inspired by the work of Archimedes of Syracuse, the Flemish and Dutch scientists Simon Stevin (1548-1620) and Christiaan Huygens (1629-1695) studied the equilibrium of floating bodies. In his work *Byvough der weeghconst* [158] of 1605 Stevin wrote (following Archimedes) that a floating body takes such a position that its centre of gravity is on the vertical centre line of gravity of the displaced fluid. Furthermore, he gives the following corollary:

> ### 1 VERVOLGH.
>
> Tis kennelick dat als des lichaems fwaerheyts middelpunt,is boven des water-
> hols fwaerheyts middelpunt,fo heeftet fulcken topfwaerheyt dat alles omkeert,
> (midts welverftaende dattet niet onderhouden en .worde) tot dat des lichaems
> fwaerheyt middellijn,is in des waterhols hanghende fwaerheyts middellijn,on-
> der des waterhols fwaerheyts middelpunt.

S. Stevin, Van de vlietende topswaerheyt, Byvough der weeghconst, p. 202.

which reads in modern English as

> *It is obvious that if the body's centre of gravity is above the centre of*
> *gravity of the body of the displaced water, it has such top-heaviness*
> *that everything turns over (provided, however, it be not supported)*
> *until the body's centre line of gravity is in the vertical centre line of*
> *gravity of the body of displaced water, below the centre of gravity of*
> *the body of displaced water.*

Apparently, Stevin thought that a floating body is stable when the centre of
gravity is above the centre of buoyancy ('the centre of gravity of the body
of the displaced water'), which is not generally true. Compared with Stevin,
the work of Huygens has a more mathematical style, but Huygens work still
relies upon geometrical methods as was common in the 17th century. In many
theorems, such as the following one, Huygens addresses stability problems of
floating bodies [79]:

> *Si corpus solidum liquido supernatans ultrò inclinetur et alium si-*
> *tum acquirat; altitudo centri gravitatis totius corporis supra centrum*
> *gravitatis partis mersae, minor erit positione corporis posteriori quam*
> *priori.* [1]
>
> CH. HUYGENS, DE IIS LIQUIDO SUPERNATANT LIBRI 3, THEO-
> REMA 6.

The theorems of Huygens have a strong similarity with the work of Stevin.
However, the novelty in the work of Huygens is, that he explicitly compares
two different positions of the system. In the 18th century, the study of the roll-
stability of ships was carried on by Daniel Bernoulli (1700-1782), Leonhard
Euler (1707-1783) and Pierre Bouguer (1698-1758). Daniel Bernoulli distinguishes between stable (which he calls 'firm') and unstable equilibria of floating bodies and writes [18, 31]

> *... quo ambo aequilibrii situs ab invicem distinguuntur; minima quidem*
> *vis quaevis corpora etiamsi in aequilibrio firmo posita aliquantillum*
> *nutare facit, sublata autem vi corpus rursus ad situm naturalem tendit,*
> *nisi nutatio certos quosdam terminos transgressa fuerit.* [2]
>
> D. BERNOULLI, COMMENTATIONES DE STATU AEQUILIBRII CORPO-
> RUM HUMIDO INSIDENTIUM, 1738, P. 148.

---

[1] Free translation: If a solid body, floating on a liquid, inclines and acquires another
position, then the height of the center of gravity of the total body over the center
of gravity of the submerged part will be smaller in the latter position than in the
former position.

[2] Translation by F. Cerulus [31]: .. by this, both positions of equilibrium are distinguished one from the other; indeed, a minimal arbitrary force makes a body
- although put in firm equilibrium - nod a little, but when the force has been
undergone [i.e. ceases to act], the body tends again to its natural position, unless
the nodding would have exceeded certain bounds.

Daniel Bernoulli speaks explicitly of the stability of an *equilibrium* and considers the couple of restoring forces when the equilibrium of the floating body is perturbed by a small amount. Similar to Huygens, Bernoulli implicitly considers two different positions of the system but their distance is small. Euler refines Daniel Bernoulli's work in his two volume treatise *Scientia Navalis* [49] and distinguishes between equilibria which are stable, unstable and indifferent:

> *Stabilitas, qua corpus aquae innatans in situ aequilibrii persuerat, aestimanda est ex momento potentiae restituentis, si corpus dato angulo infinite paruo ex situ aequilibrii fuerit declinatum.* [3]
>
> L. EULER, SCIENTIA NAVALIS, PART 1, CHAPTER 3, PROPOSITION 19, P. 86.

Euler uses the word 'stability' and associates stability with the response on an *infinitely small* disturbance from the equilibrium position. The idea of infinitely small disturbances will later play a role in the work of Lagrange. Bouguer introduced the term 'metacentric height', which became the modern expression to determine the roll-stability of ships, see [125].

The theory of elastic stability in statics began with the work of Euler on the critical buckling load of columns. Daniel Bernoulli suggested in a letter to Euler (Oct. 20th 1742) that the differential equation of the *elastica* could be found by minimising the integral of the square of the curvature along the rod, being proportional to the elastic strain energy. Euler acted on this suggestion in his 'Additamentum' *de curvas elasticis* of his work *Methodus inveniendi lineas curvas maximi minive proprietate gaudentes* [48] on the calculus of variations. Euler found a certain length which a column must attain to be bent by its own weight or an applied weight, and concluded that for shorter lengths it will simply be compressed, while for greater lengths it will be bent, i.e. buckle. Although Euler initiated the analysis of the elastic stability of the static equilibrium, he tacitly left the notion of stability undefined. In this context, we have to remark that stability is in essence a concept of dynamical systems as time plays an essential role.

The development of a stability concept in dynamics was continued by J. L. Lagrange (1736-1813), who formalised the axiom of Torricelli for conservative dynamical systems employing the concept of potential energy. Lagrange wrote in his work *Méchanique Analytique* [91]:

> *On vient de voir que la fonction Φ [the potential energy] est un minimum ou un maximum, lorsque la position du système est celle de l'équilibre; nous allons maintenant démontrer que si cette fonction est un minimum, l'équilibre aura de stabilité; ensorte que le système étant d'abord supposé dans l'état de l'équilibre, & venant ensuite à être tant*

---

[3] Free translation: The stability of a floating body in equilibrium is determined by the restoring moment arising when the body has been displaced from equilibrium by an infinitesimally small angle.

soit peu déplacé de cet état, il tendra de lui-même à s'y remettre, en faisant des oscillations infiniment petites.[4]

J. L. LAGRANGE, MÉCHANIQUE ANALYTIQUE,
PART 1, SECTION 3, NO. 16.

In other words, Lagrange posed the theorem that, if the system is conservative, a state corresponding to zero kinetic energy and minimum potential energy is a stable equilibrium point. Moreover, Lagrange gave a definition of stability of an equilibrium. Clearly, Lagrange meant that an equilibrium is stable when neighbouring solutions remain close to the equilibrium, which agrees with our modern concept of stability in the sense of Lyapunov. Lagrange speaks of infinitely small oscillations around an equilibrium because a stable equilibrium in a conservative system is necessarily a centre. Using a Taylor series approximation of the potential energy up to second-order terms, Lagrange proved that the equilibrium is indeed stable when the first-order terms vanish and the second-order terms are positive, corresponding to a minimum of the potential energy. J. P. G. Lejeune Dirichlet (1805-1859) added a note [105] to the theorem of Lagrange, arguing that a minimum of the potential energy might also be caused by fourth or higher order terms in the Taylor series, but a minimum of the potential energy is sufficient to prove stability. The theorem is in literature therefore referred to as the Lagrange-Dirichlet stability theorem and plays an important role in elasto-statics. An equilibrium in elasto-statics is called stable when it corresponds to a minimum of the potential energy. Hence, the Lagrange-Dirichlet stability theorem is used in elasto-statics as the *definition* of stability (instead of a condition for stability). The reason for this is that the notions of time, velocity and solution are non-existent in statics and a definition of stability based on those notions can not be given.

Celestial mechanics has greatly influenced the terminology of modern stability theory. P. S. Laplace (1749-1827) studied the celestial three-body problem with a perturbation analysis neglecting terms in the mass of second-order and higher and assuming small values of eccentricity [93]. He concluded that, under these assumptions, the variation of the semi-major axis of the orbits is periodic with a constant amplitude, i.e. of the form $A\sin(\alpha t + \beta)$. Laplace speaks of the *stabilité du système du monde*, the stability of the world system, and uses the same notion of stability as employed by Lagrange. Following the work of Laplace, Lagrange extended this result for arbitrary eccentricities, still neglecting terms in the mass of second-order and higher [90]. S. D. Poisson (1781-1840) extended the perturbation analysis to second as well as to third-order terms in the mass. Poisson showed that the variation of the semi-major

---

[4] Free translation: We have shown that the [potential energy] function Φ is in a minimum or maximum, when the configuration of the system is one of equilibrium; we are now going to demonstrate that if this function is in a minimum then the equilibrium will be stable, such that the system, being assumed in equilibrium and displaced by a small amount, will tend to return to it by itself while making infinitely small oscillations.

axis only contains terms of the form $A\sin(\alpha t + \beta)$ if terms in the mass of second-order are taken into account [145]. Based on this analysis, the motion of the earth and moon with respect to the sun therefore remains bounded. Similar to Laplace, Poisson associates the boundedness of the variation of the semi-major axis with the concept of stability. The extension of Poisson to third-order terms in the perturbation analysis reveals in addition terms of the form $At\sin(\alpha t + \beta)$ in the variation of the semi-major axis. The amplitude of these oscillatory terms grows unboundedly with time. An analysis based on third-order terms therefore implies that the motion of the planets does not remain bounded but the planets come arbitrarily close to their original position infinitely many times. Also C. G. J. Jacobi (1804-1851) contemplates whether the results of Laplace, Lagrange and Poisson are the proof of the *Stabilität des Weltsystems*, i.e. the *stability* of the world-system [80]. Jacobi, like Laplace and Poisson, associates the word stability with periodic functions which do not grow unboundedly. In 1887, King Oscar II of Sweden sponsored a mathematical competition with a prize for a resolution of the question of how stable the solar system is, a variation of the three-body problem. H. Poincaré (1854-1912) won the prize which was the beginning of his work on the stability of the solar system which accumulated in his work *Les Méthodes Nouvelles de la Mécanique Céleste* [144]. Poincaré states that the term 'stability' is used in different ways and he continues to discuss the differences between the results of Lagrange and Poisson on the variation of the semi-major axis. According to Poincaré, Lagrange found that the variation of the semi-major axis is governed by terms of the form $A\sin(\alpha t + \beta)$, whereas Poisson found that there are in addition terms of the form $At\sin(\alpha t + \beta)$. He concludes that the word stability does not have the same meaning for Lagrange and for Poisson. With respect to the solar system, Poincaré speaks of the stability in the sense of Lagrange, with which he means bounded behaviour of the planetary orbits, and stability in the sense of Poisson, for which the planets come arbitrarily close to their original position infinitely many times. Hence, Poincaré gives the false impression that Lagrange associates his work on the semi-major axis with the concept of stability. Moreover, Poincaré gives the false impression that Poisson's notion of stability is related to non-periodic terms which grow unboundedly. These two errors have persisted in modern times and have led to misnomers. Nowadays, the terms 'Lagrange-stability' and 'Poisson-stability', inherited from Poincaré, are used to denote boundedness and recurrence of solutions of arbitrary dynamical systems. The modern concept of Lagrange stability is therefore very different from the concept of stability as expressed by Lagrange himself in *Méchanique Analytique*. The work of Poincaré on periodic orbits led to the modern concept of Poincaré stability, which is also called orbital stability.

In the nineteenth century, the development of regulators for steam engines and water turbines led to the stability analysis of machines. J. C. Maxwell (1831-1879) analysed in a paper of 1868 [112] the stability of Watt's flyball governor. His technique was to linearise the differential equations of motion to find the characteristic equation of the system. He studied the effect of the system parameters on stability and showed that the system is stable if the roots of the characteristic equation have negative real parts. In 1877, E. J. Routh (1831-1907) provided an algorithm for determining when a characteristic equation has stable roots. The Russian I. I. Vishnegradsky analysed in 1877 the stability of regulators using differential equations independently of Maxwell and studied the stability of the Watt governor in more detail. In 1893, A. B. Stodola studied the regulation of a water turbine using the techniques of Vishnegradsky. Stodola modelled the actuator dynamics and included the delay of the actuating mechanism in his analysis and was the first to mention the notion of the system time constant. Unaware of the work of Maxwell and Routh, Stodola posed the problem of determining the stability of the characteristic equation to A. Hurwitz, who solved it independently.

An exact definition of stability for a dynamical system, as well as general stability theorems for nonlinear systems, were first formulated in 1892 by the Russian mathematician and engineer A.M. Lyapunov (1857-1918). Lyapunov defined an equilibrium to be stable when for each ε-neighbourhood one can find a δ-neighbourhood of initial conditions, such that their solutions remain within the ε-neighbourhood. Loosely speaking, this means that neighbouring solutions remain close to the equilibrium, which is essentially the same as what Lagrange understood under the term stability (see the above citation). Lyapunov proved stability using two distinct methods. In the first method, known as Lyapunov's first method or Lyapunov's indirect method, the local stability of an equilibrium is studied through linearisation. The second method, also called the direct method of Lyapunov, is far more general. The fundamental idea behind the direct method of Lyapunov is the stability theorem of Lagrange-Dirichlet, which is based on the mechanical energy. The direct method of Lyapunov is able to prove the stability of equilibria of nonlinear differential equations using a generalised notion of energy functions. Unfortunately, though his work was applied and continued in Russia, the time was not ripe in the West for his elegant theory, and it remained unknown there until its French translation in 1907 [106] (reproduced in [109]). Its importance was finally recognised in the 1960's with the emergence of control theory.

1.3 Non-smooth Dynamical Systems

In this section we try to clarify the terminology around non-smooth dynamical systems and briefly discuss various mathematical frameworks to describe non-smooth dynamical systems.

Loosely speaking, a dynamical system is a system whose state evolves in time. The time-evolution is governed by a set of rules (usually equations). We distinguish between continuous-time dynamical systems and discrete-time dynamical systems. The state of a continuous-time dynamical system may change at every time-instance, either continuously or discontinuously. The evolution of continuous-time systems is usually expressed by a set of ordinary differential equations (and sometimes algebraic equations if equality constraints are involved). The state of a discrete-time system, for which the evolution is governed by a discrete mapping, is only defined and can change at certain discrete time-instances. In the following, we will mean with the term 'system' a dynamical system in continuous-time if not stated otherwise. Note that the state of a continuous-time system can still vary discontinuously in time, but time itself is a continuum.

Lagrangian mechanical systems are described by second-order differential equations in the generalised positions $q(t)$. A second-order differential equation can always be written as a system of first-order differential equations of the form $\dot{x}(t) = f(t, x(t))$. We refer to the function $f(t, x(t))$ as the right-hand side of the system which induces a vector field in the state-space.

The term 'non-smooth dynamical system' or 'discontinuous dynamical system' is often used in literature without stating explicitly which properties of the system are considered to be 'non-smooth'. Non-smooth (continuous-time) dynamical systems in first-order form can be divided in three types according to their degree of non-smoothness:

1. Non-smooth continuous systems which are described by a Lipschitz continuous right-hand side $f(t, x(t))$ which is non-differentiable with respect to x at certain hyper-surfaces in the state-space. For example, consider a mechanical oscillator with a one-sided elastic support and external excitation. If the position of the mass m is described by q, the spring stiffness by k and the support stiffness k_{sup}, then the system is described by the second-order differential equation

$$m\ddot{q} + kq = f_0 \cos(\omega t) - f_{\mathrm{sup}}(q), \qquad f_{\mathrm{sup}}(q) = \begin{cases} 0, & q \geq 0 \\ k_{\mathrm{sup}}q, & q < 0 \end{cases}, \qquad (1.1)$$

which can be put in the first-order form

$$\dot{x}(t) = f(t, x(t)) = \begin{bmatrix} \dot{q}(t) \\ -\frac{k}{m}q(t) + \frac{f_0}{m}\cos(\omega t) - \frac{1}{m}f_{\mathrm{sup}}(q(t)) \end{bmatrix}, \qquad (1.2)$$

where $x(t) = \begin{bmatrix} q(t) & \dot{q}(t) \end{bmatrix}^{\mathrm{T}}$ and $f(t, x(t))$ is a right-hand side which is continuous in x but non-smooth at the hyper surface $\Sigma = \{x \in \mathbb{R}^2 \mid q = 0\}$.

2. Systems described by differential equations $\dot{x}(t) = f(t, x(t))$ with a right-hand side $f(t, x(t))$ which is bounded but discontinuous on certain hyper-surfaces Σ_i in the state-space. The solution of such a discontinuous differential equation has to be understood in the sense of Filippov [51] (i.e. the differential equation is extended to a differential inclusion). The discontinuities in $f(t, x(t))$ cause the time-derivative $\dot{x}(t)$ not to be defined for all t, but the state $x(t)$ remains time-continuous. Examples are mechanical systems with visco-elastic supports or dry friction.

3. Systems which expose discontinuities (or jumps) in the state, such as systems with impulsive effects. The state $x(t)$ is not defined on such discontinuity points. Examples are mechanical systems with velocity jumps due to impact.

Non-smooth dynamical systems are studied by various scientific communities using different mathematical frameworks. Examples of such mathematical frameworks are:

1. *Singular perturbations.* The non-smooth system is replaced by a singularly perturbed smooth system. The resulting ordinary differential equation is extremely stiff and hardly suited for numerical integration. More importantly, possible stationary states, which exist in the non-smooth dynamical system, may be lost due to the smoothening. For instance, the stationary state of a block on a rough slope can not be described with a smooth friction model for which the friction force vanishes at zero relative velocity (the block will therefore always slide). Therefore, this framework can generally not be used to study stability properties of non-smooth systems.

2. *Switched systems* or *Hybrid systems.* A kind of switching is involved in non-smooth systems and these systems are therefore often called 'switched systems' or 'differential equations with switching conditions' [46, 107]. In the field of systems and control theory, the term *hybrid system* is frequently used for systems composed of continuous differential equations and discrete event parts [23]. Nowadays, the term 'hybrid system' is used for any system which exposes a mixed continuous and discrete nature [71, 111, 174]. The switched/hybrid system concept switches between differential equations with possible state re-initialisations and is not able to describe accumulation points, e.g. infinitely many impacts which occur in a finite time interval such as a bouncing ball coming to rest on a table.

3. *Complementarity systems.* Non-smooth systems can in some cases be considered as dynamical extensions of (non)linear complementarity problems, which gives the complementarity systems concept [72].

4. *Differential inclusions* or *Filippov systems.* Systems described by differential equations with a discontinuous right-hand side, but with a time-continuous state, can be extended to differential inclusions with a set-valued right-hand side. The work of Filippov [51] proved to be seminal in the extension of differential equations with discontinuous right-hand sides to differential inclusions with set-valued right-hand sides and these systems are therefore often called Filippov systems (Section 4.2). The differential inclusion concept gives a simultaneous description of the dynamics in terms of a single inclusion, which avoids the need to switch between different differential equations. Moreover, this framework is able to describe accumulation points of time-instances at which the time-derivative of the state jumps.

5. *Measure differential inclusions.* Systems which expose discontinuities in the state and/or vector field can be described by measure differential

inclusions. The differential measure of the state does not only consist of a part with a density with respect to the Lebesgue measure, but is also allowed to contain an atomic part. The dynamics of the system is described by an inclusion of the differential measure of the state to a state-dependent set (similar to the concept of differential inclusions). Consequently, the measure differential inclusion concept describes the continuous dynamics as well as the impulse dynamics with a single statement in terms of an inclusion and is able to describe accumulation phenomena. Moreover, the framework of measure differential inclusions leads directly to a numerical discretisation, called the time-stepping method, which is a robust algorithm to simulate the dynamics of non-smooth systems.

The framework of measure differential inclusions allows us to describe systems with state discontinuities and this framework is therefore more general than differential inclusions. However, the great advantage of this framework over other frameworks is, that physical interaction laws, such as friction and impact in mechanics or diode characteristics in electronics, can be formulated as set-valued force laws and be seamlessly incorporated in the formulation. We will therefore use the framework of measure differential inclusions in this work to study stability properties of non-smooth systems in general and mechanical systems with unilateral constraints, such as Coulomb friction, unilateral contact and impact, in particular.

1.4 Stability and Convergence

Stability theory plays today a dominant role in system and control theory. Various stability concepts exist as has been already mentioned in the historical overview of Section 1.2. The most important stability concept is stability in the sense of Lyapunov. An equilibrium point of a dynamical system is stable in the sense of Lyapunov if all solutions which start nearby remain in the neighbourhood of the equilibrium point for all future times. An equilibrium point which is not stable is called unstable. If all solutions starting within some neighbourhood of the equilibrium point converge towards the equilibrium point as time goes on, then the equilibrium point is called attractive. Stability and attractivity are two different notions. An equilibrium point can for instance be stable without being attractive, or it can be unstable and at the same time attractive. An equilibrium point which is stable and attractive is usually called asymptotically stable. This terminology will turn out to be cumbersome when dealing with non-smooth systems. The notions of stability and attractivity can be generalised to positively invariant sets such as periodic solutions, quasi-periodic solutions etc.

One of the main issues in stability theory is the question of how to determine whether an equilibrium point is stable and/or attractive. The so-called Lyapunov theorems give sufficient conditions for the stability (and sometimes attractivity) of equilibrium points (see e.g. [85] for a recent overview). Furthermore, LaSalle's Invariance Principle can give sufficient conditions for attractivity [89].

The title of this monograph preludes that this work also deals with the so-called convergence property. Lyapunov stability and attractivity concepts are used to characterise stability properties of equilibrium points, or in the more general case, of positively invariant sets. In contrast, convergence is a property of the dynamic system. A system, which is excited by an input, is called (uniformly) convergent if it has a unique solution that is bounded on the whole time axis and this solution is globally attractive and stable. Obviously, if such a solution does exist, then all other solutions converge to this solution, regardless of their initial conditions. Moreover, such a solution can be considered as a steady-state solution. The property of convergence can be beneficial from several points of view. Firstly, in many control problems it is required that controllers are designed in such a way that all solutions of the corresponding closed-loop system "forget" their initial conditions. Actually, one of the main tasks of feedback is to eliminate the dependency of solutions on initial conditions. In this case, all solutions converge to some steady-state solution that is determined only by the input of the closed-loop system. This input can be, for example, a command signal or a signal generated by a feedforward part of the controller or, as in the observer design problem, it can be the measured signal from the observed system. Such a convergence property of a system plays an important role in many nonlinear control problems including tracking, synchronisation, observer design, and the output regulation problem, see [131, 132, 169]. Secondly, from a dynamics point of view, convergence is an interesting property because it excludes the possibility of different coexisting steady-state solutions: namely, a convergent system excited by a periodic input has a *unique* globally attractive and stable periodic solution with the same period time. Moreover, the notion of convergence is a powerful tool for the analysis of time-varying systems. This tool can be used, for example, for performance analysis of nonlinear control systems [73, 171]. Namely, as shown in [133], convergent systems allow for the definition of so-called generalised frequency response functions. For linear systems, frequency domain analysis is based on frequency response functions and has been crucial for the performance analysis of linear control systems. The recent extension to generalised frequency response functions for the class of (nonlinear) convergent systems may open up a route towards more performance-based control design techniques for nonlinear (non-smooth) control systems. Here, we aim to provide conditions for convergence for a class of measure differential inclusions, making such systems accessible for analysis and control synthesis results for convergent systems.

1.5 Literature Survey

In this section a short literature survey will be given, which may help the reader to obtain some background knowledge or to direct the reader to further references, but the survey is by no means complete.

First, a brief review will be given of the mathematical literature on non-smooth analysis and (measure) differential inclusions, which forms a prerequisite for the study of non-smooth mechanical systems. Subsequently, the contributions from the field of non-smooth mechanics are discussed. Finally, the literature on stability in non-smooth dynamical systems and convergent dynamics is reviewed. Clearly, some publications contribute to more than one topic.

Non-smooth Analysis

The book [147] by ROCKAFELLAR is regarded to be the classical reference on Convex Analysis. Meanwhile, many other books on Convex Analysis have been published, such as the books [77,78] by HIRIART-URRUTY and LEMARÉCHAL, the book [149] by ROCKAFELLAR and WETS and the short overview in the encyclopedia [56] of GAMKRELIDZE by THIKOMIROV.

The generalised gradient, which is used extensively here, is introduced in the book [36] by CLARKE and is also presented in the book [37] by CLARKE et al. The books [10,12] by AUBIN and co-workers are extensive references on set-valued functions.

Differential Measures and (Measure) Differential Inclusions

The fundamental work of FILIPPOV [50,51,96,152] extends a discontinuous differential equation to a differential inclusion (see Chapter 4). More results on differential inclusions can be found in the book [10] by AUBIN and CELLINA, the book [43] by DEIMLING and in the book [37] by CLARKE et al. UTKIN introduced in [165–168] the concept of 'equivalent control', which is similar to Filippov's convex method, and is popular in control theory.

The classical theory of measures and integration can be found in the textbooks of ELSTRODT [47] and RUDIN [150]. MOREAU [122,124] directly introduces the so-called differential measure, which can be related to real measures and (signed) measures known from classical measure theory. Discontinuous evolution problems with state re-initialisations can be described by measure differential inclusions, see MOREAU [122,124], GLOCKER [63], MONTEIRO MARQUES [116], BALLARD [14] and STEWART [159].

Stochastic measure differential inclusions are treated by BERNARDIN, SCHATZMAN and LAMARQUE [17].

Non-smooth Mechanics

The number of publications on non-smooth systems in the field of Non-smooth Mechanics is vast. The book [23] by BROGLIATO gives an extensive review as well as the article of BROGLIATO et al. [28]. Only the main publications and those which are of relevance for the next chapters will be briefly reviewed here.

The work of MOREAU and JEAN [81, 121, 123] and of PANAGIOTOPUO-LOS [126, 127] has fulfilled a pioneering role in the Non-smooth Mechanics community. The work of PANAGIOTOPOULOS focusses on variational formulations and on the formulation of contact laws in elastostatics (see also the work of GOELEVEN et al. [67,68]). MOREAU and JEAN developed a framework to describe frictional impact and unilateral contact using Convex Analysis to its full extent. The framework of MOREAU treats the dynamics of systems with impact on velocity–impulse level in terms of measure differential inclusions. This framework is directly related to the framework of complementarity systems, see ACARY et al. [1, 25]. A school of researchers (mostly French) is actively extending the work of MOREAU.

MONTEIRO MARQUES [116] and SCHATZMAN [153] prove the existence of solutions of measure differential inclusions describing a mechanical system with a single unilateral constraint. Uniqueness of solutions of frictionless impact problems has been studied by SCHATZMAN [154], PERCIVALE [135, 136] and BALLARD [14]. Non-existence and non-uniqueness of solutions of mechanical systems with frictional unilateral contact is discussed in GÉNOT and BROGLIATO [23, 55] and LEINE, BROGLIATO and NIJMEIJER [98].

Much effort has been put in the formulation of frictional contact problems by means of set-valued force laws. GLOCKER, PFEIFFER and co-workers applied the theory on linear complementarity problems and Convex Analysis on rigid multibody systems with impact and friction. In GLOCKER [57] and the book of PFEIFFER and GLOCKER [139], unilateral contact laws and Coulomb friction are formulated as linear complementarity problems on acceleration level and velocity level and simulation is performed using an event-driven integration method. The theory and methods have been applied to a variety of engineering applications, see e.g. [142, 160, 177] and the overview in [139]. Later, spatial frictional contact problems have been formulated as solvable nonlinear complementarity problems in GLOCKER [60, 63]. The theory of set-valued force laws (i.e. set-valued constitutive laws) was put in a framework of non-smooth potential theory in the book [63] by GLOCKER. The book [63] discusses non-smooth extensions to the classical variational principles in dynamics, e.g. Gauß, Jourdain, d'Alembert/Lagrange (see also GLOCKER [58,59]). A set-valued force law for spatial Coulomb-Contensou friction (combined sliding and pivoting friction) can be found in LEINE and GLOCKER [99].

ALART and CURNIER [5] presented an augmented Lagrangian approach to the solution of static frictional contact problems (see also the work of SIMO and LAURSEN [94, 155] and KLARBRING [86]). The application of the augmented Lagrangian method to dynamic frictional contact problems is presented in LEINE and GLOCKER [99] and LEINE and NIJMEIJER [101].

The developed approach for the description and numerical treatment of non-smooth Lagrangian mechanical systems can very well be equally applied to other types of engineering systems. For example, electrical circuits are treated as measure differential inclusions with set-valued force laws by GLOCKER [64].

Stability in Non-smooth Systems

The literature on stability theory for non-smooth systems is vast and rapidly growing, providing dedicated results for specific mathematical formalisms. To name a few, see e.g. [82] for results on piece-wise affine systems, [74, 107] for results on switched systems, [29, 138, 151, 180] for results on hybrid systems, [13, 92] for results on impulsive dynamical systems, and [30] for results on complementarity systems. For the sake of brevity, in the remainder of this overview we will mainly focus on the existing literature on the stability of (measure) differential inclusions (and, to a lesser extent, hybrid systems also allowing for state jumps) and results for the specific case of non-smooth mechanical systems. Herein, it is notable that most results in literature focus on the stability of isolated equilibria and few results for equilibrium sets exist (a topic receiving considerable attention in the current work).

YAKUBOVICH, LEONOV and GELIG [179] discuss the stability of equilibrium sets and the dichotomy property (meaning that solutions either converge to an equilibrium or grow unboundedly) in differential inclusions. It is explained how the construction of Lyapunov functions, which guarantee global stability, leads to algebraic problems on the solution of matrix inequalities.

ADLY, GOELEVEN and BROGLIATO [2, 3, 66] and VAN DE WOUW and LEINE [170] study stability properties of equilibrium sets of differential inclusions describing mechanical systems with friction. It is assumed that the non-smoothness is stemming from a maximal monotone operator. Existence and uniqueness of solutions is therefore always fulfilled. A basic Lyapunov theorem for stability and attractivity is given in [3, 66] for first-order differential inclusions and in [2] for linear second-order differential inclusions with maximal monotone operators. The results are applied to linear mechanical systems with friction. It is assumed in [3, 170] that the relative sliding velocity of the frictional contacts depends linearly on the generalised velocities and conditions for the attractivity of an equilibrium set are given. The results are generalised in [3] to conservative systems with an arbitrary potential energy function. The paper [170] gives many illustrating examples and a useful (and intuitive) condition for attractivity of an equilibrium set for linear mechanical systems. A numerical analysis of the stability properties of equilibrium sets in a simple mechanical system with frictional unilateral contact is given in [16].

The stability of hybrid systems with state-discontinuities is addressed by a vast number of researchers in the field of control theory. The books of

LAKSHMIKANTHAM, BAINOV and SIMEONOV [13, 92] focus on systems with impulsive effects and give many useful Russian references. Lyapunov stability theorems, instability theorems and theorems for boundedness are given by YE, MICHEL and HOU [180]. PETTERSSON and LENNARTSON [138] propose stability theorems using multiple Lyapunov functions. By using piecewise quadratic Lyapunov function candidates and replacing the regions where the different stability conditions have to be valid by quadratic inequality functions, the problem of verifying stability is turned into a Linear Matrix Inequality (LMI) problem. PEREIRA and SILVA [137] study the stability of equilibrium points of measure differential inclusions, but only using continuous Lyapunov functions. See also the review article of DAVRAZOS and KOUSSOULAS [42].

Many publications focus on the control of mechanical systems with frictionless unilateral contacts by means of Lyapunov functions. See for instance BROGLIATO et al. [27] and TORNAMBÈ [162] and the book [23] for further references.

The Lagrange-Dirichlet stability theorem is extended by BROGLIATO [24] to measure differential inclusions describing mechanical systems with frictionless impact. The idea to use Lyapunov functions involving indicator functions associated with unilateral constraints is most probably due to [24]. More generally, the work of CHAREYRON and WIEBER [32–34] is concerned with a Lyapunov stability framework for measure differential inclusions describing mechanical systems with frictionless impact. It is clearly explained in [34] why the Lyapunov function in the presence of state-discontinuities has to be globally positive definite, in order to prove local stability properties (when no further assumptions on the system or the form of the Lyapunov function are made). The importance of this condition has also been stated in [13, 180] for hybrid systems and in [27, 162] for mechanical systems with frictionless unilateral constraints. Moreover, LaSalle's invariance principle is generalised in [26] to differential inclusions and in [33, 34] to measure differential inclusions describing mechanical systems with frictionless impact. The proof of LaSalle's invariance principle strongly relies on the positive invariance of limit sets. It is assumed in [33, 34] that the system enjoys continuity of the solution with respect to the initial condition which is a sufficient condition for positive invariance of limit sets. In [34], an extension of LaSalle invariance principle to systems with unilateral constraints is presented (more specifically it is applied to mechanical systems with frictionless unilateral contacts). In [26] an extension of LaSalle invariance principle for a class of unilateral dynamical systems, so-called evolution variational inequalities, is presented.

Instability results for finite dimensional variational inequalities can be found in the work of GOELEVEN and BROGLIATO [65, 66], whereas QUITTNER [146] gives instability results for a class of parabolic variational inequalities in Hilbert space.

Closely related to stability problems are bifurcations of equilibria and periodic solutions. Bifurcations in non-smooth systems do recently receive much attention in literature (see the work of LEINE [96, 97, 100, 101] for a literature survey).

Convergent Dynamics

The first conditions for the convergence property have been formulated by PLISS [143] and DEMIDOVICH [44] for smooth nonlinear systems, where Pliss focussed on systems with periodic right-hand sides. YAKUBOVICH [178] considered Lur'e-type systems, possibly with discontinuities, and proposed sufficient convergence conditions based on the circle criterion. More recently, the work of PAVLOV, VAN DE WOUW and NIJMEIJER [130] has given sufficient conditions for continuous (though non-smooth) piece-wise affine (PWA) systems in terms of the existence of a common quadratic Lyapunov function for all affine systems constituting the PWA system. It is shown that the existence of such a common quadratic Lyapunov function is by no means sufficient for convergence of discontinuous PWA systems and sufficient conditions for convergence of discontinuous PWA systems are proposed. Results on the convergence property of a class of switched linear systems are proposed in [173].

Similar notions describing the property of solutions converging to each other are studied in literature. The notion of contraction has been introduced by LOHMILLER and SLOTINE [108] (see also references therein). An operator-based approach towards studying the property that all solutions of a system converge to each other is pursued by FROMION and co-workers [52, 53]. In the work of ANGELI [7], a Lyapunov-based approach has been developed to study the global uniform asymptotic stability of all solutions of a system (in [7], this property is called incremental stability) as well as the so-called incremental input-to-state stability property, which is compatible with the input-to-state stability approach (see e.g. [157]). In the same spirit, the convergence property has been extended to the so-called input-to-state convergence property [132].

Convergence properties of measure differential inclusions have been studied by [104], see also Chapter 8.

1.6 Objective and Scope

As stated in Section 1.1, there exists a need to redefine the notion of stability and related properties in the light of non-smooth systems as well as to formulate stability results providing stability properties in the redefined sense. The objective of this research monograph is, in one phrase, to partially address this need within the setting of measure differential inclusions and with an emphasis on mechanical (Lagrangian) systems with frictional unilateral constraints. More specifically the aims are:

- to introduce the reader to the formulation of finite-dimensional Lagrangian mechanical systems with unilateral constraints as measure differential inclusions, as well as to present various friction models as set-valued force laws,

- to precisely define stability, attractivity and related terminology for measure differential inclusions,
- to present the Lyapunov stability theory (Lyapunov's direct method, LaSalle's invariance principle, instability theorems) for measure differential inclusions,
- to apply and specialise the Lyapunov stability theory to Lagrangian mechanical systems with frictional unilateral constraints,
- to study the convergence property for measure differential inclusions and to give sufficient conditions for convergence for systems which enjoy certain monotonicity properties,
- to show the use of the convergence property in solving the tracking control problem for certain non-smooth mechanical systems.

Special attention will be paid to the stability/attractivity of equilibrium sets of measure differential inclusions.

1.7 Outline

The first few chapters of this monograph (Chapter 2–5) introduce the mathematical framework to describe non-smooth systems and deal with the formulation of mechanical systems with frictional unilateral constraints within this framework. Chapters 6 and 7 are devoted to the stability of non-smooth (mechanical) systems and form the core of the work. Finally, Chapter 8 revolves around the convergence property for measure differential inclusions.

Chapter 2 presents some basic mathematical theory from non-smooth analysis and convex analysis. The notions of generalised derivatives, normal cones and indicator functions will prove to be essential when dealing with non-smooth systems and in particular with set-valued force laws derived from non-smooth potentials for the description of frictional unilateral contact in mechanical systems.

The theory of measures forms the basis for the modern integration theory and both are quintessential for the understanding of measure differential inclusions. In Chapter 3 we give a short overview of measure and integration theory and relate the differential measure (which is pivotal in the definition of measure differential inclusions) to real measures and (signed) measures.

Chapter 4 is concerned with non-smooth dynamical systems and their solution concepts. A brief introduction to differential equations is followed by a discussion of differential equations with a discontinuous right-hand side. The requirement for the existence of a solution leads to the need to fill in the graph of the discontinuous right-hand side (Filippov's convex method). The resulting set-valued right-hand side brings forth a differential inclusion. Subsequently, the differential measure of the state, which classically contains a density with respect to the Lebesgue measure, is extended with an atomic part, which leads to a measure differential inclusion.

An important class of measure differential inclusions is formed by Lagrangian mechanical systems with frictional unilateral constraints. Chapter 5 deals with the mathematical formulation of Lagrangian mechanical systems with unilateral contact and friction modelled with set-valued force laws. First, it is shown how set-valued force laws can be derived from non-smooth potentials. Subsequently we treat the contact laws for unilateral contact and various types of friction within the setting of non-smooth potential theory. This leads to a unified approach with which all set-valued forces can be formulated. Finally, the set-valued forces are incorporated as Lagrangian multipliers in the Newton-Euler equations.

Chapter 6 is devoted to the Lyapunov stability theory for measure differential inclusions. Definitions of stability, attractivity and related terminology are given, that make sense for non-smooth dynamical systems with possible discontinuities in the state and loss of uniqueness of solutions. Generalisations of the direct method of Lyapunov, LaSalle's invariance principle and Chetaev's instability theorem are given.

The stability results of Chapter 6 are applied in Chapter 7 to Lagrangian mechanical systems with frictional unilateral constraints. The special structure of mechanical systems allows for a refinement of the stability results and the formulation of intuitive stability conditions.

In Chapter 8 the convergence property of measure differential inclusions is studied and results are presented which give sufficient conditions for the convergence of measure differential inclusions with certain maximal monotonicity properties. It is shown how these convergence results for measure differential inclusions can be exploited to solve tracking control problems for certain classes of non-smooth mechanical systems with friction and one-way clutches. Illustrative examples of convergent mechanical systems are discussed in detail.

Finally, some concluding remarks are given in Chapter 9, which put the presented results in a broader perspective.

2

Non-smooth Analysis

This chapter presents some basic mathematical theory from non-smooth analysis [10, 12, 37, 56, 77, 78, 147, 149]. The aim of this chapter is not to give a real introduction to non-smooth analysis as the above textbooks are much better suited for that task. Instead, the primary aim of the chapter is to make the reader is familiar with the terminology and notation used in this monograph. Moreover, it provides a compendium on non-smooth and convex analysis which is useful when reading the following chapters. The reader might want to look up how a mathematical term is exactly defined, making use of the index in combination with this chapter.

We begin with a brief introduction to sets (Section 2.1). The notion of continuity of functions is relaxed in Section 2.2 to semi-continuity and the notion of the classical derivative of smooth functions is extended to generalised differentials for non-smooth functions in Section 2.3. Subsequently, we discuss set-valued functions in Section 2.4. Topics from convex analysis are reviewed in Section 2.5 and the subderivative is discussed in Section 2.6.

2.1 Sets

A number of properties of sets and set-valued functions will be briefly reviewed. Let C be a subset of the normed space $\mathbb{R}^n$, equipped with the Euclidean norm $\| \cdot \|$.

Definition 2.1 (Closed Set). *A set $C \subset \mathbb{R}^n$ is closed if it contains all its limit points. Every limit point of a set C is the limit of some sequence $\{x_k\}$ with $x_k \in C$ for all $k \in \mathbb{N}$.*

The boundary of a set C, denoted by bdry C, is the set of points which can be approached both from C and from the outside of C. We define the closure of a set C as the smallest closed set containing C, i.e. $\overline{C} = C \cup \text{bdry}\, C$. Furthermore, those points in C which are not on the boundary form the interior of C, int $C = C \setminus \text{bdry}\, C$. We can uniquely decompose the closure of a set in its boundary and its interior: $\overline{C} = \text{bdry}\, C \cup \text{int}\, C$. A set is called *open*, if it does not contain any of its boundary points, i.e. $C \cap \text{bdry}\, C = \emptyset$. It holds that $\text{int}\, C$ is an open set.

Definition 2.2 (Bounded Set). *A set $C \subset \mathbb{R}^n$ is bounded if there exists a point $\boldsymbol{y} \in \mathbb{R}^n$ and a finite number $c > 0$ such that $\|\boldsymbol{x} - \boldsymbol{y}\| < c$ for all $\boldsymbol{x} \in C$.*

A set is bounded if it is contained in a ball of finite radius.

Definition 2.3 (Compact Set). *A set $C \subset \mathbb{R}^n$ is compact if it is closed and bounded.*

An important property in Non-smooth Analysis is the convexity of sets.

Definition 2.4 (Convex Set). *A set $C \subset \mathbb{R}^n$ is convex if for each $\boldsymbol{x} \in C$ and $\boldsymbol{y} \in C$ also $(1 - q)\boldsymbol{x} + q\boldsymbol{y} \in C$ for arbitrary q with $0 \leq q \leq 1$.*

It follows that a convex set contains all line segments between any two points in the set.

Definition 2.5 (Convex Hull). *The convex hull of a set $C \subset \mathbb{R}^n$, denoted by $\text{co}(C)$, is the smallest convex set that contains all the points of C.*

The convex hull of a set C is therefore the intersection of all the convex sets containing C. Consequently, the closed convex hull of $\{\boldsymbol{x}, \boldsymbol{y}\} \in \mathbb{R}^n$ is the line segment between $\boldsymbol{x}$ and $\boldsymbol{y}$, i.e. the smallest closed convex set containing $\boldsymbol{x}$ and $\boldsymbol{y}$

$$\overline{\text{co}}\{\boldsymbol{x}, \boldsymbol{y}\} = \{(1 - q)\boldsymbol{x} + q\boldsymbol{y}, \forall q \in [0, 1]\}. \tag{2.1}$$

Figure 2.1 illustrates the notion of convexity and the convex hull.

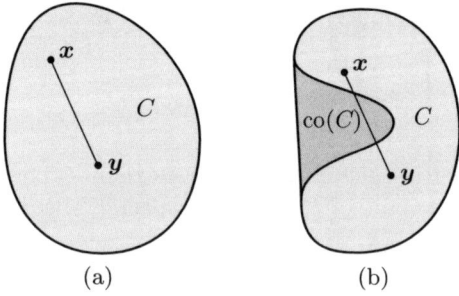

(a)   (b)

Fig. 2.1. Convex set (a) and non-convex set with its convex hull (b).

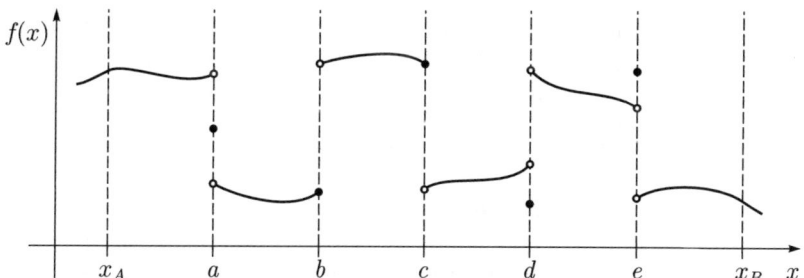

Fig. 2.2. Upper and lower semi-continuity of functions.

2.2 Functions and Continuity

A function $f : \mathbb{R} \to \mathbb{R}$ associates to any element x of its domain a single function value $f(x)$. A function is therefore single-valued which excludes vertical lines and loops on its graph, where

$$\text{graph}(f) = \{(x, f(x)) \mid x \in \mathbb{R}\}. \tag{2.2}$$

We define the epigraph of f to be the set above the graph of f

$$\text{epi}(f) = \{(x, y) \mid y \geq f(x); x \in \mathbb{R}\}. \tag{2.3}$$

Definition 2.6 (Convex Function). *A function $f : \mathbb{R} \to (-\infty, \infty]$ is convex when $\text{epi}(f)$ is a convex set.*

For a differentiable function $f : \mathbb{R}^n \to \mathbb{R}$, it follows that $f(\boldsymbol{x})$ is convex if

$$f(\boldsymbol{x}^*) \geq f(\boldsymbol{x}) + \nabla f(\boldsymbol{x})^{\mathrm{T}}(\boldsymbol{x}^* - \boldsymbol{x}), \quad \forall \boldsymbol{x}^* \text{ and } \boldsymbol{x}. \tag{2.4}$$

Let B_δ denote the open ball with radius δ centred at the origin. A function $\boldsymbol{f}(\boldsymbol{x}) : \mathbb{R}^n \to \mathbb{R}^n$ is continuous at $\boldsymbol{x} \in X$, $X \subset \mathbb{R}^n$, provided that for all $\varepsilon > 0$, there exists a $\delta > 0$ such that $\boldsymbol{y} \in \boldsymbol{x} + B_\delta \subset X$ implies $\|\boldsymbol{f}(\boldsymbol{x}) - \boldsymbol{f}(\boldsymbol{y})\| < \varepsilon$. For single-valued functions this means that we can draw the graph of the function without taking the pencil of the paper. In the following chapters we will often use the word 'smooth'.

Definition 2.7 (Smooth function). *A function $f(\boldsymbol{x}) : \mathbb{R}^n \to \mathbb{R}$ is called smooth if it is continuously differentiable up to any order in $\boldsymbol{x}$. A vector-valued function $\boldsymbol{f}(\boldsymbol{x}) : \mathbb{R}^n \to \mathbb{R}$ is smooth if each element $f_i(\boldsymbol{x})$ is a smooth function.*

For a function $f : \mathbb{R}^n \to (-\infty, +\infty]$ we can define semi-continuity:

Definition 2.8 (Upper Semi-continuity). *A function $f(\boldsymbol{x})$ is upper semi-continuous if*

$$\limsup_{\boldsymbol{y} \to \boldsymbol{x}} f(\boldsymbol{y}) \leq f(\boldsymbol{x}) \qquad \forall \boldsymbol{x} \in \mathbb{R}^n.$$

We say that f is upper semi-continuous at $\boldsymbol{x}$ if for every $\varepsilon > 0$ there exists a neighbourhood U of $\boldsymbol{x}$ such that $f(\boldsymbol{y}) < f(\boldsymbol{x}) + \varepsilon$ for all $\boldsymbol{y} \in U$. Similarly, we introduce lower semi-continuity of a real-valued function:

Definition 2.9 (Lower Semi-continuity). *A function $f(\boldsymbol{x})$ is lower semi-continuous if*

$$\liminf_{\boldsymbol{y} \to \boldsymbol{x}} f(\boldsymbol{y}) \geq f(\boldsymbol{x}) \qquad \forall \boldsymbol{x} \in \mathbb{R}^n.$$

A function $f(\boldsymbol{x})$ is lower semi-continuous, when $-f(\boldsymbol{x})$ is upper semi-continuous. A function which is both lower and upper semi-continuous is called continuous. The function f in Figure 2.2 is continuous on $(x_A, x_B) \backslash \{a, b, c, d, e\}$. It is lower semi-continuous on $(x_A, x_B) \backslash \{a, c, e\}$ and upper semi-continuous on $(x_A, x_B) \backslash \{a, b, d\}$. Function values on discontinuities are depicted in Figure 2.2 as bullets ($\bullet$). Left and right limits, which do not agree with the function value, are depicted as circles ($\circ$).

A stronger concept than continuity is absolute continuity:

Definition 2.10 (Absolute Continuity). *A function $\boldsymbol{f} : I \to \mathbb{R}^n$ is absolutely continuous on $I \subset \mathbb{R}$ if for every $\varepsilon > 0$ there exists a $\delta(\varepsilon) > 0$ such that*

$$\sum_{i=1}^{n} \| \boldsymbol{f}(b_i) - \boldsymbol{f}(a_i) \| < \varepsilon$$

for any n and any disjoint collection of intervals $[a_i, b_i] \in I$ satisfying

$$\sum_{i=1}^{n} (b_i - a_i) < \delta.$$

Absolutely continuous functions have the property that they can be obtained from integration of their derivative, which exists almost everywhere. A function $\boldsymbol{f} : \mathbb{R}^n \to \mathbb{R}^n$ is said to satisfy a Lipschitz condition with constant K provided that $\boldsymbol{f}$ is finite and satisfies

$$\| \boldsymbol{f}(\boldsymbol{x}) - \boldsymbol{f}(\boldsymbol{y}) \| \leq K \| \boldsymbol{x} - \boldsymbol{y} \| \quad \forall \boldsymbol{x}, \boldsymbol{y} \in \mathbb{R}^n. \tag{2.5}$$

Every Lipschitz-continuous function $\boldsymbol{f} : I \to \mathbb{R}^n$ is absolutely continuous.

Let I be a real interval and X be a Euclidean space with the norm $\| \cdot \|$.

Definition 2.11 (Variation). *Let $\boldsymbol{f} : I \to X$ and let $[a, b]$ be a subinterval of I. The variation of $\boldsymbol{f}$ on $[a, b]$ is the nonnegative extended real number*

$$\mathrm{var}(\boldsymbol{f}, [a, b]) = \sup \sum_{i=1}^{n} \| \boldsymbol{f}(x_i) - \boldsymbol{f}(x_{i-1}) \|, \tag{2.6}$$

where the supremum is taken over all strictly increasing finite sequences $x_1 < x_2 < \ldots < x_n$ of points on $[a, b]$.

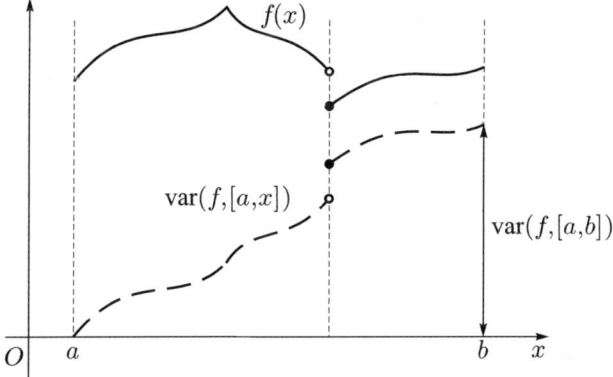

Fig. 2.3. The function $f(x)$ and its associated variation function $\mathrm{var}(f, [a, x])$ [63].

The variation of a constant function is therefore zero. Moreover, it holds that

$$A \subset B \subset I \Rightarrow \mathrm{var}(\boldsymbol{f}, A) \leq \mathrm{var}(\boldsymbol{f}, B) \tag{2.7}$$

and

$$\mathrm{var}(\boldsymbol{f}, [a, c]) = \mathrm{var}(\boldsymbol{f}, [a, b]) + \mathrm{var}(\boldsymbol{f}, [b, c]), \qquad \forall a \leq b \leq c \in I. \tag{2.8}$$

Figure 2.3 illustrates the concept of the variation of a function $f(x)$. The associated variation function $\mathrm{var}(f, [a, x])$ is an increasing function which has kinks and discontinuities in its graph on the same location where f has kinks and discontinuities.

Definition 2.12 (Locally Bounded Variation). *The function $\boldsymbol{f} : I \to X$ is said to be of locally bounded variation, $\boldsymbol{f} \in \mathrm{lbv}(I, X)$, if and only if*

$$\mathrm{var}(\boldsymbol{f}, [a, b]) < \infty \tag{2.9}$$

for every compact subinterval $[a, b]$ of I.

If $\boldsymbol{f} \in \mathrm{lbv}(I, X)$, then we can define at each $x \in I$ a right limit $\boldsymbol{f}^+(x)$ and a left limit $\boldsymbol{f}^-(x)$ of $\boldsymbol{f}$:

$$\boldsymbol{f}^+(x) = \lim_{y \downarrow x} \boldsymbol{f}(y), \qquad \boldsymbol{f}^-(x) = \lim_{y \uparrow x} \boldsymbol{f}(y). \tag{2.10}$$

If $\boldsymbol{f}(x)$ is locally continuous at x, then it holds that $\boldsymbol{f}(x) = \boldsymbol{f}^-(x) = \boldsymbol{f}^+(x)$. The Cantor function [47,63] is an example of a function that is continuous and of locally bounded variation, but is not absolutely continuous. It can be proven that every absolutely continuous function is of locally bounded variation and is differentiable almost everywhere (see [87], Theorem 2, p.337). To give an example of a function which is continuous but not of locally bounded variation, consider the function in Figure 2.4 (see [8])

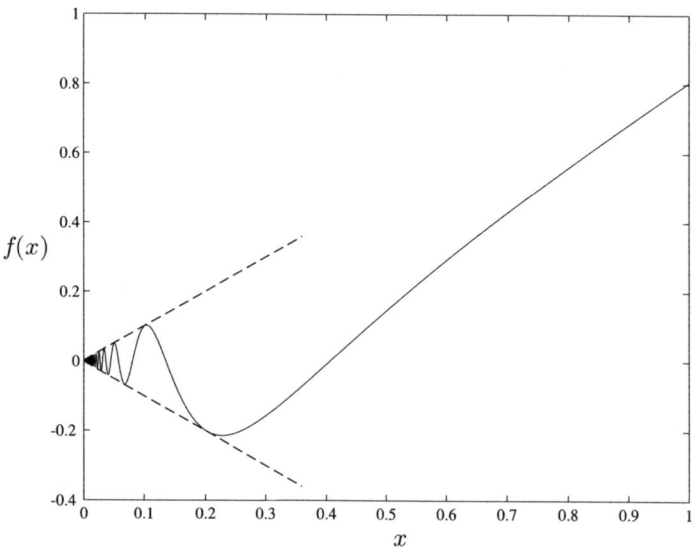

Fig. 2.4. Function (2.11) is continuous but not of bounded variation and not absolutely continuous.

$$f(x) = x \cos\left(\frac{\pi}{2x}\right) \quad \text{for } x \in (0,1] \text{ and } f(0) = 0. \tag{2.11}$$

Clearly, the function f is continuous on $[0,1]$. For the partition

$$x_0 = 0, \quad x_1 = \frac{1}{2n}, \quad x_2 = \frac{1}{2n-1}, \quad \ldots, \quad x_{2n} = 1$$

one calculates

$$\sum_{i=0}^{2n-1} |f(x_{i+1}) - f(x_i)| = x_1 + x_1 + x_3 + x_3 + \cdots + x_{2n-1} + x_{2n-1} = \sum_{k=1}^{n} \frac{1}{k}$$

and hence the supremum over all possible partitions equals $\sum_{k=1}^{\infty} \frac{1}{k} = \infty$. The function $f(x)$ is therefore not of bounded variation on $[0,1]$ and therefore also not absolutely continuous.

2.3 Generalised Derivatives

The classical derivative of smooth continuous functions will be extended in this section to the generalised derivative (and differential) of Clarke for non-smooth lower semi-continuous functions.

Consider a Lipschitz-continuous piecewise differentiable function $f(x)$ with a kink (i.e. non-smooth point) at one value of x, such as $f(x) = |x|$ (Figure 2.5). The derivative $f'(x)$ is defined by the tangent line to the graph of f when the graph is smooth at x

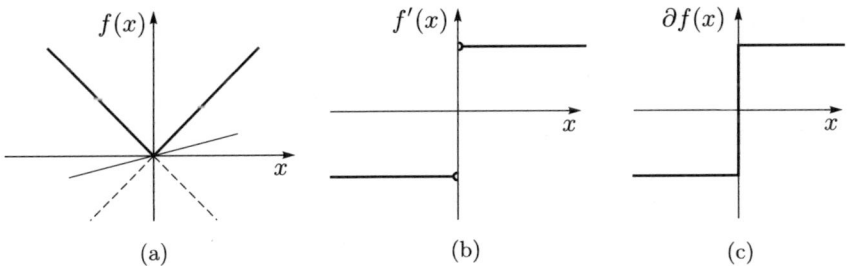

Fig. 2.5. Function (a), classical derivative (b) and generalised derivative (c).

$$f'(x) = \frac{\mathrm{d}f}{\mathrm{d}x}(x) = \lim_{y \to x} \frac{f(y) - f(x)}{y - x}. \tag{2.12}$$

Although the function is not differentiable at every point x, it possesses at each x a left and right derivative defined as

$$f'_-(x) = \lim_{y \uparrow x} \frac{f(y) - f(x)}{y - x}, \qquad f'_+(x) = \lim_{y \downarrow x} \frac{f(y) - f(x)}{y - x}. \tag{2.13}$$

The generalised derivative of f at x is declared as *any* value $f'_q(x)$ included between its left and right derivatives [37, 41]. Such an intermediate value can be expressed as a convex combination of the left and right derivatives:

$$f'_q(x) = (1 - q)f'_-(x) + qf'_+(x), \qquad 0 \leq q \leq 1. \tag{2.14}$$

Geometrically, a generalised derivative is the slope of *any* line which is tangent to the graph at the point $(x, f(x))$. Such a line is necessarily between the left and right tangent lines (drawn by dashed lines in Figure 2.5a). The set of all the generalised derivatives of f at x, more generally the convex hull of the derivative extremes, is called the generalised differential of f at x:

$$\begin{aligned} \partial f(x) &= \overline{\mathrm{co}}\{f'_-(x), f'_+(x)\} \\ &= \{f'_q(x) \mid f'_q(x) = (1 - q)f'_-(x) + qf'_+(x), 0 \leq q \leq 1\}. \end{aligned} \tag{2.15}$$

The generalised differential of Clarke at x is the set of the slopes of all the lines included in the cone bounded by the left and right tangent lines and is a closed convex set (Figure 2.5b,c). In non-smooth analysis, the generalised differential is for instance used to give a necessary condition for a local extremum of f at x by $0 \in \partial f$, which is the generalised form of $f'(x) = 0$ in smooth analysis [37, 56].

Infinitely many directional derivatives exist for functions in $\mathbb{R}^n$, whereas only two directional derivatives exist for scalar functions (the left and right derivative). For $f : \mathbb{R}^n \to \mathbb{R}$, Lipschitz-continuous and differentiable almost everywhere, the generalised differential of Clarke is defined as [37]

$$\partial f(\boldsymbol{x}) = \overline{\mathrm{co}}\{ \lim_{i \to \infty} \nabla f(\boldsymbol{x}_i) : \boldsymbol{x}_i \to \boldsymbol{x}, \nabla f(\boldsymbol{x}_i) \text{ exists}\} \quad \subset \mathbb{R}^n, \tag{2.16}$$

with the gradient

$$\nabla f(\boldsymbol{x}) = \left(\frac{\partial f(\boldsymbol{x})}{\partial \boldsymbol{x}}\right)^{\mathrm{T}} \subset \mathbb{R}^n. \tag{2.17}$$

Equivalently, the generalised differential can be defined as [56]

$$\partial f(\boldsymbol{x}) = \bigcap_{\delta > 0} \overline{\mathrm{co}}\{\nabla f(\boldsymbol{y}) \mid \boldsymbol{y} \in \boldsymbol{x} + B_\delta\} \quad \subset \mathbb{R}^n. \tag{2.18}$$

The generalised differential (2.16) simplifies to (2.15) for the scalar case. Note that $f(\boldsymbol{x})$ can be convex or non-convex in the above definitions. For continuous functions, the image of $\boldsymbol{x}$ under the generalised differential $\partial f(\cdot)$ is always a closed convex set if it exists. If $f(\boldsymbol{x})$ is a continuous convex function, then the following relation holds

$$\partial f(\boldsymbol{x}) = \{\boldsymbol{y} \mid f(\boldsymbol{x}^*) \geq f(\boldsymbol{x}) + \boldsymbol{y}^{\mathrm{T}}(\boldsymbol{x}^* - \boldsymbol{x}); \forall \boldsymbol{x}^*\} \quad \subset \mathbb{R}^n. \tag{2.19}$$

The generalised differential of a continuous convex function, defined by (2.19), agrees with the subdifferential. The subdifferential is not restricted to continuous convex functions, but to lower semi-continuous convex functions $f(\boldsymbol{x}) > -\infty$ with $f(\boldsymbol{x}) < \infty$ for at least one $\boldsymbol{x} \in \mathbb{R}^n$.

Definition 2.13 (Subdifferential). *For a lower semi-continuous convex function $f : \mathbb{R}^n \to (-\infty, +\infty]$, with the property*

$$\exists \boldsymbol{x} \in \mathbb{R}^n, \quad f(\boldsymbol{x}) < +\infty,$$

we define the subdifferential of f at $\boldsymbol{x} \in \mathbb{R}^n$, denoted as $\partial f(\boldsymbol{x})$, by

$$\partial f(\boldsymbol{x}) = \{\boldsymbol{y} \mid f(\boldsymbol{x}^*) \geq f(\boldsymbol{x}) + \boldsymbol{y}^{\mathrm{T}}(\boldsymbol{x}^* - \boldsymbol{x}); \forall \boldsymbol{x}^*\} \quad \subset \mathbb{R}^n.$$

2.4 Set-valued Functions

In this section we consider the main notions concerning set-valued functions and their properties. We first define what we mean by a set-valued function.

Definition 2.14 (Set-valued Function). *A set-valued function $\mathcal{F} : \mathbb{R}^n \to \mathbb{R}^n$ is a map that associates with any $\boldsymbol{x} \in \mathbb{R}^n$ a set $\mathcal{F}(\boldsymbol{x}) \subset \mathbb{R}^n$.*

A set-valued function can therefore contain vertical segments on its graph

$$\mathrm{Graph}(\mathcal{F}) = \{(\boldsymbol{x}, \boldsymbol{y}) \in \mathbb{R}^n \times \mathbb{R}^n \mid \boldsymbol{x} \in \mathbb{R}^n, \boldsymbol{y} \in \mathcal{F}(\boldsymbol{x})\}. \tag{2.20}$$

The image of $\boldsymbol{x}$ under $\mathcal{F}$, being a set, is in general not closed or convex. Convexity of a function should not be confused with convexity of the image of $\boldsymbol{x}$ under $\mathcal{F}$.

Definition 2.15 (Convex Image). *A set-valued function $\mathcal{F}$ has a convex image on $X \subset \mathbb{R}^n$ if the image of x under $\mathcal{F}$ is a convex set for all fixed values $x \in X$.*

We will reserve lowercase letters for single-valued functions, $f(x) \in \mathbb{R}^n$, and uppercase calligraphic letters or uppercase Greek letters for set-valued functions, i.e. $\mathcal{F}(x) \subset \mathbb{R}^n$. The term multi-valued function or multifunction is sometimes used instead of set-valued function [77].

We use the graph to define monotonicity of a set-valued function [12].

Definition 2.16 (Monotone Set-valued Function). *A set-valued map $\mathcal{F}(x)$ is called monotone if its graph is monotone in the sense that*

$$\forall (x, y) \in \mathrm{Graph}(\mathcal{F}), \quad \forall (x^*, y^*) \in \mathrm{Graph}(\mathcal{F}), \quad (y - y^*)^{\mathrm{T}}(x - x^*) \geq 0.$$

In addition, if

$$(y - y^*)^{\mathrm{T}}(x - x^*) \geq \alpha \|x - x^*\|^2$$

for some $\alpha > 0$, then the set-valued map is strictly monotone.

For example, set-valued function

$$\mathcal{F}(x) = \begin{cases} -1 & x < 0 \\ \{-1, +1\} & x = 0 \\ +1 & x > 0 \end{cases} \tag{2.21}$$

is a monotone set-valued function, whereas $\mathcal{F}(x) + cx$ is strictly monotone for $c > 0$. The subdifferential ∂f of a lower semi-continuous convex function $f(x)$ is monotone. Indeed, if $y_1 \in \partial f(x_1)$ and $y_2 \in \partial f(x_2)$, then the definition of the subdifferential (Definition 2.13) gives $f(x_2) \geq f(x_1) + y_1^{\mathrm{T}}(x_2 - x_1)$ and $f(x_1) \geq f(x_2) + y_2^{\mathrm{T}}(x_1 - x_2)$, which yields $(y_1 - y_2)^{\mathrm{T}}(x_1 - x_2) \geq 0$ after substitution. The set-valued function

$$\mathcal{F}(x) = \partial |x| = \begin{cases} -1 & x < 0 \\ [-1, +1] & x = 0 \\ +1 & x > 0 \end{cases} \tag{2.22}$$

is also monotone, but is even maximal monotone.

Definition 2.17 (Maximal Monotone Set-valued Function). *A monotone set-valued function $\mathcal{F}(x)$ is called maximal monotone if there exists no other monotone set-valued function whose graph strictly contains the graph of $\mathcal{F}$. If $\mathcal{F}$ is strictly monotone and maximal, then it is called strictly maximal monotone.*

A necessary and sufficient condition for a set-valued function $\mathcal{F}$ to be maximal monotone is that the property

$$\forall (x, y) \in \mathrm{Graph}(\mathcal{F}), \quad (y - y^*)^{\mathrm{T}}(x - x^*) \geq 0 \tag{2.23}$$

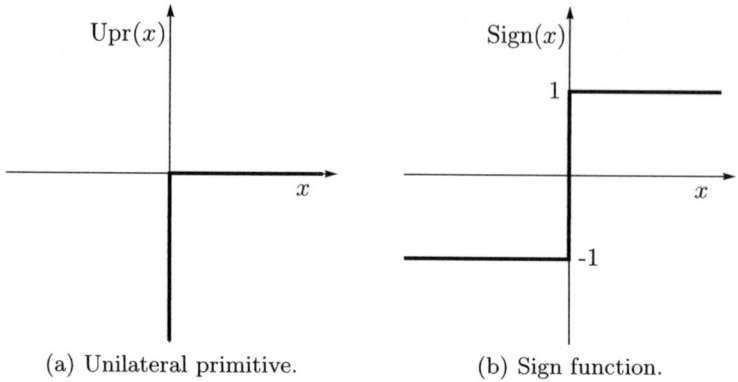

(a) Unilateral primitive.   (b) Sign function.

Fig. 2.6. Maximal monotone set-valued functions.

is equivalent to $\boldsymbol{y}^* \in \mathcal{F}(\boldsymbol{x}^*)$, see [12].

Two maximal monotone set-valued functions, which are depicted in Figure 2.6, play a very dominant role in non-smooth analysis. The first one is the *unilateral primitive* [63]

$$\mathrm{Upr}(x) = \partial \Psi_{\mathbb{R}^+} = \begin{cases} 0 & x > 0, \\ (-\infty, 0] & x = 0, \\ \emptyset & x < 0, \end{cases} \tag{2.24}$$

which is the subdifferential of the indicator function $\Psi_{\mathbb{R}^+}$ on the set $\mathbb{R}^+$ which will be defined in Section 2.5. The second maximal monotone set-valued, which is of prime importance, is the set-valued Sign-function

$$\mathrm{Sign}(x) = \partial |x| = \begin{cases} -1 & x < 0, \\ [-1, +1] & x = 0, \\ +1 & x > 0, \end{cases} \tag{2.25}$$

which we already met in (2.22).

It follows from Definition 2.17 that if $\mathcal{F}$ is maximal monotone, then the image of $\boldsymbol{x}$ under $\mathcal{F}$ is closed and convex for each $\boldsymbol{x} \in \mathbb{R}^n$. It can be shown, that the subdifferential ∂f of a lower semi-continuous convex function $f(\boldsymbol{x})$ is not only monotone but also maximal monotone (see Corollary 31.5.1 in [147]). Furthermore, the closure of the effective domain $\mathrm{dom}(\mathcal{F}) = \{\boldsymbol{x} \mid \mathcal{F}(\boldsymbol{x}) \neq \emptyset\}$ of a maximal monotone set-valued function is a convex set (Theorem 2.2 in [22]).

Proposition 2.18 (Convexity of the effective domain [22]). *If $\mathcal{F}(\boldsymbol{x})$ is a maximal monotone set-valued function, then $\overline{\mathrm{dom}}(\mathcal{F})$ is convex.*

Proof: Consider $\boldsymbol{x} \in \mathbb{R}^n$ and take $\boldsymbol{x}_\lambda$ and $\boldsymbol{y}_\lambda = \frac{\boldsymbol{x} - \boldsymbol{x}_\lambda}{\lambda}$ such that $\boldsymbol{y}_\lambda \in \mathcal{F}(\boldsymbol{x}_\lambda)$ for $\lambda > 0$. The monotonicity of $\mathcal{F}$ requires that

$$(y_\lambda - y^*)^{\mathrm{T}} (x_\lambda - x^*) \geq 0$$

for all $(x^*, y^*) \in \mathrm{Graph}(\mathcal{F})$. Hence, it must hold that

$$\lambda \left(y_\lambda^{\mathrm{T}}(x_\lambda - x^*) - y^{*\mathrm{T}}(x_\lambda - x^*) \right) \geq 0$$

or, using $y_\lambda = \frac{x - x_\lambda}{\lambda}$,

$$-(x_\lambda - x)^{\mathrm{T}} (x_\lambda - x^*) - \lambda y^{*\mathrm{T}} (x_\lambda - x^*) \geq 0,$$

which simplifies to

$$\|x_\lambda\|^2 \leq x^{\mathrm{T}}(x_\lambda - x^*) + x_\lambda^{\mathrm{T}} x^* - \lambda y^{*\mathrm{T}}(x_\lambda - x^*).$$

It holds that x_λ remains bounded for $\lambda \to 0$ and that $\mathrm{dom}\,\mathcal{F}$ is dense in $\mathbb{R}^n$ [22]. Consider a sequence $\lambda_n \to 0$ such that $x_{\lambda_n} \to x_0$. Using this sequence, we obtain the inequality

$$\|x_0\|^2 \leq x^{\mathrm{T}}(x_0 - x^*) + x_0^{\mathrm{T}} x^* \qquad \forall x^* \in \mathrm{dom}(\mathcal{F}).$$

In particular, for $x_1^*, x_2^* \in \mathrm{dom}(\mathcal{F})$ it holds that

$$\|x_0\|^2 \leq x^{\mathrm{T}}(x_0 - x_1^*) + x_0^{\mathrm{T}} x_1^*$$
$$\|x_0\|^2 \leq x^{\mathrm{T}}(x_0 - x_2^*) + x_0^{\mathrm{T}} x_2^*$$

A convex combination $x_\alpha^* = \alpha x_1^* + (1 - \alpha)x_2^*$ for all $\alpha \in [0, 1]$ gives

$$\|x_0\|^2 \leq x^{\mathrm{T}}(x_0 - x_\alpha^*) + x_0^{\mathrm{T}} x_\alpha^*$$

which is equivalent to $\|x_0\|^2 \leq x^{\mathrm{T}}(x_0 - x_\alpha^*) + x_0^{\mathrm{T}} x_\alpha^*$ for all $x_\alpha^* \in C := \overline{\mathrm{co}}\,\mathrm{dom}(\mathcal{F})$, in which we exploited the fact that the inequality contains its boundary. A further reformulation gives $(x - x_0)^{\mathrm{T}}(x_\alpha^* - x_0) \leq 0$ for all $x_\alpha^* \in C$, from which we deduce that $x_0 = \mathrm{prox}_C(x)$ (see Definition 2.31). It therefore follows that $x_\lambda \to \mathrm{prox}_C(x)$ for $\lambda \downarrow 0$. Taking arbitrary values of $x \in \mathbb{R}^n$, we immediately see that $x_\lambda \in \mathrm{dom}(\mathcal{F})$ can attain every point in C. Consequently, it must hold that $C = \overline{\mathrm{dom}}(\mathcal{F})$. $\qquad \square$

The notion of semi-continuity of functions can be extended to set-valued functions [37].

Definition 2.19 (Upper Semi-continuity of Set-valued Functions).
A set-valued function $\mathcal{F}(x)$ is upper semi-continuous in x if

$$\lim_{y \to x} \left(\sup_{a \in \mathcal{F}(y)} \inf_{b \in \mathcal{F}(x)} \|a - b\| \right) \to 0.$$

This condition is equivalent to the condition that for all $\varepsilon > 0$ there exists a $\delta > 0$ such that $\|x - y\| < \delta \Rightarrow \mathcal{F}(y) \subset \mathcal{F}(x) + B_\varepsilon$. If a function $\mathcal{F}$ is set-valued at a distinct x, then the graph of the function in a neighbourhood around x is connected to the set $\mathcal{F}(x)$. Upper semi-continuity does not imply convexity of the image.

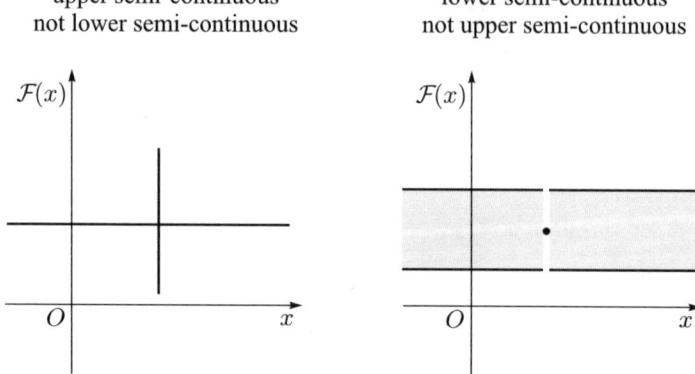

Fig. 2.7. Upper semi-continuity and lower semi-continuity of a set-valued function [12].

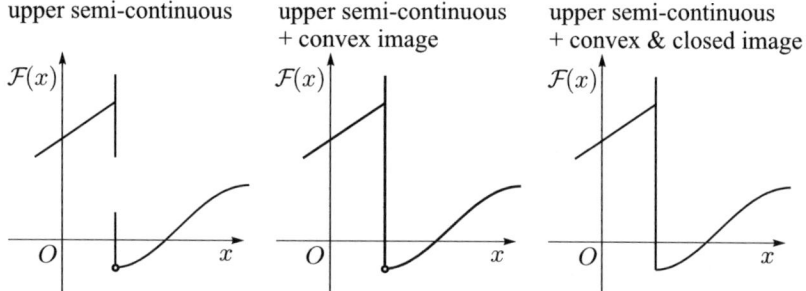

Fig. 2.8. Upper semi-continuity, convexity and closedness of a set-valued function.

Definition 2.20 (Lower Semi-continuity of Set-valued Functions).
A set-valued function $\mathcal{F}(x)$ is lower semi-continuous in x if

$$\lim_{y \to x} \left(\inf_{a \in \mathcal{F}(y)} \sup_{b \in \mathcal{F}(x)} \|a - b\| \right) \to 0.$$

This condition is equivalent to the condition that for all $\varepsilon > 0$ there exists a $\delta > 0$ such that $\|x - y\| < \delta \Rightarrow \mathcal{F}(x) \subset \mathcal{F}(y) + B_\varepsilon$. A finite-valued function $\mathcal{F}(x)$, that is both upper and lower semi-continuous is continuous. Upper and lower semi-continuity of set-valued functions are sometimes called outer and inner semi-continuity [77].

The difference between upper and lower semi-continuity is illustrated in Figure 2.7 and the notions of upper semi-continuity, convexity and closedness of (the images of) set-valued functions are illustrated in Figure 2.8. Left and right limits, which do not agree with the function value, are depicted as circles (∘) in Figure 2.8.

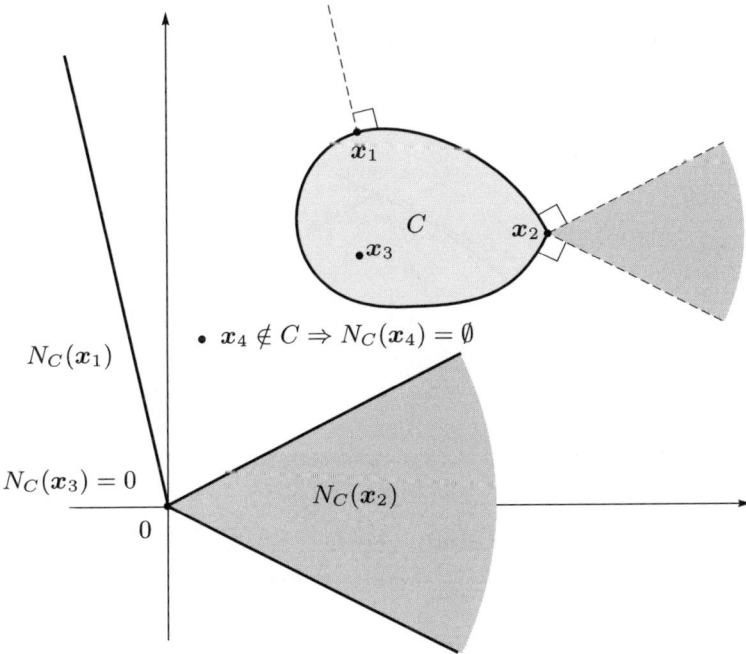

Fig. 2.9. Normal cone at different points of a convex set [57].

2.5 Definitions from Convex Analysis

In this section, the basic concepts of normal cone, indicator function, support function, proximal point and distance are introduced. We will use the notion of the subdifferential to reveal the relation between the indicator function and the normal cone.

A cone is a subset of $\mathbb{R}^n$ consisting of rays (half lines emanating from the origin).

Definition 2.21 (Cone). *A subset $C \subset \mathbb{R}^n$ is called a cone if for any $x \in C$ and $\lambda > 0$ also $\lambda x \in C$ holds. A cone is convex if the subset $C \subset \mathbb{R}^n$ is convex.*

A vector y is a normal vector of C at x if y does not make an acute angle with every line segment in C starting from x.

Definition 2.22 (Normal Vector). *Let $C \subset \mathbb{R}^n$ be a convex set and $x \in C$. A vector $y \in \mathbb{R}^n$ is a normal vector of C at $x \in \mathbb{R}^n$ if*

$$y^{\mathrm{T}}(x^* - x) \leq 0, \quad x \in C, \forall x^* \in C.$$

The normal cone of a set C at x is the set of rays that are normals of C at x (Figure 2.9).

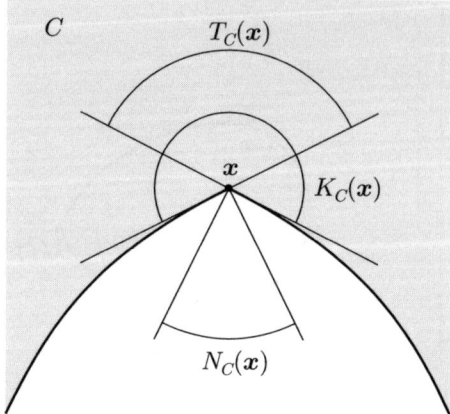

Fig. 2.10. Tangent cone and contingent cone to a non-convex set C [63].

Definition 2.23 (Normal Cone). *Let $C \subset \mathbb{R}^n$ be a convex set and $x \in C$. The set of vectors $y \in \mathbb{R}^n$ that are normal vectors of C at $x \in C$ form the normal cone of C in x*

$$N_C(x) = \{y \mid y^T(x^* - x) \leq 0, \quad x \in C, \forall x^* \in C\}.$$

If x is in the interior of C then $N_C(x) = 0$. If $x \notin C$ then $N_C(x) = \emptyset$.

Definition 2.24 (Polar Cone). *Let $K \subset \mathbb{R}^n$ be a convex cone. The set of normal vectors y to 0 with respect to K form the polar cone of K*

$$K^\circ = \{y \mid x^T y \leq 0, \forall x \in K\}.$$

It therefore holds that $K^\circ = N_K(0)$.

There exist many ways in which one can express the concept of tangency to a set C at a point $x \in C$. This leads to various cones such as the tangent cone and contingent[1] cone [12]. The contingent cone is the cone which touches the boundary of C with its sides (see Figure 2.10).

Definition 2.25 (Contingent Cone [148]). *Let C be a closed set and let x be a point in C. The contingent cone is defined as*

$$K_C(x) = \{y \mid \exists t_k \downarrow 0, \ y_k \rightarrow y, \ with \ x + t_k y_k \in C\}.$$

Loosely speaking, the contingent cone is the cone which one obtains if one zooms in on the point x after a shift of the set C such that x is at the origin. The contingent cone, which is also called the 'tangent cone' of Bouligand, is closed but not necessarily convex. If C is convex, then the contingent cone agrees with the tangent cone.

---

[1] from the Latin *contingere*, which means to touch on all sides.

Definition 2.26 (Tangent Cone [148]). *Let C be a closed set and let $\boldsymbol{x}$ be a point in C. The tangent cone is defined as*

$$T_C(\boldsymbol{x}) = \{\boldsymbol{y} \mid \forall t_k \downarrow 0,\, \boldsymbol{x}_k \to \boldsymbol{x} \text{ with } \boldsymbol{x}_k \in C,\, \exists \boldsymbol{y}_k \to \boldsymbol{y},\, \text{with } \boldsymbol{x} + t_k \boldsymbol{y}_k \in C\}.$$

The tangent cone, which is sometimes called the circatangent cone or tangent cone of Clarke [12], is a closed convex cone and a subset of the contingent cone, i.e. $T_C(\boldsymbol{x}) \subset K_C(\boldsymbol{x})$. The normal cone to an arbitrary set C, which is not necessarily convex, can be defined as $N_C(\boldsymbol{x}) = T_C^\circ(\boldsymbol{x})$.

One often needs to distinguish between points which belong to a set and points which are outside. A useful tool to describe this distinction is the indicator function.

Definition 2.27 (Indicator Function). *Let C be a set. The indicator function of C is defined as*

$$\Psi_C(\boldsymbol{x}) = \begin{cases} 0, & \boldsymbol{x} \in C, \\ +\infty, & \boldsymbol{x} \notin C. \end{cases}$$

The indicator function is a convex function if C is convex. With the Definition 2.13 of the subdifferential and the indicator function it follows for a convex set C that

$$\begin{aligned}
\partial \Psi_C(\boldsymbol{x}) &= \{\boldsymbol{y} \mid \Psi_C(\boldsymbol{x}^*) \geq \Psi_C(\boldsymbol{x}) + \boldsymbol{y}^\mathrm{T}(\boldsymbol{x}^* - \boldsymbol{x}),\quad \boldsymbol{x} \in C, \forall \boldsymbol{x}^*\} \\
&= \{\boldsymbol{y} \mid 0 \geq \boldsymbol{y}^\mathrm{T}(\boldsymbol{x}^* - \boldsymbol{x}),\quad \boldsymbol{x} \in C, \forall \boldsymbol{x}^* \in C\}.
\end{aligned} \tag{2.26}$$

This is exactly the definition of the normal cone at C. The subdifferential of the indicator function at $\boldsymbol{x} \in C$ is therefore the normal cone of C at $\boldsymbol{x}$,

$$\partial \Psi_C(\boldsymbol{x}) = N_C(\boldsymbol{x}). \tag{2.27}$$

Definition 2.28 (Conjugate Function). *Let f be a convex function. The function f^* is called the conjugate function of f and is defined as*

$$f^*(\boldsymbol{x}^*) = \sup_{\boldsymbol{x}} \{\boldsymbol{x}^\mathrm{T} \boldsymbol{x}^* - f(\boldsymbol{x})\}.$$

The conjugate f^* of a convex function is again convex. If we take the conjugate of the conjugate function f^*, then we retrieve the original function.

Theorem 2.29 (Fenchel-Moreau). *If f is a lower semi-continuous convex function, then it holds that $f^{**} = f$.*

Proof: See Theorem 11.1 in [149] and Theorem 4.4.2 in [11].   □

It therefore holds that

$$f(\boldsymbol{x}) = \sup_{\boldsymbol{x}^*} \{\boldsymbol{x}^\mathrm{T} \boldsymbol{x}^* - f^*(\boldsymbol{x}^*)\}. \tag{2.28}$$

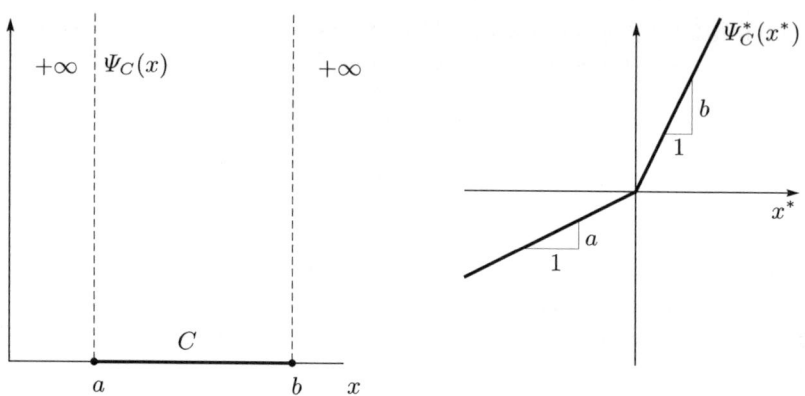

Fig. 2.11. Indicator function and support function [57].

From (2.28) follows Fenchel's inequality

$$\boldsymbol{x}^{\mathrm{T}}\boldsymbol{x}^* \leq f(\boldsymbol{x}) + f^*(\boldsymbol{x}^*). \tag{2.29}$$

The equality holds when $\boldsymbol{x}^*$ is a subgradient of f at $\boldsymbol{x}$, i.e.

$$\boldsymbol{x}^{\mathrm{T}}\boldsymbol{x}^* = f(\boldsymbol{x}) + f^*(\boldsymbol{x}^*) \Longleftrightarrow \boldsymbol{x}^* \in \partial f(\boldsymbol{x}) \Longleftrightarrow \boldsymbol{x} \in \partial f^*(\boldsymbol{x}^*). \tag{2.30}$$

Of special interest is the conjugate of the indicator function, which is called the support function.

Definition 2.30 (Support Function). *Let C be a convex set. The conjugate function of the indicator function Ψ_C is called support function,*

$$\Psi_C^*(\boldsymbol{x}^*) = \sup_{\boldsymbol{x}}\{\boldsymbol{x}^{\mathrm{T}}\boldsymbol{x}^* - \Psi_C(\boldsymbol{x})\}$$

$$= \sup_{\boldsymbol{x}}\{\boldsymbol{x}^{\mathrm{T}}\boldsymbol{x}^* \mid \boldsymbol{x} \in C\}.$$

The support function is positively homogeneous in the sense that

$$\Psi_C^*(a\boldsymbol{x}^*) = a\Psi_C^*(\boldsymbol{x}^*) \quad \forall a > 0. \tag{2.31}$$

Using (2.30), we see that if $\boldsymbol{x} \in \partial\Psi_C^*(\boldsymbol{x}^*)$ then it holds that

$$\boldsymbol{x}^{\mathrm{T}}\boldsymbol{x}^* = \Psi_C(\boldsymbol{x}) + \Psi_C^*(\boldsymbol{x}^*). \tag{2.32}$$

However, $\boldsymbol{x} \in \partial\Psi_C^*(\boldsymbol{x}^*)$ implies $\boldsymbol{x} \in C$ and $\Psi_C(\boldsymbol{x}) = 0$. Hence, equation (2.32) simplifies to

$$\boldsymbol{x}^{\mathrm{T}}\boldsymbol{x}^* = \Psi_C^*(\boldsymbol{x}^*). \tag{2.33}$$

Moreover, applying (2.30) on the function $\Psi_C(\boldsymbol{x})$ with $\boldsymbol{x}^* = \boldsymbol{0}$, we obtain

$$\boldsymbol{0} \in \partial\Psi_C(\boldsymbol{x}) \Longleftrightarrow \boldsymbol{x} \in \partial\Psi_C^*(\boldsymbol{0}) \tag{2.34}$$

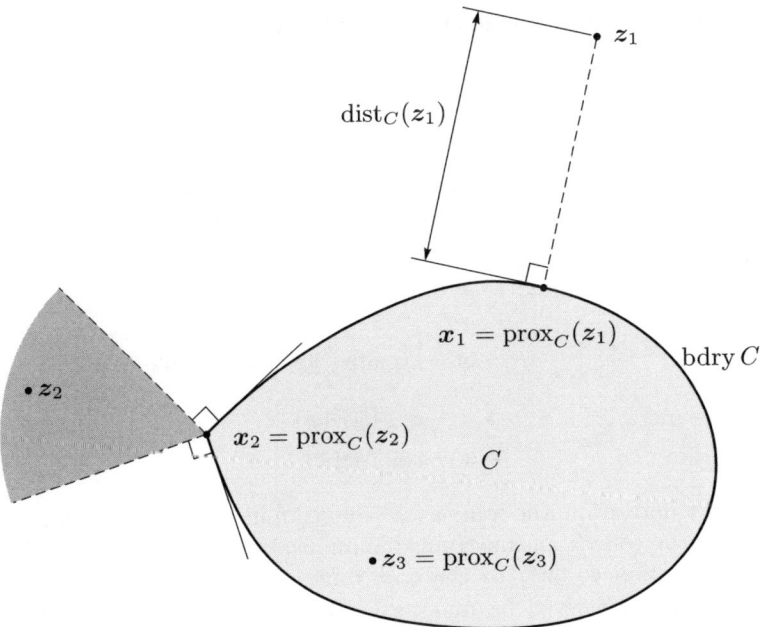

Fig. 2.12. Normal cone, proximal point and distance to a convex set, $z_1, z_2 \notin C$, $z_3 \in C$.

or

$$0 \in N_C(x) \Longleftrightarrow x \in \partial \Psi_C^*(0) \qquad (2.35)$$

and therefore

$$x \in C \Longleftrightarrow x \in \partial \Psi_C^*(0). \qquad (2.36)$$

Consequently, it follows that $\partial \Psi_C^*(0) = C$. The support function $\Psi_C^*(x^*)$ is a convex function with $\Psi_C^*(0) = 0$. Hence, if $0 \in C$, then $0 \in \partial \Psi_C^*(0)$ from which follows that $\Psi_C^*(x^*)$ attains a minimum at $x^* = 0$, i.e.

$$0 \in C \Longrightarrow \Psi_C^*(x^*) \geq 0. \qquad (2.37)$$

Moreover, if $0 \subset \text{int } C$ then it follows that $\Psi_C^*(x^*)$ attains a global minimum at $x^* = 0$, i.e.

$$0 \in \text{int } C \Longrightarrow \Psi_C^*(x^*) > 0 \ \forall x^* \neq 0. \qquad (2.38)$$

Definition 2.31 (Proximal Point). *The proximal point of a closed convex set C to a point z is the closest point in C to z:*

$$\text{prox}_C(z) = \underset{x^* \in C}{\text{argmin}} \, \|z - x^*\|, \quad z \in \mathbb{R}^n. \qquad (2.39)$$

Let $x = \mathrm{prox}_C(z)$ then it holds that $x \in C$ and

$$\begin{cases} x = z, & \text{if } z \in C, \\ x \in \mathrm{bdry}\, C, & \text{if } z \notin C, \end{cases} \tag{2.40}$$

where $\mathrm{bdry}\, C$ denotes the boundary of C. The vector $z - \mathrm{prox}_C(z)$ is an element of the normal cone of C at the proximal point:

$$x = \mathrm{prox}_C(z) \Leftrightarrow x = \underset{x^* \in C}{\mathrm{argmin}}\, \frac{1}{2}\|z - x^*\|^2$$

$$\Leftrightarrow x = \underset{x^*}{\mathrm{argmin}}\, \frac{1}{2}\|z - x^*\|^2 + \Psi_C(x^*) \tag{2.41}$$

$$\Leftrightarrow z - x \in \partial \Psi_C(x)$$

$$\Leftrightarrow z - x \in N_C(x).$$

In the above derivation, the convex constrained minimisation problem is transformed into a convex unconstrained minimisation problem by adding the indicator function $\Psi_C(x^*)$ to the cost function. Subsequently, the stationarity condition is derived by making use of the subdifferential of the indicator function which equals the normal cone. Note that the solution of a convex minimisation problem is unique and the implication therefore holds in both directions. Furthermore, by substituting $z = x - ry$, and using $-y \in N_C(x) \Leftrightarrow -ry \in N_C(x)$ for $r > 0$ we obtain

$$x = \mathrm{prox}_C(x - ry),\ r > 0 \quad \Longleftrightarrow \quad -y \in N_C(x). \tag{2.42}$$

Definition 2.32 (Distance). *The distance from a point $z \in \mathbb{R}^n$ to a closed convex set C is the Euclidian distance from the point to its proximal point in C*

$$\mathrm{dist}_C(z) = \|z - \mathrm{prox}_C(z)\|. \tag{2.43}$$

It therefore holds that $\mathrm{prox}_C(x) = x$ and $\mathrm{dist}_C(x) = 0$ for all $x \in C$.

Proposition 2.33. *Let C be a closed convex set. It holds that*

$$\nabla \frac{1}{2}\, \mathrm{dist}_C^2(z) = z - \mathrm{prox}_C(z). \tag{2.44}$$

Proof: Consider the convex function $f(z) = \frac{1}{2}\,\mathrm{dist}_C^2(z) = \frac{1}{2}\|z - \mathrm{prox}_C(z)\|^2$. Definition 2.13 of the subdifferential applied on $f(z)$ yields the condition

$$\partial f(z)^{\mathrm{T}}(z^* - z) \le f(z^*) - f(z), \quad \forall z^*. \tag{2.45}$$

Moreover, we deduce from Definition 2.31 the result

$$\|z^* - \mathrm{prox}_C(z^*)\| = \min_{z^0 \in C} \|z^* - z^0\| \le \|z^* - z^0\|, \quad z^0 \in C. \tag{2.46}$$

We now take any point z such that $z^0 = \text{prox}_C(z) \in C$ in (2.46) and substitute the result in (2.45)

$$
\begin{aligned}
\partial f(z)^{\mathrm{T}}(z^* - z) &\le \frac{1}{2}\|z^* - \text{prox}_C(z^*)\|^2 - \frac{1}{2}\|z - \text{prox}_C(z)\|^2 \\
&\le \frac{1}{2}\|z^* - \text{prox}_C(z)\|^2 - \frac{1}{2}\|z - \text{prox}_C(z)\|^2 \\
&= \frac{1}{2}\|z^* - z + z - \text{prox}_C(z)\|^2 - \frac{1}{2}\|z - \text{prox}_C(z)\|^2 \\
&= \frac{1}{2}\|z^* - z\|^2 + (z^* - z)^{\mathrm{T}}(z - \text{prox}_C(z)), \quad \forall z^*.
\end{aligned}
\tag{2.47}
$$

Dividing by $\|z^* - z\|$ for $z^* \ne z$ yields

$$
\partial f(z)^{\mathrm{T}} \frac{z^* - z}{\|z^* - z\|} \le \frac{1}{2}\|z^* - z\| + (z - \text{prox}_C(z))^{\mathrm{T}} \frac{(z^* - z)}{\|z^* - z\|}, \quad \forall z^* \ne z. \tag{2.48}
$$

Consider an arbitrary sequence z_n^* with $\|z_n^* - z\| \xrightarrow{n \to \infty} 0$ and denote the following limit as

$$
e := \lim_{n \to \infty} \frac{(z_n^* - z)}{\|z_n^* - z\|}. \tag{2.49}
$$

Applying the sequence z_n^* to (2.48) and taking the limit $n \to \infty$ yields

$$
\partial f(z)^{\mathrm{T}} e \le (z - \text{prox}_C(z))^{\mathrm{T}} e, \quad \forall e \ne 0. \tag{2.50}
$$

Consequently, it must hold that $\partial f(z) = z - \text{prox}_C(z)$. $\qquad \square$

The concepts of proximal point, distance and normal cone are illustrated in Figure 2.12.

2.6 Subderivative

The differentiability of a function $f : \mathbb{R}^n \to \mathbb{R}$ at a point x is connected with the existence of a tangent hyperplane to the graph of f at the point $(x, f(x))$ [148]. The concept of differentiability can be generalised by considering the contingent cone (Definition 2.25) to the epigraph of f instead. In this section, we consider a lower semi-continuous extended function $f : \mathbb{R}^n \to \mathbb{R} \cup \{-\infty, \infty\}$ whose domain $\text{dom}(f) = \{x \in \mathbb{R}^n \mid f(x) \ne \pm\infty\}$ is non-empty (i.e. the function is not trivial). The epigraph of the function f is closed, because f is lower semi-continuous. Various generalised notions of gradients exist, but the subderivative is the most natural object to focus on and is often called the contingent epiderivative [12] or epicontingent derivative [11].

Definition 2.34 (Subderivative [149]). *We define the function*

$$
df(x)(v) = \liminf_{t \downarrow 0, \, v' \to v} \frac{f(x + tv') - f(x)}{t}
$$

as the subderivative of f at x in the direction v.

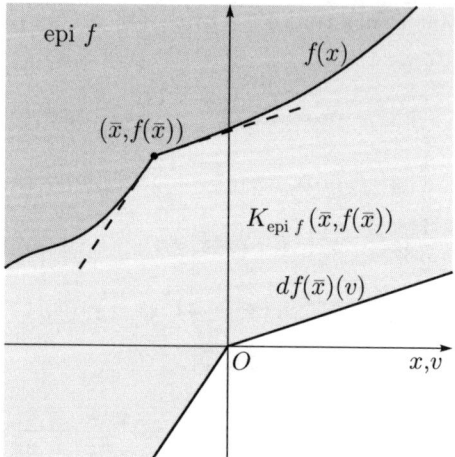

Fig. 2.13. Subderivative of f and its relation to the contingent cone.

The epigraph of $df(\boldsymbol{x})(\cdot)$, as a function of the second argument, is the contingent cone at the epigraph of f at $(\boldsymbol{x}, f(\boldsymbol{x}))$

$$\mathrm{epi}\, df(\boldsymbol{x})(\cdot) = K_{\mathrm{epi}\, f}\left((\boldsymbol{x}, f(\boldsymbol{x}))\right). \tag{2.51}$$

We observe that $df(\boldsymbol{x})(\boldsymbol{0}) = 0$. If f is differentiable at $\boldsymbol{x}$, then it holds that $df(\boldsymbol{x})(\boldsymbol{v}) = -df(\boldsymbol{x})(-\boldsymbol{v}) = (\nabla f(\boldsymbol{x}))^{\mathrm{T}}\boldsymbol{v}$ which agrees with the Lie derivative $L_{\boldsymbol{v}}f(\boldsymbol{x})$. The epigraph of the subderivative is a cone (the contingent cone) and the subderivative $df(\boldsymbol{x})(\cdot)$ is therefore positively homogeneous

$$df(\boldsymbol{x})(a\boldsymbol{v}) = a\, df(\boldsymbol{x})(\boldsymbol{v}), \qquad \forall a \geq 0. \tag{2.52}$$

If the function f is convex, then we can express the subdifferential (Definition 2.13) as

$$\partial f(\boldsymbol{x}) = \{\boldsymbol{y} \mid df(\boldsymbol{x})(\boldsymbol{v}) \geq \boldsymbol{v}^{\mathrm{T}}\boldsymbol{y}\}. \tag{2.53}$$

Of special interest is the subderivative of an indicator function $\Psi_C(\boldsymbol{x})$,

$$d\Psi_C(\boldsymbol{x})(\boldsymbol{v}) = \Psi_{K_C(\boldsymbol{x})}(\boldsymbol{v}), \tag{2.54}$$

where $K_C(\boldsymbol{x})$ is the contingent cone to C at the point $\boldsymbol{x}$.

2.7 Summary

An overwhelming amount of definitions and terminology has been introduced in this chapter. Yet, virtually all concepts of non-smooth analysis which have briefly been exposed here will be used in the following chapters. Set-valued functions will be used in Chapter 4 to construct (measure) differential inclusions. A maximal monotonicity property of (measure) differential inclusion will appear to lead to a contracting behaviour in those systems. Set-valued force laws to describe frictional unilateral contact will be cast in Chapter 5 as an inclusion to a normal cone on a convex set of admissible contact forces. The normal cone formulation of set-valued force laws brings the constitutive description in relation with non-smooth potential theory, in which indicator functions, conjugate functions and subdifferentials play an important role. Moreover, non-smooth potentials and their associated mathematical apparatus will be used in Chapters 6 and 7 to set up Lyapunov functions to prove stability properties of measure differential inclusions.

3

Measure and Integration Theory

The theory of measures forms the basis for the modern integration theory and both are quintessential for the understanding of measure differential inclusions which will be dealt with in Chapter 4. Moreau, who introduced the notion measure differential inclusions in [117–120, 122, 123], directly introduces the so-called differential measure. In Chapter 4, it will be explained in more detail that a measure differential inclusion actually describes how the differential measure of the state relates to the state and time in analogy with the fact that a (first-order) differential inclusion (or equation) describes how the time derivative of the state depends on the state and time. In this chapter we give a brief overview of measure and integration theory and relate the differential measure used by Moreau to real measures and (signed) measures. The theory presented in this chapter is based on standard textbooks about measure and integration theory [47,150], various publications of Moreau and others [63,116, 122,124] as well as the work of P. Ballard communicated through lecture notes of a summer course on Non-smooth Dynamical Systems (2003, Praz-sur-Arly).

This chapter may be very demanding and a reader on the verge of despair should keep the key result of this chapter in mind: every function of locally bounded variation can be decomposed in an absolutely continuous function, a step function and a singular function. As will appear in Chapter 4, solutions of measure differential inclusions are defined as functions of locally bounded variation. The latter fact clearly implies that the solutions of measure differential inclusions may include the contribution of step functions, which can account for jumps in the state evolution (such as e.g. velocity jumps in mechanical systems with impacts).

3.1 Measures

The theory of measures associates with subsets $A, B \ldots$ of a set M nonnegative numbers $\mu(A)$, $\mu(B)$, $\ldots$ We can think of $\mu(A)$ as the 'volume' or 'mass' of A and $\mu(A)$ is called the measure of A. We now recall what we understand by a σ-algebra and we define a measure μ on a σ-algebra $\mathfrak{A}$:

Definition 3.1 (σ-Algebra). *Let M be a set. A non-empty system $\mathfrak{A}$ of subsets of M is called a σ-algebra if and only if:*

1. *$M \in \mathfrak{A}$ and $\emptyset \in \mathfrak{A}$,*
2. *$A, B \in \mathfrak{A}$ implies $A \cup B \in \mathfrak{A}$, $A \cap B \in \mathfrak{A}$ and $A \backslash B \in \mathfrak{A}$,*
3. *$A_1, A_2 \cdots \in \mathfrak{A}$ implies $\bigcup\limits_{k=1}^{\infty} A_k \in \mathfrak{A}$ and $\bigcap\limits_{k=1}^{\infty} A_k \in \mathfrak{A}$.*

Definition 3.2 (Measure). *Let $\mathfrak{A}$ be a σ-algebra. A map $\mu : \mathfrak{A} \to [0, \infty]$, which maps each $A \in \mathfrak{A}$ to a non-negative number $\mu(A)$, such that $\mu(\emptyset) = 0$, is called a measure μ if*

1. *$\mu(A \cup B) = \mu(A) + \mu(B)$ if A and B are disjoint subsets, i.e. $A \cap B = \emptyset$.*
2. *$\mu\left(\bigcup\limits_{k=1}^{\infty} A_k\right) = \sum\limits_{k=1}^{\infty} \mu(A_k)$ if $A_1, A_2 \ldots$ are disjoint.*

This definition allows $\mu(A) = \infty$ for some $A \in \mathfrak{A}$. For $a \in \mathbb{R}$ we define: $-\infty < a < \infty$, $\infty + a = \infty$ and $\infty - a = \infty$. An element $\{a\}$ for which $\mu(\{a\}) > 0$ is called an atom[1] of μ. A measure μ without any atoms is called diffuse. The sets in $\mathfrak{A}$, for which we can define a measure μ, are called measurable. Moreover, we call $(\mu, M, \mathfrak{A})$ a measure space. A measurable set M is called a (μ-)null-set if $\mu(M) = 0$. A measure μ is called finite if $\mu(M) < \infty$ for all $M \in \mathfrak{A}$ and is called σ-finite if for all $X \in \mathfrak{A}$ there exists a sequence of disjoint sets $A_n \in \mathfrak{A}$ such that $X = \bigcup_{n=1}^{\infty} A_n$ and $\mu(A_n) < \infty$. In the following, we will mostly consider measures on subsets of the real line $\mathbb{R}$.

The most elementary measure is the Lebesgue measure on $\mathbb{R}$, which assigns the value $b - a$ to the set $(a, b]$, i.e. $\mu((a, b]) = b - a$ for $b \geq a$. The Lebesgue measure is defined for half-open intervals, because every half-open interval $(a, c]$ can be split in two other half-open intervals $(a, b]$ and $(b, c]$, $a \leq b \leq c$, such that

$$\mu((a, c]) = \mu((a, b]) + \mu((b, c]), \tag{3.1}$$

but the same holds of course for open and closed intervals because the Lebesgue measure of a single element is zero, i.e. $\mu(\{a\}) = 0$ (i.e. the Lebesgue measure is diffuse). A set M, with $\mu(M) = 0$ and therefore a null-set with respect to the Lebesgue measure, is called Lebesgue negligible.

Of much importance are measures which measure the variation of functions on intervals. Let $f : \mathbb{R} \to \mathbb{R}$ be a nondecreasing function of locally bounded variation, i.e. $f : \text{lbv}(I, \mathbb{R})$ with $I \subset \mathbb{R}$. Then f has an at most countable[2] number of discontinuities on I and for each $t \in I$ we can define $f^+(t)$ and $f^-(t)$ (see (2.10)). Let $\mathfrak{A}$ be a σ-algebra on I. We now define the Lebesgue-Stieltjes measure of f as the measure $\mu_f : \mathfrak{A} \to [0, \infty]$ for which holds that

---

[1] The word atom comes from the Greek ἄτομον, which means 'that which is indi-visible/uncuttable'.

[2] A set $\mathcal{M}$ is *at most countable* if there exists an injective map $f : \mathcal{M} \to \mathbb{N}$.

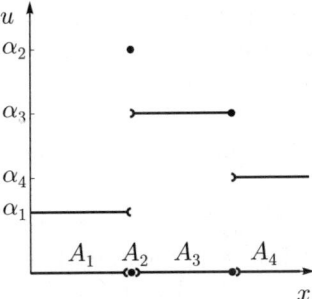

Fig. 3.1. Step function.

$$\begin{aligned}
\mu_f([a,b]) &= f^+(b) - f^-(a),\\
\mu_f((a,b]) &= f^+(b) - f^+(a),\\
\mu_f([a,b)) &= f^-(b) - f^-(a),\\
\mu_f((a,b)) &= f^-(b) - f^+(a),
\end{aligned} \qquad (3.2)$$

with $[a,b] \subset I$. If f is discontinuous at a, then μ_f has an atom at $\{a\}$, i.e.

$$\mu_f(\{a\}) = f^+(a) - f^-(a) > 0. \qquad (3.3)$$

The Lebesgue measure is a special case of the Lebesgue-Stieltjes measure μ_f with $f(x) = x$ and is, as mentioned above, a diffuse measure.

3.2 The Lebesgue Integral

The characteristic function $\chi : I \to \mathbb{R}$ of a set $A \subset I$ is defined as

$$\chi_A(x) = \begin{cases} 1 & x \in A,\\ 0 & x \in I\backslash A. \end{cases} \qquad (3.4)$$

A function $u : I \to \mathbb{R}$, which takes a finite number of values on the domain I is called a step function. Let $u : I \to \mathbb{R}^+$ be a non-negative step function such that

$$u(x) = \sum_{j=1}^{m} \alpha_j \chi_{A_j}(x), \qquad (3.5)$$

with $\alpha_j \geq 0$ and $A_j \in \mathfrak{A}$ for $j = 1 \ldots m$ (see Figure 3.1).

The Lebesgue integral of u over I is defined as

$$\int_I u\,d\mu := \sum_{j=1}^{m} \alpha_j \mu(A_j), \qquad (3.6)$$

where $\mu(A_j)$ denotes the Lebesgue measure of the set A_j. Every function of bounded variation can be approximated by a step function. Let $f^+ \in$

$\mathrm{lbv}(I, \mathbb{R}^+)$ be a non-negative function of bounded variation, such that it is the uniform limit of non-negative step functions, i.e. $u_n \uparrow f^+$ for $n \to \infty$. We define the Lebesgue integral of f^+ over I as

$$\int_I f^+ \mathrm{d}\mu := \lim_{n \to \infty} \int_I u_n \mathrm{d}\mu. \tag{3.7}$$

Using the decomposition $f = f^+ - f^-$ with $f^+ = \max(f, 0)$ and $f^- = \max(-f, 0)$ we define the Lebesgue integral of an arbitrary function of locally bounded variation $f \in \mathrm{lbv}(I, \mathbb{R})$ over I as

$$\int_I f \mathrm{d}\mu := \int_I f^+ \mathrm{d}\mu - \int_I f^- \mathrm{d}\mu. \tag{3.8}$$

The function f is called quasi-integrable if f is measurable and if at least one of the integrals $\int_I f^+ \mathrm{d}\mu$ and $\int_I f^- \mathrm{d}\mu$ is finite. The Lebesgue integral (3.8) can be applied to arbitrary measures μ. If we plug the Lebesgue-Stieltjes measure μ_f in the integral (3.8), then we obtain the Lebesgue-Stieltjes integral.

3.3 Signed Measures

Signed measures, as opposed to measures, can also take negative values.

Definition 3.3 (Signed Measure). *Let $\mathfrak{A}$ be a σ-algebra. A map $\nu : \mathfrak{A} \to (-\infty, \infty]$, which maps each $A \in \mathfrak{A}$ to a real number $\nu(A)$, such that $\nu(\emptyset) = 0$, is called a signed measure ν if*

1. $\nu(A \cup B) = \nu(A) + \nu(B)$ if A and B are disjoint subsets, i.e. $A \cap B = \emptyset$.

2. $\nu\left(\bigcup_{k=1}^{\infty} A_k\right) = \sum_{k=1}^{\infty} \nu(A_k)$ if $A_1, A_2 \ldots$ are disjoint.

For instance, let $\nu : \mathfrak{A} \to (-\infty, \infty]$ be a signed measure, which maps intervals to a real number and let $[a, c] \in \mathfrak{A}$ with $a \leq b \leq c$. Then the application of a signed measure ν on an interval can be decomposed in the application of ν on disjoint subsets:

$$\begin{aligned} \nu([a, b]) &= \nu(\{a\} + \nu((a, b)) + \nu(\{b\}), \\ \nu((a, c)) &= \nu((a, b)) + \nu(\{b\}) + \nu((b, c)), \\ \nu((a, b]) &= \nu([a, b]) - \nu(\{a\}), \\ \nu([a, b)) &= \nu([a, b]) - \nu(\{b\}). \end{aligned} \tag{3.9}$$

If $\nu : \mathfrak{A} \to (-\infty, \infty]$ is a signed measure, then the set $P \in \mathfrak{A}$ is called ν-positive, if $\nu(A) \geq 0$ for all $A \in \mathfrak{A}$ with $A \subset P$. Similarly, the set $N \in \mathfrak{A}$ is called ν-negative, if $\nu(A) \leq 0$ for all $A \in \mathfrak{A}$ with $A \subset N$. The set $Q \in \mathfrak{A}$ is called a ν-null-set if $\nu(A) = 0$ for all $A \subset Q$. The disjoint decomposition of ν

into a ν-positive and a ν-negative set (called Hahn decomposition) is unique apart from a ν-null-set. The measures $\nu^+, \nu^- : \mathfrak{A} \to \overline{\mathbb{R}}$, with $\overline{\mathbb{R}} := \mathbb{R} \cup \{+\infty\}$,

$$\nu^+(A) := \nu(A \cap P), \qquad \nu^-(A) := -\nu(A \cap N) \tag{3.10}$$

are called the positive and negative variation of ν and lead to the measure $|\nu| : \mathfrak{A} \to \overline{\mathbb{R}}$

$$|\nu|(A) := \nu^+(A) + \nu^-(A), \tag{3.11}$$

which is called the variation of ν. At least one of the measures ν^+ and ν^- is finite and it holds that

$$\nu(A) = \nu^+(A) - \nu^-(A). \tag{3.12}$$

Every function $f \in \mathrm{lbv}(I, \mathbb{R})$ of locally bounded variation equals the difference of two nondecreasing functions f_p and f_n such that

$$f = f_p - f_n. \tag{3.13}$$

Using the Lebesgue-Stieltjes measures μ_{f_p} and μ_{f_n}, we can therefore define the signed Lebesgue-Stieltjes measure ν_f as

$$\nu_f(A) = \mu_{f_p}(A) - \mu_{f_n}(A), \tag{3.14}$$

which gives the Hahn decomposition $\nu_f^+(A) = \mu_{f_p}(A)$ and $\nu_f^-(A) = \mu_{f_n}(A)$. Consequently, the relations (3.2) also hold for the signed Lebesgue-Stieltjes measure:

$$\begin{aligned} \nu_f([a, b]) &= f^+(b) - f^-(a), \\ \nu_f((a, b]) &= f^+(b) - f^+(a), \\ \nu_f([a, b)) &= f^-(b) - f^-(a), \\ \nu_f((a, b)) &= f^-(b) - f^+(a). \end{aligned} \tag{3.15}$$

If $\mu : \mathfrak{A} \to [0, \infty]$ is a measure and $f : X \to \overline{\mathbb{R}}$ is quasiintegrable, then

$$\nu(A) := \int_A f \mathrm{d}\mu \tag{3.16}$$

is a signed measure, which we call the signed measure with density f with respect to μ denoted by $\nu = f \odot \mu$. Signed measures which obey a density with respect to some measure play an important role in integration theory. We now introduce the notions of continuity and orthogonality (singularity) of measures and come to the theorem of Radon-Nykodým:

Definition 3.4 (μ-Continuity). *Let μ and ν be signed measures on $\mathfrak{A}$. The signed measure ν is called μ-continuous, or absolutely continuous with respect to μ, if every μ-null-set is a ν-null-set, which we denote by $\nu \ll \mu$.*

Definition 3.5 (Orthogonality). *Two (signed) measures $\nu, \rho : \mathfrak{A} \to \overline{\mathbb{R}}$ are called orthogonal (or mutually singular), denoted $\nu \perp \rho$, if there exists a decomposition $X = A \cup B$, $A \cap B = \emptyset$, $A, B \in \mathfrak{A}$, such that A is a ν-null-set and B is a ρ-null-set.*

The theorem of Radon-Nykodým states that if ν is absolutely continuous with respect to μ, then there must exist a density function f such that (3.16) holds.

Theorem 3.6 (Radon-Nykodým). *Let μ be a σ-finite measure and $\nu \ll \mu$ be a signed measure on $\mathfrak{A}$. Then it holds that ν has a density function with respect to μ, i.e. there exists a quasiintegrable function $f : I \to \overline{\mathbb{R}}$ such that $\nu = f \odot \mu$.*

Proof: See Elstrodt [47], p. 279.   □

If μ and ν are two finite (signed) measures on $\mathfrak{A}$, then ν is not necessarily μ-continuous, i.e. ν does not necessarily have a density with respect to μ. Now we can try to separate from ν a part $f \odot \mu$, such that $\nu - f \odot \mu$ is as 'small' as possible.

Let μ be a finite measure and ν be a finite signed measure on $\mathfrak{A}$. Then there exists a unique decomposition

$$\nu = \nu_1 + \nu_2, \tag{3.17}$$

into two signed measures ν_1 and ν_2, such that $\nu_1 \ll \mu$ and $\nu_2 \perp \mu$. This decomposition is called the Lebesgue decomposition. The signed measure ν_1 is therefore μ-continuous and admits a density function f with respect to μ, i.e. $\nu_1 = f \odot \mu$. We now use the Lebesgue decomposition to decompose the signed Lebesgue-Stieltjes measure ν_f of a function f, being of locally bounded variation.

Theorem 3.7 (Lebesgue Decomposition of the Lebesgue-Stieltjes Measure). *For ever function $f \in \mathrm{lbv}(I, \mathbb{R})$ there exists a decomposition*

$$f = f_{\mathrm{abs}} + f_{\mathrm{s}} + f_{\mathrm{sing}} \tag{3.18}$$

in functions of locally bounded variation and a decomposition of the signed Lebesgue-Stieltjes measure

$$\nu_f = \nu_{\mathrm{abs}} + \nu_{\mathrm{s}} + \nu_{\mathrm{sing}} \tag{3.19}$$

in signed measures such that

1. *f_{abs} is absolutely continuous with respect to the Lebesgue measure μ:*

$$\nu_{\mathrm{abs}} \ll \mu, \qquad \nu_{\mathrm{abs}} = f'_{\mathrm{abs}} \odot \mu. \tag{3.20}$$

   The signed measure ν_{abs} is diffuse.
2. *f_{s} is a step function, which is constant almost everywhere and ν_{s} is a purely atomic signed measure.*
3. *f_{sing} is singular, i.e. ν_{sing} is a diffuse measure which is orthogonal to the Lebesgue measure μ, i.e. $\nu_{\mathrm{sing}} \perp \mu$. f_{sing} is continuous and constant almost everywhere.*

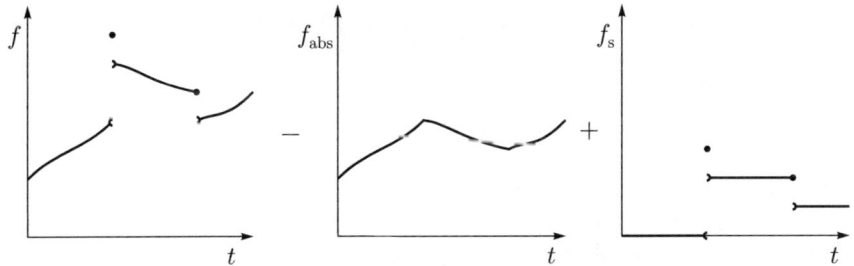

Fig. 3.2. Decomposition of f in $f_{\text{abs}} + f_{\text{s}}$ if $f_{\text{sing}} = 0$.

Proof: We decompose f in a continuous part $f_{\text{c}} = f_{\text{abs}} + f_{\text{sing}}$ and a step function f_{s}. Subsequently, we use the Lebesgue decomposition to decompose f_{c} in an absolutely continuous part and a singular part such that $\nu_{\text{abs}} \ll \mu$ and $\nu_{\text{sing}} \perp \mu$. □

The jumps in f are therefore solely described by the step function f_{s}

$$\nu_{\text{s}}(\{a\}) = f^{+}(a) - f^{-}(a) = \nu_f(\{a\}) \tag{3.21}$$

and $f' = f'_{\text{abs}}$ almost everywhere (see Figure 3.2). The Cantor function [47,63] is an example of a function which is purely singular.

3.4 Real Measures

Measures and signed measures have been introduced above as mappings from sets to the real numbers. We now introduce the so-called real measure which maps a function to a real number.

We denote by $C^0(I, \mathbb{R})$ the vector space of continuous functions on the real closed interval I, equipped with the norm of uniform convergence (supremum norm)

$$\|g\|_{C^0} = \sup_{t \in I} |g(t)|. \tag{3.22}$$

Definition 3.8 (Real Measure). *We call any linear form $v : C^0(I, \mathbb{R}) \to \mathbb{R}$ which is continuous, i.e.*

$$\exists M \in \mathbb{R} \quad \forall g \in C^0(I, \mathbb{R}) \quad |v(g)| \leq M \sup_{t \in I} |g(t)|, \tag{3.23}$$

a real measure on I.

The linearity of the linear form v implies that $v(g - h) = v(g) - v(h)$ for all $g, h \in C^0(I, \mathbb{R})$. Condition (3.23) states that

$$\exists L \in \mathbb{R} \quad \forall g, h \in C^0(I, \mathbb{R}) \quad |v(g) - v(h)| \leq L \sup_{t \in I} |g(t) - h(t)|, \tag{3.24}$$

which is a Lipschitz continuity condition. A real measure is therefore a Lipschitz continuous linear functional on the vector space $C^0(I, \mathbb{R})$. We use the following notation

$$v(g) =: \int_I g \, v, \qquad (3.25)$$

which involves artificially an integral sign and will be related in the next section to a true Lebesgue integral if v is a differential measure. A real measure is said to be positive if $v(g) \geq 0$ for all $g \in C^0(I, \mathbb{R}^+)$. For any function $g \in C^0(I, \mathbb{R})$ we define its positive part $g^+(t) = \max(g(t), 0)$ and negative part $g^-(t) = \max(-g(t), 0)$, with which we obtain the relations

$$g = g^+ - g^-, \qquad |g| = g^+ + g^-. \qquad (3.26)$$

Any function $g(t)$ can therefore be decomposed into two nonnegative functions $g^+(t)$ and $g^-(t)$. We now define the modulus measure $|v|$ for nonnegative functions $h(t) \in C^0(I, \mathbb{R}^+)$:

$$|v|(h) := \sup_{\substack{f \in C^0(I, \mathbb{R}) \\ |f(t)| \leq h(t)}} \int_I f \, v. \qquad (3.27)$$

It holds that $|v|(g^+) \geq v(g^+)$ and $|v|(g^-) \geq v(g^-)$. For instance, if $v = dt$, then $|v|(g^\pm) = v(g^\pm)$ and if $v = -dt$, then $|v|(g^\pm) = v(-g^\pm)$. The modulus measure $|v|$ for arbitrary functions $g(t) \in C^0(I, \mathbb{R})$ is defined as

$$|v|(g) := |v|(g^+) - |v|(g^-). \qquad (3.28)$$

The modulus measure $|v|$ is a positive real measure. If $v = dt$, then $|v|(g) = v(g)$ and if $v = -dt$, then $|v|(g) = v(-g)$. For a real measure v we define the positive real measures

$$v^+ = \frac{1}{2}(|v| + v), \qquad v^- = \frac{1}{2}(|v| - v). \qquad (3.29)$$

We now define the linear operators $\int_{(a,b)} v$ and $\int_{[a,b]} v$ on a real measure v. Using the decomposition

$$v = v^+ - v^-, \qquad (3.30)$$

which is known as the Jordan decomposition, we only need to define these linear operators for positive real measures. The main idea is to relate $\int_{(a,b)} v$ and $\int_{[a,b]} v$ to the already defined integral $v(g) =: \int_I g \, v$ through the characteristic functions $\chi_{(a,b)}$ and $\chi_{[a,b]}$ respectively. Unfortunately, we are not able to apply a real measure v on a characteristic function χ_A, because a characteristic function is not continuous. We therefore need to approach χ_A from below if A is an open interval or from above if A is a closed interval (see Figure 3.3). Let v be a positive real measure; then, we define

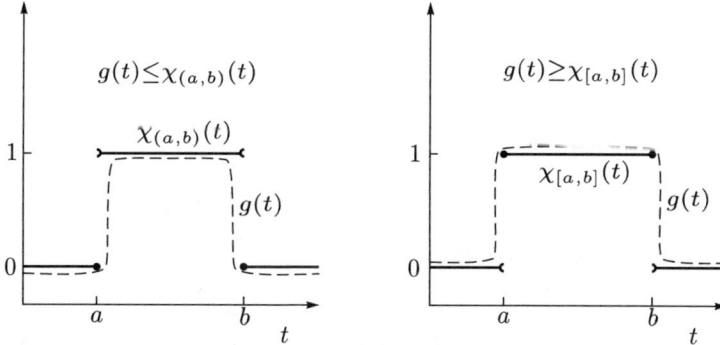

Fig. 3.3. Supremum for $g(t) \leq \chi_{(a,b)}(t)$ and infimum for $g(t) \geq \chi_{[a,b]}(t)$ with $g \in C^0(I, \mathbb{R})$.

$$\int_{(a,b)} v := \sup_{\substack{g \in C^0(I,\mathbb{R}) \\ g(t) \leq \chi_{(a,b)}(t) \, \forall t}} v(g), \qquad \int_{[a,b]} v := \inf_{\substack{g \in C^0(I,\mathbb{R}) \\ g(t) \geq \chi_{[a,b]}(t) \, \forall t}} v(g). \qquad (3.31)$$

If v is an arbitrary real measure, then the Jordan decomposition (3.30) yields

$$\int_A v = \int_A v^+ - \int_A v^- \qquad (3.32)$$

for any interval $A \in \mathfrak{A}$, which forms a signed measure $\nu : \mathfrak{A} \to \overline{\mathbb{R}}$

$$\nu(A) := \int_A v. \qquad (3.33)$$

Using the relations (3.9) for the decomposition of a signed measure on an interval in the sum of signed measures on disjoint subsets it follows that

$$\int_{[a,b]} v = \int_{\{a\}} v + \int_{(a,b)} v + \int_{\{b\}} v$$

$$\int_{(a,c)} v = \int_{(a,b)} v + \int_{\{b\}} v + \int_{(b,c)} v$$

$$\int_{(a,b]} v = \int_{[a,b]} v - \int_{\{a\}} v \qquad (3.34)$$

$$\int_{[a,b)} v = \int_{[a,b]} v - \int_{\{b\}} v.$$

If in turn the signed measure ν, which is associated with the real measure v, has a density function f with respect to a measure μ, i.e. $\nu = f \odot \mu$, then the real measure v is associated with a measure μ

$$\int_A v =: \nu(A) := \int_A f \mathrm{d}\mu. \qquad (3.35)$$

Consequently, if $\nu = f \odot \mu$ then we can write

$$v = fd\mu \tag{3.36}$$

and we call such a real measure v a differential measure with a density f with respect to μ.

3.5 Differential Measures

Let $\boldsymbol{f} : I \rightarrow \mathbb{R}^n$ be a vector function of locally bounded variation, $\boldsymbol{f} \in \mathrm{lbv}(I, \mathbb{R}^n)$ (i.e. each $f_i \in \mathrm{lbv}(I, \mathbb{R})$), and $g : I \rightarrow \mathbb{R}$ be a continuous function, $g \in C^0(I, \mathbb{R})$, with a compact support on the real interval I. Furthermore, let $[a, b]$ be a compact subinterval of I and $\mathcal{S}$ be the set of finite subsets of $[a, b]$. Consider a discretisation $S_m \in \mathcal{S}$ of $[a, b]$ with $m + 1$ points

$$S_m : a = t_0 < t_1 < t_2 \ldots t_{m-1} < t_m = b. \tag{3.37}$$

Between every two neighbouring points t_{i-1} and t_i ($1 \leq i \leq m$) of S_m we choose an intercalator $\vartheta_i^{S_m} \in [t_{i-1}, t_i]$. We now construct the Riemann-Stieltjes sum

$$\boldsymbol{H}(S_m, \vartheta^{S_m}, g, \boldsymbol{f}) = \sum_{i=1}^{m} g(\vartheta_i^{S_m}) \left(\boldsymbol{f}(t_i) - \boldsymbol{f}(t_{i-1}) \right). \tag{3.38}$$

It has been proven in [122] that the mapping $S_m \rightarrow \boldsymbol{H}(S_m, \vartheta^{S_m}, g, \boldsymbol{f})$ converges to a limit

$$\lim_{m \rightarrow \infty, S_m \in \mathcal{S}} \boldsymbol{H}(S_m, \vartheta^{S_m}, g, \boldsymbol{f}), \tag{3.39}$$

which is independent on the choice of the intercalator ϑ^{S_m}. This limit will be denoted by

$$\int_{[a,b]} g \, \mathrm{d}\boldsymbol{f} := \lim_{m \rightarrow \infty, S_m \in \mathcal{S}} \boldsymbol{H}(S_m, \vartheta^{S_m}, g, \boldsymbol{f}), \tag{3.40}$$

which is the Riemann-Stieltjes integral of g with respect to $\mathrm{d}\boldsymbol{f}$. The set of discontinuity points of $\boldsymbol{f}$ is Lebesgue negligible. The Riemann-Stieltjes integral on a bounded domain I agrees therefore with the Lebesgue-Stieltjes integral. The linear mapping

$$\mathrm{d}\boldsymbol{f} : g \rightarrow \int g \, \mathrm{d}\boldsymbol{f} \tag{3.41}$$

constitutes a real vector measure $\mathrm{d}\boldsymbol{f} : C^0(I, \mathbb{R}) \rightarrow \mathbb{R}^n$ on I. We call $\mathrm{d}\boldsymbol{f}$ the differential measure (or Stieltjes measure) of $\boldsymbol{f}$. For example, if $f(t) = t$, then the differential measure $\mathrm{d}f$ equals the (differential) Lebesgue measure $\mathrm{d}t$. Moreover, if f is the right-continuous unit step function

$$f(t) = \begin{cases} 0 & t < t_i, \\ 1 & t \geq t_i, \end{cases} \tag{3.42}$$

then the differential measure df equals the differential Dirac point measure $d\delta_{t_i}$, which is such that

$$\int_{[a,b]} \gamma(t)d\delta_{t_i} = \gamma(t_i), \quad \text{for } t_i \in [a,b], \tag{3.43}$$

or

$$\delta_{t_i}([a,b]) = \int_{[a,b]} d\delta_{t_i} = \begin{cases} 1 & t_i \in [a,b], \\ 0 & t_i \notin [a,b]. \end{cases} \tag{3.44}$$

Let $\boldsymbol{f}_s \in \mathrm{lbv}(I,\mathbb{R}^n)$ be a right-continuous step function with p discontinuities on the nodes τ_p with

$$a < \tau_1 < \tau_2 \ldots \tau_p < b. \tag{3.45}$$

The step function $\boldsymbol{f}_s(\tau)$ is constant for $t \neq \tau_j$. Now we choose the points t_i of $S \in \mathcal{S}$ such that every open interval (t_{i-1}, t_i) contains at most one node τ_j and every node falls in an open interval (t_{i-1}, t_i). As intercalator we choose $\bar{\vartheta} = t_i$. The only non-zero terms in the Riemann-Stieltjes integral (3.38) originate from intervals (t_{i-1}, t_i) which contain a node τ_j and yield $g(t_i)\left(\boldsymbol{f}_s(t_i) - \boldsymbol{f}_s(t_{i-1})\right)$. In the limit, when $t_i - t_{i-1}$ tends to zero, this term tends to $g(\tau_j)\left(\boldsymbol{f}_s^+(\tau_j) - \boldsymbol{f}_s^-(\tau_j)\right)$. The integral (3.40) is in this case

$$\int_{[a,b]} g\, d\boldsymbol{f}_s = \sum_{j=1}^{p} g(\tau_j)\left(\boldsymbol{f}_s^+(\tau_j) - \boldsymbol{f}_s^-(\tau_j)\right). \tag{3.46}$$

Consequently, the differential measure $d\boldsymbol{f}_s$ is the weighted sum of a number of Dirac point measures

$$d\boldsymbol{f}_s = \sum_{j=1}^{p}\left(\boldsymbol{f}_s^+(\tau_j) - \boldsymbol{f}_s^-(\tau_j)\right) d\delta_{\tau_j}. \tag{3.47}$$

If $g(t) = 1$, then the integral (3.40) is given by

$$\begin{aligned}
\int_{[a,b]} d\boldsymbol{f}_s &= \int_{[a,b]} \sum_{j=1}^{p}\left(\boldsymbol{f}_s^+(\tau_j) - \boldsymbol{f}_s^-(\tau_j)\right) d\delta_{\tau_j} \\
&= \sum_{j=1}^{p}\left(\boldsymbol{f}_s^+(\tau_j) - \boldsymbol{f}_s^-(\tau_j)\right) \int_{[a,b]} d\delta_{\tau_j} \\
&= \sum_{j=1}^{p}\left(\boldsymbol{f}_s^+(\tau_j) - \boldsymbol{f}_s^-(\tau_j)\right) \\
&= \boldsymbol{f}_s^+(\tau_p) - \boldsymbol{f}_s^-(\tau_1) = \boldsymbol{f}_s^+(b) - \boldsymbol{f}_s^-(a).
\end{aligned} \tag{3.48}$$

Every right-continuous locally bounded function $\boldsymbol{f}$ can be approximated by a right-continuous local step function $\boldsymbol{f}_s$ and their difference can be made arbitrarily small by refining the partition of the local step function. The result

$$\int_{[a,b]} \mathrm{d}\boldsymbol{f} = \boldsymbol{f}^+(b) - \boldsymbol{f}^-(a) \tag{3.49}$$

does therefore not only hold for local step functions, but holds for arbitrary locally bounded functions $\boldsymbol{f}$ as has been proven rigourously in [122]. We can shrink the interval $[a, b]$ to a singleton $\{a\}$ such that

$$\int_{\{a\}} \mathrm{d}\boldsymbol{f} = \boldsymbol{f}^+(a) - \boldsymbol{f}^-(a), \tag{3.50}$$

which yields a non-zero result if the differential measure $\mathrm{d}\boldsymbol{f}$ has an atom at $\{a\}$. Note that there is no one-to-one relationship between $\boldsymbol{f}$ and $\mathrm{d}\boldsymbol{f}$: the differential measure $\mathrm{d}\boldsymbol{f}$ does not depend on the value $\boldsymbol{f}(a)$ but only on its right and left value $\boldsymbol{f}^\pm(a)$. Using the integral relations (3.34) it follows for $\boldsymbol{f} \in \mathrm{lbv}(I, \mathbb{R}^n)$ that

$$\int_{(a,b)} \mathrm{d}\boldsymbol{f} = \boldsymbol{f}^-(b) - \boldsymbol{f}^+(a),$$

$$\int_{(a,b]} \mathrm{d}\boldsymbol{f} = \boldsymbol{f}^+(b) - \boldsymbol{f}^+(a), \tag{3.51}$$

$$\int_{[a,b)} \mathrm{d}\boldsymbol{f} = \boldsymbol{f}^-(b) - \boldsymbol{f}^-(a).$$

Hence, the integral of the differential measure $\mathrm{d}\boldsymbol{f}$ is precisely the signed Lebesgue-Stieltjes vector measure $\nu_{\boldsymbol{f}}$ of $\boldsymbol{f}$

$$\nu_{\boldsymbol{f}}(A) = \int_A \mathrm{d}\boldsymbol{f}. \tag{3.52}$$

Using the Lebesgue decomposition of the Lebesgue-Stieltjes measure (Theorem 3.7), we decompose the right-continuous function $\boldsymbol{f} \in \mathrm{lbv}(I, \mathbb{R}^n)$ in functions of locally bounded variation such that

$$\boldsymbol{f}(t) = \boldsymbol{f}_{\mathrm{abs}}(t) + \boldsymbol{f}_{\mathrm{s}}(t) + \boldsymbol{f}_{\mathrm{sing}}(t) \tag{3.53}$$

and decompose the differential measure correspondingly

$$\mathrm{d}\boldsymbol{f} = \mathrm{d}\boldsymbol{f}_{\mathrm{abs}} + \mathrm{d}\boldsymbol{f}_{\mathrm{s}} + \mathrm{d}\boldsymbol{f}_{\mathrm{sing}}. \tag{3.54}$$

The function $\boldsymbol{f}_{\mathrm{abs}}$ is absolutely continuous with respect to the Lebesgue measure and admits the density function $\boldsymbol{f}'_{\mathrm{abs}}$. Its differential measure $\mathrm{d}\boldsymbol{f}_{\mathrm{abs}}$ is the product of the density function $\boldsymbol{f}'_{\mathrm{abs}}$ and the Lebesgue measure $\mathrm{d}t$

$$\mathrm{d}\boldsymbol{f}_{\mathrm{abs}} = \boldsymbol{f}'_{\mathrm{abs}}\mathrm{d}t, \tag{3.55}$$

such that

$$\int_{[a,b]} \mathrm{d}\boldsymbol{f}_{\mathrm{abs}} = \int_a^b \boldsymbol{f}'_{\mathrm{abs}}\mathrm{d}t = \boldsymbol{f}_{\mathrm{abs}}(b) - \boldsymbol{f}_{\mathrm{abs}}(a). \tag{3.56}$$

The right-continuous step function $\boldsymbol{f}_{\mathrm{s}}$ takes the discontinuities in $\boldsymbol{f}$ into account and is almost everywhere constant. The corresponding signed Lebesgue-Stieltjes measure ν_{s} is purely atomic. The corresponding differential measure $\mathrm{d}\boldsymbol{f}_{\mathrm{s}}$ equals the weighted sum of Dirac point measures $\mathrm{d}\delta_{t_i}$, defined by (3.43), with jump heights $\boldsymbol{f}^{+}(t_i) - \boldsymbol{f}^{-}(t_i)$

$$\mathrm{d}\boldsymbol{f}_{\mathrm{s}} = \sum_i (\boldsymbol{f}^{+}(t_i) - \boldsymbol{f}^{-}(t_i))\mathrm{d}\delta_{t_i}. \qquad (3.57)$$

For general step functions $\boldsymbol{f}_{\mathrm{s}}(t)$, we say that $\boldsymbol{f}_{\mathrm{s}}$ admits a density function $\boldsymbol{f}'_{\eta}$ with respect to the differential measure $\mathrm{d}\eta$ (which we call the atomic measure) and, consequently,

$$\int \mathrm{d}\boldsymbol{f}_{\mathrm{s}} = \int \boldsymbol{f}'_{\eta}\mathrm{d}\eta. \qquad (3.58)$$

Now, using (3.57), the meaning of the atomic measure $\mathrm{d}\eta$ becomes clear (in an integral sense) from

$$\int_{[a,b]} \sum_i (\boldsymbol{f}^{+}(t_i) - \boldsymbol{f}^{-}(t_i))\mathrm{d}\delta_{t_i} = \int_{[a,b]} \boldsymbol{f}'_{\eta}\mathrm{d}\eta. \qquad (3.59)$$

More conveniently, it is proposed in [63] to write (3.57) as

$$\mathrm{d}\boldsymbol{f}_{\mathrm{s}} = (\boldsymbol{f}^{+} - \boldsymbol{f}^{-})\mathrm{d}\eta, \qquad \mathrm{d}\eta = \sum_i \mathrm{d}\delta_{t_i} \qquad (3.60)$$

and $\mathrm{d}\boldsymbol{f}_{\mathrm{s}}$ is called the atomic differential measure of $\boldsymbol{f}$. Here, we will use the notation (3.60), keeping in mind that differential measures always need to be understood in an integral sense.

The singular term $\boldsymbol{f}_{\mathrm{sing}}$ is continuous (but not absolutely continuous) and constant for almost all t. The corresponding signed Lebesgue-Stieltjes measure ν_{sing} is non-atomic (diffuse) and is orthogonal to the Lebesgue measure. Its differential measure $\mathrm{d}\boldsymbol{f}_{\mathrm{sing}}$ is therefore said to be singular with respect to the Lebesgue measure $\mathrm{d}t$.

The Lebesgue decomposition (3.53) is possible for right-continuous functions which are of locally bounded variation. A function $\boldsymbol{f} \in \mathrm{lbv}(I, \mathbb{R}^n)$ which is not defined on its discontinuity points can be decomposed in an absolutely continuous function, a step function and a singular function for almost all t. The decomposition of measures (3.54) remains valid as measures have to be understood in the sense of integration. Moreover, if the function $\boldsymbol{f} \in \mathrm{lbv}(I, \mathbb{R}^n)$ is not right-continuous, then we can still apply the Lebesgue decomposition (3.53) on the function $\boldsymbol{f}^{+} \in \mathrm{lbv}(I, \mathbb{R}^n)$, which is again right-continuous.

3.6 The Differential Measure of a Bilinear Form

In this section, which is based primarily on [122], we study the differential measure of a bilinear form. For instance, in mechanics the kinetic energy

$T(\boldsymbol{u}) = F(\boldsymbol{u}, \boldsymbol{u})$, with $F(\boldsymbol{u}, \boldsymbol{v}) = \frac{1}{2}\boldsymbol{u}^{\mathrm{T}}\boldsymbol{M}\boldsymbol{v}$, is a symmetric bilinear form in the generalised velocities.

Consider the bounded variation functions $\boldsymbol{x} \in \mathrm{lbv}(I, X)$ and $\boldsymbol{y} \in \mathrm{lbv}(I, Y)$ and the function $t \rightarrow F(\boldsymbol{x}(t), \boldsymbol{y}(t))$, being a continuous bilinear form $F : X \times Y \rightarrow \mathbb{R}$, denoted by $F(\boldsymbol{x}, \boldsymbol{y})$ in short. First assume that $\boldsymbol{x}$ and $\boldsymbol{y}$ are local step functions, each having their own set of discontinuity points. The set of discontinuity points of $F(\boldsymbol{x}, \boldsymbol{y})$ is the union of the discontinuity points of $\boldsymbol{x}$ and of $\boldsymbol{y}$. Construct a sequence of nodes $t_1 < t_t < \ldots < t_n$ on the discontinuity points of $F(\boldsymbol{x}, \boldsymbol{y})$ on a subinterval $[a, b]$ of I. The functions $\boldsymbol{x}(t)$, $\boldsymbol{y}(t)$ and $F(\boldsymbol{x}, \boldsymbol{y})$ are therefore constant on each open subinterval (t_i, t_{i+1}). The differential measure $\mathrm{d}F(\boldsymbol{x}, \boldsymbol{y})$ equals the sum of a locally finite collection of point measures placed at the discontinuity points of F

$$\int_{[a,b]} \mathrm{d}F(\boldsymbol{x}, \boldsymbol{y}) = \sum_{i=1}^{n} \left(F(\boldsymbol{x}^+(t_i), \boldsymbol{y}^+(t_i)) - F(\boldsymbol{x}^-(t_i), \boldsymbol{y}^-(t_i)) \right). \qquad (3.61)$$

Similarly, $\mathrm{d}\boldsymbol{x}$ equals the sum of point measures placed at the nodes t_i with values $\boldsymbol{x}^+(t_i) - \boldsymbol{x}^-(t_i)$ (and the same applies for $\mathrm{d}\boldsymbol{y}$). It therefore holds that

$$\begin{aligned} \int_{[a,b]} F(\mathrm{d}\boldsymbol{x}, \boldsymbol{y}^-) &= \sum_{i=1}^{n} F(\boldsymbol{x}^+(t_i) - \boldsymbol{x}^-(t_i), \boldsymbol{y}^-(t_i)), \\ \int_{[a,b]} F(\boldsymbol{x}^+, \mathrm{d}\boldsymbol{y}) &= \sum_{i=1}^{n} F(\boldsymbol{x}^+(t_i), \boldsymbol{y}^+(t_i) - \boldsymbol{y}^-(t_i)). \end{aligned} \qquad (3.62)$$

Exploiting the bilinearity of F yields

$$\int_{[a,b]} \mathrm{d}F(\boldsymbol{x}, \boldsymbol{y}) = \int_{[a,b]} \left(F(\mathrm{d}\boldsymbol{x}, \boldsymbol{y}^-) + F(\boldsymbol{x}^+, \mathrm{d}\boldsymbol{y}) \right). \qquad (3.63)$$

Using again the fact that every locally bounded function can be approximated by a local step function we conclude that equation (3.63) holds for arbitrary locally bounded functions $\boldsymbol{x}$ and $\boldsymbol{y}$ (see [122]).

Consider now a symmetric quadratic form $G(\boldsymbol{x}) = F(\boldsymbol{x}, \boldsymbol{x}) = \boldsymbol{x}^{\mathrm{T}}\boldsymbol{A}\boldsymbol{x}$, with $\boldsymbol{A} = \boldsymbol{A}^{\mathrm{T}} \in \mathbb{R}^{n \times n}$. We deduce from (3.63) that

$$\begin{aligned} \int_{[a,b]} \mathrm{d}G(\boldsymbol{x}) &= \int_{[a,b]} \left(F(\mathrm{d}\boldsymbol{x}, \boldsymbol{x}^-) + F(\boldsymbol{x}^+, \mathrm{d}\boldsymbol{x}) \right) \\ &= \int_{[a,b]} F(\boldsymbol{x}^+ + \boldsymbol{x}^-, \mathrm{d}\boldsymbol{x}) \\ &= \int_{[a,b]} (\boldsymbol{x}^+ + \boldsymbol{x}^-)^{\mathrm{T}} \boldsymbol{A} \mathrm{d}\boldsymbol{x} \end{aligned} \qquad (3.64)$$

or simply $\mathrm{d}G = (\boldsymbol{x}^+ + \boldsymbol{x}^-)^{\mathrm{T}} \boldsymbol{A} \mathrm{d}\boldsymbol{x}$.

3.7 Summary

An overview of measure and integration theory has been given in this chapter. A measure assigns a non-negative real number to a set, while signed measures can also take negative values. If a signed measure is absolutely continuous with respect to a measure, then it admits a density function with respect to that measure (Radon-Nykodým). A real measure is a continuous linear functional which assigns a real number to a continuous function and is related to a signed measure through the characteristic function. The differential measure is a real measure, which assigns a real number to a continuous function by means of the limit of Riemann-Stieltjes sums.

Of utmost importance for the next chapters is the fact that every (right-continuous) function of locally bounded variation can be decomposed in an absolutely continuous function, a (right-continuous) step function and a singular function (3.53)–(3.60). The differential measures can be split correspondingly. The singular function is assumed to vanish.

Differential measures will be used in Chapter 4 to formulate measure differential inclusions, which are generalisations of differential inclusions and equations. The solution concept of measure differential inclusions allows the solution to be a function of locally bounded variation, and therefore to undergo discontinuities in time (such as e.g. velocity jumps in mechanical systems with impacts).

4

Non-smooth Dynamical Systems

A dynamical system is a system whose state evolves with time. The evolution is governed by a set of rules, and is usually put in the form of equations. Continuous-time systems are dynamical systems of which the state is allowed to change (continuously or discontinuously) at all times t. Continuous-time systems are usually described by ordinary differential equations. Discrete-time systems are dynamical systems of which the state can only change at discrete time-instances $t_1 < t_2 < t_3 \ldots$. In this chapter we will consider continuous-time dynamical systems with a non-smooth (and possibly discontinuous) time-evolution of the state.

After a brief introduction to differential equations, we will consider differential equations with a discontinuous right-hand side. The requirement for the existence of a solution leads to the need to fill in the graph of the discontinuous right-hand side (Filippov's convex method). The resulting set-valued right-hand side brings forth a differential inclusion, which will be discussed in detail in Section 4.2. Subsequently, the differential measure of the state, which classically contains a density with respect to the Lebesgue measure, is extended in Section 4.3 with an atomic part. This leads to a measure differential inclusion, being the mathematical framework used in this work to describe dynamical systems with state evolutions with discontinuities (state jumps). In Chapters 6 and 8, we will present stability and convergence results for measure differential inclusions.

4.1 Differential Equations

We consider a dynamical system described by a set of ordinary differential equations

$$\dot{\boldsymbol{x}}(t) = \boldsymbol{f}(t, \boldsymbol{x}(t)), \qquad \boldsymbol{x}(t) \in \mathbb{R}^n, \tag{4.1}$$

where $\boldsymbol{x}$ is the state vector and $\boldsymbol{f}(t, \boldsymbol{x}(t))$ is the right-hand side vector, called the vector field, describing the time derivative of the state vector. We will assume that $f(t, x)$ is linearly bounded [37], i.e. there exist positive constants γ and c such that

$$\|f(t, x)\| \leq \gamma \|x\| + c, \quad \forall (t, x). \tag{4.2}$$

If the vector field is smooth (see Definition 2.7), then a solution $x(t)$ of the system (4.1) exists for any given initial condition and is unique for all $t \in \mathbb{R}$, i.e. solutions of (4.1) are defined globally (in time). In fact, smoothness of the vector field is not a necessary condition for existence and uniqueness of the solution, as can be concluded from the following theorem (see [37], theorem 1.1, page 178):

Theorem 4.1 (Existence and Uniqueness for Continuous Systems).
Suppose that $f(t, x)$ is continuous, and let $(t_0, x_0) \in \mathbb{R} \times \mathbb{R}^n$ be given. Then the following holds:

1. *There exists a solution of system (4.1) on an open interval $(t_0 - \delta, t_0 + \delta)$, for $\delta > 0$, satisfying $x(t_0) = x_0$.*
2. *If in addition we assume that $f(t, x)$ is linearly bounded, so that (4.2) holds, then there exists a solution of system (4.1) on $(-\infty, \infty)$ such that $x(t_0) = x_0$.*
3. *We now add the hypothesis that $f(t, x)$ is locally Lipschitz at x, i.e. there exists a constant $L(x) > 0$ and $r > 0$ such that*

$$\|f(t, x_1) - f(t, x_2)\| \leq L(x)\|x_1 - x_2\|, \forall x_1, x_2 \in B_r + x.$$

Then there exists a unique solution of (4.1) on $(-\infty, \infty)$ such that $x(t_0) = x_0$.

Note that above theorem deals with systems with a continuous vector field $f(t, x)$ but the vector field is allowed to be *non-smooth*, i.e. the right-hand side f may not be differentiable everywhere with respect to x.

4.2 Differential Inclusions

In this section we introduce the concept of a solution to a differential inclusion

$$\dot{x}(t) \in \mathcal{F}(t, x(t)), \tag{4.3}$$

where $\mathcal{F} : \mathbb{R} \times \mathbb{R}^n \to \mathbb{R}^n$ is a set-valued map.

Existence and uniqueness of solutions of continuous dynamical systems have been discussed in Section 4.1. However, differential equations stemming from systems with switches may be discontinuous, i.e. the right-hand side f can be piecewise continuous such that it is discontinuous in x on certain hyper-surfaces in the state-space. The theory of Filippov [50,51,96,152] gives a generalised definition of the solution of differential equations which incorporates systems with a discontinuous right-hand side. The solution $x(t)$ in the sense of Filippov of a differential equation with a discontinuous right-hand side (also called Filippov systems) is absolutely continuous in time, i.e. there are no discontinuities in the state $x(t)$. Systems with discontinuities in the state $x(t)$ can be described by measure differential inclusions (see Section 4.3). Filippov's theory will be briefly outlined in this section.

In order to make things as clear as possible, we first look at a very simple one-dimensional example (see [88]). Consider the following differential equation with a discontinuous right-hand side

$$\dot{x} = f(x) = 1 - 2\,\mathrm{sign}(x) = \begin{cases} 3, & x < 0, \\ 1, & x = 0, \\ -1, & x > 0, \end{cases} \tag{4.4}$$

with

$$\mathrm{sign}(x) = \begin{cases} -1, & x < 0, \\ 0, & x = 0, \\ 1, & x > 0. \end{cases} \tag{4.5}$$

For a given initial condition $x(0) \neq 0$ we can obtain a solution of the initial value problem

$$x(t) = \begin{cases} 3t + C_1, & x < 0, \\ -t + C_2, & x > 0, \end{cases} \tag{4.6}$$

with constants C_1 and C_2 being determined by the initial condition. Each solution reaches $x = 0$ in finite time. If the solution arrives at $x = 0$, it can not leave $x = 0$, because $\dot{x} > 0$ for $x < 0$ and $\dot{x} < 0$ for $x > 0$. The solution will therefore stay at $x = 0$, which implies $\dot{x}(t) = 0$. Note that $x(t) = 0$ with $\dot{x}(t) = 0$ is not a solution of the problem since $0 \neq 1 - 2\,\mathrm{sign}(0)$. Hence, a natural idea to extend the notion of solution is to replace the right-hand side $f(x)$ by a *set-valued* function $\mathcal{F}(x)$ such that $\mathcal{F}(x) = f(x)$ for all x for which f is continuous in x. At the points for which f is discontinuous in x a suitable choice of $\mathcal{F}(x)$ is required. The differential equation is then replaced by the differential inclusion [50, 51]

$$\dot{x} \in \mathcal{F}(x), \tag{4.7}$$

with the set-valued function

$$\mathcal{F}(x) = 1 - 2\,\mathrm{Sign}(x). \tag{4.8}$$

Here, $\mathrm{Sign}(x)$ denotes the set-valued Sign function (2.25), which is set-valued at $x = 0$, and which is the generalised differential of $|x|$. With this definition $x(t) = 0$ is a unique solution in *forward time* of the differential inclusion

$$\dot{x} \in 1 - 2\,\mathrm{Sign}(x), \tag{4.9}$$

with initial condition $x(0) = 0$. Note that the solution of (4.9) is non-unique in *backward time*. For instance, the solutions of (4.9) with initial condition

$x(-1) = 1$ and initial condition $x(-1) = -3$ evolve both to $x(0) = 0$. In fact, there exist infinitely many initial conditions at $t = -1$ which evolve to $x(0) = 0$. Knowledge about the solution for $t \in (-\infty, 0)$ is therefore lost by the non-uniqueness in backward time.

Let us also give a one-dimensional example of non-uniqueness in forward time. Consider the one-dimensional differential inclusion

$$\dot{x}(t) \in \text{Sign}(x(t)), \quad x(0) = 0. \tag{4.10}$$

This initial value problem has three solutions on the domain $[0, \infty)$:

$$x_1(t) = -t, \quad x_2(t) = 0, \quad x_3(t) = t. \tag{4.11}$$

The solution of (4.10) is non-unique in forward time but unique in backward time, i.e. $x(t) = 0$ for $t \in (-\infty, 0]$.

The above examples are one-dimensional. We now need to define a differential equation with a discontinuous right-hand side in a more general sense for any dimension n. We restrict ourselves to differential equations with a right-hand side that is discontinuous at one hyper-surface, but this restriction is not essential. The state-space $\mathbb{R}^n$ is split into two subspaces $\mathcal{V}_-$ and $\mathcal{V}_+$ by a hyper-surface Σ such that $\mathbb{R}^n = \mathcal{V}_- \cup \Sigma \cup \mathcal{V}_+$. The hyper-surface Σ is called the *switching boundary* and is defined by a scalar switching boundary function $h(\boldsymbol{x})$. The state $\boldsymbol{x}$ is in Σ when

$$h(\boldsymbol{x}) = 0 \Longleftrightarrow \boldsymbol{x} \in \Sigma. \tag{4.12}$$

The subspaces $\mathcal{V}_-$ and $\mathcal{V}_+$ and switching boundary Σ can be formulated as

$$\begin{aligned} \mathcal{V}_- &= \{\boldsymbol{x} \in \mathbb{R}^n \mid h(\boldsymbol{x}) < 0\}, \\ \Sigma &= \{\boldsymbol{x} \in \mathbb{R}^n \mid h(\boldsymbol{x}) = 0\}, \\ \mathcal{V}_+ &= \{\boldsymbol{x} \in \mathbb{R}^n \mid h(\boldsymbol{x}) > 0\}. \end{aligned} \tag{4.13}$$

The right-hand side of the dynamics $\dot{\boldsymbol{x}} = \boldsymbol{f}(t, \boldsymbol{x})$ is assumed to be locally continuous, smooth and linearly bounded for all $\boldsymbol{x} \notin \Sigma$. From this assumption it follows that the solution $\boldsymbol{x}(t)$ within each subspace $\mathcal{V}_-$ and $\mathcal{V}_+$ exists and is unique (see Theorem 4.1).

The set-valued extension of $\boldsymbol{f}(t, \boldsymbol{x})$ of (4.1) for $\boldsymbol{x} \in \Sigma$ is given by the closed convex hull of all the limits

$$\mathcal{F}(t, \boldsymbol{x}) = \overline{\text{co}}\{\boldsymbol{y} \in \mathbb{R}^n \mid \boldsymbol{y} = \lim_{\tilde{\boldsymbol{x}} \to \boldsymbol{x}} \boldsymbol{f}(t, \tilde{\boldsymbol{x}}), \tilde{\boldsymbol{x}} \in \mathbb{R}^n \backslash \Sigma\}, \tag{4.14}$$

where the notation (2.1) has been used. All the limits exist because $\boldsymbol{f}(t, \boldsymbol{x})$ is assumed to be locally continuous, smooth and linearly bounded for all $\boldsymbol{x} \notin \Sigma$.

We are now able to consider the following n-dimensional nonlinear system with discontinuous right-hand side

$$\dot{\boldsymbol{x}}(t) = \boldsymbol{f}(t, \boldsymbol{x}(t)) = \begin{cases} \boldsymbol{f}_-(t, \boldsymbol{x}(t)), & \boldsymbol{x} \in \mathcal{V}_-, \\ \boldsymbol{f}_+(t, \boldsymbol{x}(t)), & \boldsymbol{x} \in \mathcal{V}_+, \end{cases} \tag{4.15}$$

with the initial condition $\boldsymbol{x}(0) = \boldsymbol{x}_0$. As mentioned before, the right-hand side $\boldsymbol{f}(t, \boldsymbol{x})$ is assumed to be piecewise continuous such that it is smooth on $\mathcal{V}_-$ and $\mathcal{V}_+$ and not yet defined on Σ. We extend the function $\boldsymbol{f}_-(t, \boldsymbol{x})$ in a continuous way to $\mathcal{V}_- \cup \Sigma$ and extend $\boldsymbol{f}_+(t, \boldsymbol{x})$ continuously to $\mathcal{V}_+ \cup \Sigma$. It is not required that $\boldsymbol{f}_-(t, \boldsymbol{x})$ and $\boldsymbol{f}_+(t, \boldsymbol{x})$ agree on Σ. Subsequently, the graph is filled-in between the values of $\boldsymbol{f}_-(t, \boldsymbol{x})$ and $\boldsymbol{f}_-(t, \boldsymbol{x})$ on Σ which gives the set-valued extension $\mathcal{F}(t, \boldsymbol{x})$

$$\dot{\boldsymbol{x}}(t) \in \mathcal{F}(t, \boldsymbol{x}(t)) = \begin{cases} \boldsymbol{f}_-(t, \boldsymbol{x}(t)), & \boldsymbol{x} \in \mathcal{V}_-, \\ \overline{\mathrm{co}}\{\boldsymbol{f}_-(t, \boldsymbol{x}(t)), \boldsymbol{f}_+(t, \boldsymbol{x}(t))\}, & \boldsymbol{x} \in \Sigma, \\ \boldsymbol{f}_+(t, \boldsymbol{x}(t)), & \boldsymbol{x} \in \mathcal{V}_+. \end{cases} \quad (4.16)$$

The convex set with the two right-hand sides $\boldsymbol{f}_-$ and $\boldsymbol{f}_+$ can be cast in the form

$$\overline{\mathrm{co}}\{\boldsymbol{f}_-, \boldsymbol{f}_+\} = \{(1 - q)\boldsymbol{f}_- + q\boldsymbol{f}_+, \forall q \in [0, 1]\}. \quad (4.17)$$

The extension (or convexification) of a discontinuous system (4.15) into a differential inclusion (4.16) with convex image is known as *Filippov's convex method*.

It was stated that the set-valued extension $\mathcal{F}$ of $\boldsymbol{f}$ should be *suitable*. If the discontinuous system (4.15) is a mathematical model of a physical system, then we are interested in a solution concept that guarantees existence of solutions. Therefore, for practical reasons we demand that the choice for $\mathcal{F}$ guarantees existence of solutions. Existence can be guaranteed with the notion of upper semi-continuity of set-valued functions. We first define a solution of the differential inclusion (4.3).

Definition 4.2 (Solution of a Differential Inclusion). *An absolutely continuous function* $\boldsymbol{x} : [0, \tau] \to \mathbb{R}^n$ *is said to be a solution of the differential inclusion (4.3) if it fulfils*

$$\dot{\boldsymbol{x}}(t) \in \mathcal{F}(t, \boldsymbol{x}(t))$$

for almost all [1] $t \in [0, \tau]$.

The following existence theorem is proven in [10] (Theorem 3, page 98; see also [156]):

Theorem 4.3 (Existence of Solution of a Differential Inclusion). *Let* $\mathcal{F}$ *be a set-valued function. We assume that* $\mathcal{F}$ *is upper semi-continuous and that the image of* $(t, \boldsymbol{x})$ *under* $\mathcal{F}$ *is closed, convex and bounded for all* $t \in \mathbb{R}$ *and* $\boldsymbol{x} \in \mathbb{R}^n$. *Then, for each* $\boldsymbol{x}_0 \in \mathbb{R}^n$ *there exists a* $\tau > 0$ *and an absolutely continuous function* $\boldsymbol{x}(t)$ *defined on* $[0, \tau]$, *which is a solution of the initial value problem*

$$\dot{\boldsymbol{x}} \in \mathcal{F}(t, \boldsymbol{x}(t)), \quad \boldsymbol{x}(0) = \boldsymbol{x}_0.$$

---

[1] *for almost all* t means except for a set t of Lebesgue measure 0.

The theorem clearly holds at values of x for which $\mathcal{F}(t, x)$ is locally single-valued and continuous, because of the boundedness restriction (see Theorem 4.1). To illustrate the theorem for set-valued $\mathcal{F}(t, x)$ we once more look at the example of (4.9)

$$\dot{x} \in 1 - a \operatorname{Sign}(x), \text{ with } |a| > 1. \tag{4.18}$$

To allow for the solution $x(t) = 0$ we must demand that $0 \in 1 - a \operatorname{Sign}(0)$. In order to guarantee the existence of a solution we therefore have to define $\operatorname{Sign}(0)$ to be the set $[-1, 1]$. This set is upper semi-continuous with the values of $\operatorname{sign}(x) = \pm 1$ for $x \neq 0$, and the set is closed, convex and bounded.

Filippov's convex method together with the above existence theorem defines the *solution in the sense of Filippov* for a discontinuous differential equation.

Definition 4.4 (Solution in the Sense of Filippov of a Discontinuous Differential Equation). *An absolutely continuous function $x : [0, \tau] \to \mathbb{R}^n$ is said to be a solution of $\dot{x}(t) = f(t, x)$ (4.15) in the sense of Filippov if for almost all $t \in [0, \tau]$ it holds that*

$$\dot{x}(t) \in \mathcal{F}(t, x(t)),$$

where $\mathcal{F}(t, x(t))$ is the closed convex hull of all the limits of $f(t, x(t))$ as in (4.14).

Remark: If $x(t)$ is in a region where the vector field is continuous, $x(t) \in \mathcal{V}$, then of course $\mathcal{F}(t, x(t)) = f(t, x(t))$ must hold. If the solution $x(t)$ slides along a switching boundary, $x(t) \in \Sigma$, then $\dot{x}(t) \in \mathcal{F}(t, x(t))$. However, $\dot{x}(t)$ is not defined at time instances t_Σ where the solution $x(t)$ arrives at a switching boundary Σ or leaves from Σ (the solution $x(t)$ arrives at or leaves from Σ if there exists an arbitrary small $\varepsilon > 0$ and $t^* \in t_\Sigma + B_\varepsilon \backslash 0$ such that $x(t^*) \notin \Sigma$ and $x(t_\Sigma) \in \Sigma$). The set of t for which the solution $x(t)$ arrives at or leaves from Σ is of Lebesgue measure zero. Hence, the set of time-instances t for which $\dot{x}(t)$ is not defined is Lebesgue negligible. Note that the absolutely continuous solution $x(t) = x(t_0) + \int_{t_0}^{t} \dot{x} dt$ does not depend on the value (or the lack of having a value) of $\dot{x}(t)$ on a Lebesgue negligible set of time-instances t.

It was assumed that $f(t, x)$ is linearly bounded for $x \notin \Sigma$. In addition, $\mathcal{F}(t, x(t))$ is assumed to be bounded at values (t, x) for which $\mathcal{F}$ is set-valued. Consequently, $\mathcal{F}(t, x(t))$ is linearly bounded, i.e. there exist positive constants γ and c such that for all $t \in [0, \infty)$ and $x \in \mathbb{R}^n$ it holds that:

$$\|\mathcal{F}(t, x)\| \leq \gamma \|x\| + c.$$

Solutions $x(t)$ to (4.16) therefore exist on $[0, \infty)$ (see [10, 37]) but uniqueness is not guaranteed.

In the following, it will sometimes be useful to explicitly write the dependence of the solution on the initial condition. We therefore denote a solution $\boldsymbol{x}(t)$ of $\dot{\boldsymbol{x}}(t) \in \boldsymbol{F}(t, \boldsymbol{x}(t))$ starting from the initial condition $\boldsymbol{x}(t_0) = \boldsymbol{x}_0$ by $\boldsymbol{\varphi}(t, t_0, \boldsymbol{x}_0)$. The solution $\boldsymbol{\varphi}(t, t_0, \boldsymbol{x}_0)$ is in general non-unique. We denote by $\mathcal{S}(\boldsymbol{F}, t_0, \boldsymbol{x}_0)$ the set of solutions curves $\boldsymbol{x}(t)$, with $t \geq t_0$, starting from the initial condition $\boldsymbol{x}(t_0) = \boldsymbol{x}_0$, i.e.

$$\boldsymbol{\varphi}(\cdot, t_0, \boldsymbol{x}_0) \in \mathcal{S}(\boldsymbol{F}, t_0, \boldsymbol{x}_0).$$

If the solution is unique, then $\mathcal{S}(\boldsymbol{F}, t_0, \boldsymbol{x}_0)$ reduces to a singleton $\{\boldsymbol{\varphi}(\cdot, t_0, \boldsymbol{x}_0)\}$.

A sufficient criterion for uniqueness of solutions can be gained from a maximal monotonicity property (Definition 2.17) of the system [22].

Theorem 4.5 (Uniqueness of Solutions through Maximal Monotonicity). *Consider a differential inclusion of the form*

$$\dot{\boldsymbol{x}} \in -\boldsymbol{A}(\boldsymbol{x}).$$

If $\boldsymbol{A}(\boldsymbol{x})$ is a maximal monotone set-valued function with a linear boundedness condition, then the differential inclusion has a unique solution $\boldsymbol{x}(t)$ for all initial conditions $\boldsymbol{x}(0) = \boldsymbol{x}_0$.

Proof: The maximal monotonicity implies that $\boldsymbol{A}(\boldsymbol{x})$ is upper semi-continuous and that the image of $\boldsymbol{x}$ under $\boldsymbol{A}$ is closed, convex and bounded for all $\boldsymbol{x} \in \mathbb{R}^n$. Existence of solutions therefore follows from a linear boundedness condition. Moreover, because $\boldsymbol{A}(\boldsymbol{x})$ is monotone, it holds that

$$(\boldsymbol{A}(\boldsymbol{u}) - \boldsymbol{A}(\boldsymbol{v}))^{\mathrm{T}} (\boldsymbol{u} - \boldsymbol{v}) \geq 0 \quad \forall \boldsymbol{u}, \boldsymbol{v}. \tag{4.19}$$

Let $\boldsymbol{u}(\cdot) \in \mathcal{S}(-\boldsymbol{A}, 0, \boldsymbol{u}_0)$ and $\boldsymbol{v}(\cdot) \in \mathcal{S}(-\boldsymbol{A}, 0, \boldsymbol{v}_0)$ be solutions of the differential inclusion. The proof of uniqueness follows from *reductio ad absurdum*. Assume that the system has non-uniqueness of solutions. Then there exists a $\boldsymbol{u}_0 = \boldsymbol{v}_0$ such that $\boldsymbol{u}(t) \neq \boldsymbol{v}(t)$. Consider the positive definite function $V \in \mathbb{R}$

$$V(t) = \frac{1}{2} \|\boldsymbol{u}(t) - \boldsymbol{v}(t)\|^2. \tag{4.20}$$

The time-derivative of V yields

$$\begin{aligned}
\dot{V} &= \|\boldsymbol{u}(t) - \boldsymbol{v}(t)\| \frac{\mathrm{d}}{\mathrm{d}t} \|\boldsymbol{u}(t) - \boldsymbol{v}(t)\| \\
&= (\boldsymbol{u}(t) - \boldsymbol{v}(t))^{\mathrm{T}} (\dot{\boldsymbol{u}}(t) - \dot{\boldsymbol{v}}(t)) \\
&\in -(\boldsymbol{u}(t) - \boldsymbol{v}(t))^{\mathrm{T}} (\boldsymbol{A}(\boldsymbol{u}(t)) - \boldsymbol{A}(\boldsymbol{v}(t))) \\
&\leq 0.
\end{aligned} \tag{4.21}$$

Hence, the function V can not increase, meaning that the distance between $\boldsymbol{u}(t)$ and $\boldsymbol{v}(t)$ can not increase, i.e.

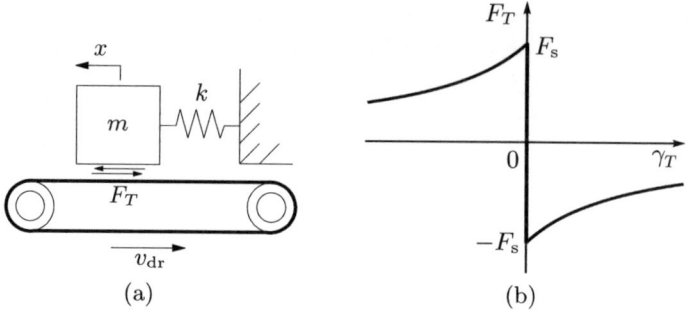

Fig. 4.1. Stick-slip system and friction curve with Stribeck effect.

$$\|\boldsymbol{u}(t) - \boldsymbol{v}(t)\| \le \|\boldsymbol{u}_0 - \boldsymbol{v}_0\|. \tag{4.22}$$

Taking $\boldsymbol{u}_0 = \boldsymbol{v}_0$ it follows that $\boldsymbol{u}(t) = \boldsymbol{v}(t)$, which is in contradiction with $\boldsymbol{u}(t) \ne \boldsymbol{v}(t)$. Consequently, the system has uniqueness of solutions. $\qquad\square$

Although mechanical systems with set-valued force laws will be described in a much more general setting in Chapter 5, we will already discuss here a small example to illustrate Filippov systems. Consider the horizontal motion of rigid mass m attached to a vertical wall by a spring k (Figure 4.1a). The mass is riding on a belt, that is moving at a constant velocity v_{dr}. We denote the horizontal position of the block by x and its velocity by $\dot{x}$. The relative velocity of the mass m with respect to the belt is denoted by $\gamma_T = \dot{x} - v_{\mathrm{dr}}$. Dry friction occurs between the mass and the belt, with a friction force F_T. The friction force $F_T(\gamma_T)$ is in the slip phase a function of the relative velocity γ_T.

$$F_T(\gamma_T) = -\mu(\gamma_T)\lambda_N \operatorname{sign}(\gamma_T), \quad \mu(\gamma_T) = \frac{\mu}{1 + \delta|\gamma_T|}, \quad \gamma_T \ne 0, \tag{4.23}$$

where $\mu(\gamma_T)$ is the velocity dependent friction coefficient, $\lambda_N = mg$ is the normal contact force. In the stick phase, the magnitude of the friction force is limited by the maximum static friction force $F_s = \mu_s \lambda_N \ge |F_T|$. The friction curve is drawn in Figure 4.1b for $\delta > 0$ and shows a Stribeck effect, i.e. the magnitude of friction force diminishes for increasing relative velocity γ_T. We now combine the description of the friction force in the slip phase, which is described by a function $F_T(\gamma_T)$, and the restriction on the friction force in the stick phase $|F_T| \le F_s$ in one set-valued force law

$$F_T \in -\mu(\gamma_T)\lambda_N \operatorname{Sign}(\gamma_T), \tag{4.24}$$

where $\operatorname{Sign}(x)$ is the set-valued sign function (2.25). Newton's second law gives the equation of motion

$$m\ddot{x}(t) + kx(t) = F_T. \tag{4.25}$$

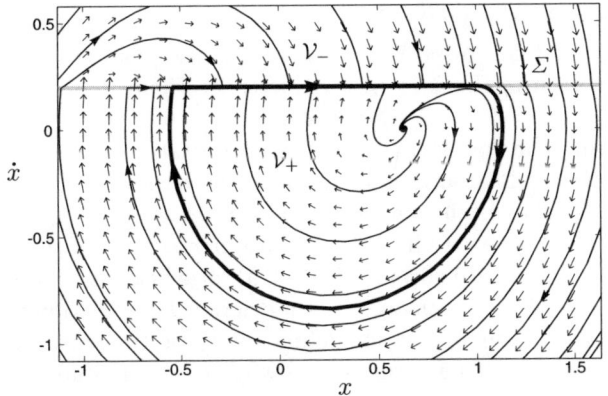

Fig. 4.2. Phase portrait of the stick-slip system.

The equation of motion together with the set-valued force law for the friction force F_T gives a second-order differential inclusion

$$m\ddot{x}(t) + kx(t) \in -\mu(\dot{x}(t) - v_{\mathrm{dr}})mg\,\mathrm{Sign}(\dot{x}(t) - v_{\mathrm{dr}}), \qquad (4.26)$$

which holds for almost all t. This second-order differential inclusion can be cast in first-order form using $\boldsymbol{x}^{\mathrm{T}} = \begin{bmatrix} x & \dot{x} \end{bmatrix}$:

$$\dot{\boldsymbol{x}}(t) \in \mathcal{F}(\boldsymbol{x}(t)) = \begin{bmatrix} \dot{x}(t) \\ -\frac{k}{m}x(t) - \mu(\dot{x}(t) - v_{\mathrm{dr}})\,g\,\mathrm{Sign}(\dot{x}(t) - v_{\mathrm{dr}}) \end{bmatrix}. \qquad (4.27)$$

The differential inclusion (4.26) or (4.27) has been obtained by considering a set-valued force law (4.24) for the friction force. Alternatively, one can also derive the differential inclusion (4.27) using 'Filippov's convex method', i.e. along the lines of equations (4.15) and (4.16). The right-hand side of the system in first-order form switches between $\boldsymbol{f}_-(\boldsymbol{x})$ and $\boldsymbol{f}_+(\boldsymbol{x})$ on a hyperplane $\Sigma = \{\boldsymbol{x} \in \mathbb{R}^n \mid h(\boldsymbol{x}) = 0\}$ with $h(\boldsymbol{x}) = \gamma_T = \dot{x} - v_{\mathrm{dr}}$ (see (4.13)) and

$$\boldsymbol{f}_-(\boldsymbol{x}) = \begin{bmatrix} \dot{x} \\ -\frac{k}{m}x + \mu(\dot{x}(t) - v_{\mathrm{dr}})\,g \end{bmatrix}, \qquad (4.28)$$

$$\boldsymbol{f}_+(\boldsymbol{x}) = \begin{bmatrix} \dot{x} \\ -\frac{k}{m}x - \mu(\dot{x}(t) - v_{\mathrm{dr}})\,g \end{bmatrix}, \qquad (4.29)$$

in the spaces $\mathcal{V}_- = \{\boldsymbol{x} \in \mathbb{R}^n \mid h(\boldsymbol{x}) < 0\}$ and $\mathcal{V}_+ = \{\boldsymbol{x} \in \mathbb{R}^n \mid h(\boldsymbol{x}) > 0\}$, respectively. The convexification of the right-hand side yields the differential inclusion in the form (4.16)

$$\dot{\boldsymbol{x}}(t) \in \mathcal{F}(\boldsymbol{x}(t)) = \begin{cases} \boldsymbol{f}_-(\boldsymbol{x}(t)), & \boldsymbol{x} \in \mathcal{V}_-, \\ \overline{\mathrm{co}}\{\boldsymbol{f}_-(\boldsymbol{x}(t)), \boldsymbol{f}_+(\boldsymbol{x}(t))\}, & \boldsymbol{x} \in \Sigma, \\ \boldsymbol{f}_+(\boldsymbol{x}(t)), & \boldsymbol{x} \in \mathcal{V}_+, \end{cases} \qquad (4.30)$$

which is equivalent to (4.27). The forward and backward slip phases are the subspaces $\mathcal{V}_+$ and $\mathcal{V}_-$ respectively, and the stick phase is contained in Σ. The phase portrait of this system is depicted in Figure 4.2 for the parameter values $m = 1$ kg, $k = 1$ N/m, $v_{\mathrm{dr}} = 0.2$ m/s, $\mu_{\mathrm{s}} = 0.1$, $g = 10$ m/s^2, $F_{\mathrm{s}} = 1$ N and $\delta = 3$ s/m. The equilibrium point is an unstable focus surrounded by a stable limit cycle, which alternates between the backward slip phase $\mathcal{V}_-$ and the stick phase. This is the so-called stick-slip motion. We can observe that the solution $(x, \dot{x})$ shows a kink in the phase portrait when the solution goes from forward to backward slip (or vice versa), or when the solution enters the stick phase. At those time-instances, the friction force F_T jumps to another value. The acceleration $\ddot{x}(t)$ is not defined at those time-instances for which such a change occurs. This is the reason why the differential inclusion (4.26) (or equivalently (4.27) or (4.30)) only describes $\dot{\boldsymbol{x}} = \begin{bmatrix} \dot{x} & \ddot{x} \end{bmatrix}^{\mathrm{T}}$ *for almost all t*, i.e. not for time instances for which the acceleration is not defined.

4.3 Measure Differential Inclusions

In this section we introduce the measure differential inclusion

$$\mathrm{d}\boldsymbol{x} \in \mathrm{d}\boldsymbol{\varGamma}(t, \boldsymbol{x}(t)) \tag{4.31}$$

as has been proposed by Moreau [117–120, 122, 123].

Differential inclusions, as studied in Section 4.2, are able to describe an absolutely continuous time-evolution $\boldsymbol{x}(t)$ which is non-differentiable on an at most countable number of time-instances. Roughly speaking, we can say that the function $\boldsymbol{x}(t)$ is allowed to have a 'kink'. Discontinuities in the time-evolution can however not be described by a differential inclusion. In the following, we will extend the concept of differential inclusions to measure differential inclusions in order to allow for discontinuities in $\boldsymbol{x}(t)$. The assumption of absolute continuity of $\boldsymbol{x}(t)$ will be relaxed to locally bounded variation in time.

Consider $\boldsymbol{x} \in \mathrm{lbv}(I, \mathcal{X})$, i.e. $\boldsymbol{x}(t)$ is a function of locally bounded variation in time. The function $\boldsymbol{x}(t)$ is therefore admitted to undergo jumps at discontinuity points $t_i \in I$, $i = 1, 2, \ldots$, but the discontinuity points are either

1. separated in time such that $t_1 < t_2 < t_3 \ldots$, leading to a finite number of discontinuities on a compact time-interval, or,
2. accumulate to an accumulation point $\lim_{i \to \infty} t_i = t^*$, leading to infinitely many discontinuities on a compact time-interval. The jump $\boldsymbol{x}^+(t_i) - \boldsymbol{x}^-(t_i)$ is assumed to decrease to zero for $i \to \infty$.

In the first case, each discontinuity point t_i is followed by a non-zero time-interval of absolutely continuous evolution. In the case of an accumulation point, the time-interval $t_{i+1} - t_i$ tends to zero as well as the discontinuities (jumps) in $\boldsymbol{x}(t)$, which guarantees that $\boldsymbol{x}(t)$ is of locally bounded variation. At a time-instance t, including the discontinuity points $t = t_i$ and accumulation points, we can therefore define a right limit $\boldsymbol{x}^+(t)$ and a left limit $\boldsymbol{x}^-(t)$ of $\boldsymbol{x}$ as one of the properties of functions of bounded variation:

$$\boldsymbol{x}^+(t) - \lim_{\tau \downarrow 0} \boldsymbol{x}(t + \tau), \qquad \boldsymbol{x}^-(t) = \lim_{\tau \uparrow 0} \boldsymbol{x}(t + \tau). \qquad (4.32)$$

If $\boldsymbol{x}$ is locally continuous at t and $\dot{\boldsymbol{x}}^+(t) = \dot{\boldsymbol{x}}^-(t)$, then $\boldsymbol{x}$ is locally differentiable (or locally smooth) at t. A function $\boldsymbol{x} : I \to \mathbb{R}^n$ is said to be smooth if it is locally smooth for all $t \in I$. A function is said to be *almost everywhere* continuous, if the set $D \subset I$ of discontinuity points $t_i \in D$, $k = 1, 2, \ldots$ is of measure zero, i.e. $\mu(D) = 0$. Similarly, a continuous function can be differentiable almost everywhere.

We want to describe with $\boldsymbol{x}(t)$ an evolution in time and therefore consider $\boldsymbol{x}(t)$ to be the result of an integration process

$$\boldsymbol{x}(t) = \boldsymbol{x}(t_0) + \int_{t_0}^{t} \mathrm{d}\boldsymbol{x}, \quad t \geq t_0, \qquad (4.33)$$

where we call $\mathrm{d}\boldsymbol{x}$ the differential measure of $\boldsymbol{x}$ [122]. If $\boldsymbol{x}(t)$ is a smooth function, then $\mathrm{d}\boldsymbol{x}$ admits a density function, say $\boldsymbol{x}'_t(t)$, with respect to the Lebesgue measure $\mathrm{d}t$

$$\mathrm{d}\boldsymbol{x} = \boldsymbol{x}'_t(t)\mathrm{d}t. \qquad (4.34)$$

We usually immediately associate the density function $\boldsymbol{x}'_t(t)$ with the derivative $\dot{\boldsymbol{x}}(t)$ and write

$$\boldsymbol{x}(t) = \boldsymbol{x}(t_0) + \int_{t_0}^{t} \dot{\boldsymbol{x}}(t)\mathrm{d}t, \quad t \geq t_0. \qquad (4.35)$$

Subsequently, we consider $\boldsymbol{x}(t)$ to be an absolutely continuous function, which is non-differentiable at the set $D \subset I$ of points $t_i \in D$. The derivative $\dot{\boldsymbol{x}}(t)$ does therefore not exist for $t = t_i$. Lebesgue integration over a singleton $\{t_i\}$, i.e. an interval with zero length, results in zero

$$\int_{\{t_i\}} \mathrm{d}\boldsymbol{x} = 0, \qquad \text{with } \mathrm{d}\boldsymbol{x} = \boldsymbol{x}'_t(t)\mathrm{d}t, \qquad (4.36)$$

even if $\dot{\boldsymbol{x}}(t)$ does not exist for $t = t_i$. The derivative $\dot{\boldsymbol{x}}(t)$ exists almost everywhere because $\boldsymbol{x}(t)$ is absolutely continuous. We say that the set D of points t_i for which $\dot{\boldsymbol{x}}(t)$ does not exist is Lebesgue negligible. Lebesgue integration over a Lebesgue negligible set results in zero. Consequently, (4.33) also holds for absolutely continuous functions, which are non-differentiable at a Lebesgue negligible set D of time-instances t_i.

Theorem 4.6 (Lebesgue-Vitali). *A function $\boldsymbol{x} : [t_l, t_k] \to \mathbb{R}^n$ is absolutely continuous if and only if it admits a representation as*

$$\boldsymbol{x}(t) = \boldsymbol{x}(t_l) + \int_{t_l}^{t} \dot{\boldsymbol{x}}(t)\mathrm{d}t, \quad t_l \le t \le t_k,$$

where $\dot{\boldsymbol{x}}$ is the derivative of $\boldsymbol{x}$, which exists at almost all points of $[t_l, t_k]$ and is Lebesgue integrable on the interval.

The proof can be found in [47].

Finally, we consider $\boldsymbol{x}(t)$ to be a function of bounded variation on the interval I, which is discontinuous at the set $D \subset I$ of points $t_i \in D$. Moreover, we assume that $\boldsymbol{x}(t)$ does not contain any singular terms, i.e. fractal-like functions such as the Cantor function. Although the function $\boldsymbol{x}(t)$ does not exist at the discontinuity points $t = t_i$, it admits a right limit $\boldsymbol{x}^+(t)$ and left limit $\boldsymbol{x}^-(t)$ at every time-instance t. Just as before, we consider $\boldsymbol{x}(t)$ to be the result of an integration process

$$\boldsymbol{x}^+(t) = \boldsymbol{x}^-(t_0) + \int_{[t_0,t]} \mathrm{d}\boldsymbol{x}, \quad t \ge t_0, \tag{4.37}$$

where the integration process takes the left limit $\boldsymbol{x}^-(t_0)$ of the initial value to the right limit $\boldsymbol{x}^+(t)$ of the final value over the closed time-interval $[t_0, t] = \{\tau \in I | t_0 \le \tau \le t\}$. The differential measure $\mathrm{d}\boldsymbol{x}$ does therefore not only contain a density $\boldsymbol{x}'_t$ with respect to the Lebesgue measure $\mathrm{d}t$ but also contains a density $\boldsymbol{x}'_\eta$ with respect to the atomic measure $\mathrm{d}\eta$, which gives a nonzero result when integrated over a singleton, such that

$$\mathrm{d}\boldsymbol{x} = \boldsymbol{x}'_t(t)\mathrm{d}t + \boldsymbol{x}'_\eta(t)\mathrm{d}\eta, \tag{4.38}$$

with

$$\int_{\{t_i\}} \mathrm{d}\eta = 1, \quad t_i \in D. \tag{4.39}$$

As discussed in Section 3.5, the atomic measure $\mathrm{d}\eta$ can be interpreted as the sum of Dirac point measures $\mathrm{d}\delta_{t_i}$,

$$\mathrm{d}\eta = \sum_i \mathrm{d}\delta_{t_i}, \quad \int_{[t_l,t_k]} \mathrm{d}\delta_{t_i} = \begin{cases} 1 & t_i \in [t_l, t_k], \\ 0 & t_i \notin [t_l, t_k], \end{cases} \tag{4.40}$$

for any interval $[t_l, t_k] \subset I$. Integration of the differential measure $\mathrm{d}\boldsymbol{x}$ over a singleton $\{t_k\}$ yields

$$\boldsymbol{x}^+(t_k) - \boldsymbol{x}^-(t_k) = \int_{\{t_k\}} \mathrm{d}\boldsymbol{x}$$

$$= \int_{\{t_k\}} \boldsymbol{x}'_\eta(t)\mathrm{d}\eta \tag{4.41}$$

$$= \boldsymbol{x}'_\eta(t_k) \int_{\{t_k\}} \mathrm{d}\eta.$$

It follows that $x^+(t_k) = x^-(t_k)$ when $t_k \notin D$, which obviously must hold for a locally continuous function at $t = t_k$. Moreover, if we choose $t_k = t_i \in D$, then we can immediately associate the density $x'_\eta(t)$ with respect to the atomic measure $d\eta$ as the jump in $x(t)$ at the discontinuity point t_i:

$$x'_\eta(t_i) = x^+(t_i) - x^-(t_i), \quad t_i \in D. \tag{4.42}$$

This reveals the fact that the interpretation of the atomic measure $d\eta$ as in (4.40) is becoming meaningful when the discontinuity points t_i are known and these discontinuity points t_i are exactly the time instances at which $x'_\eta(t)$ is non-zero. Considering (4.42), we therefore usually write the differential measure (4.38) as

$$dx = \dot{x}(t)dt + (x^+(t) - x^-(t))d\eta. \tag{4.43}$$

Consequently, using the differential measure (4.43) we are able to describe a locally absolutely continuously varying time-evolution (using the Lebesgue measurable part of dx) together with discontinuities at time-instances $t_i \subset D$ (using the atomic part). Integration of the differential measure dx therefore gives the total increment over the interval under consideration

$$x^+(t_k) - x^-(t_l) = \int_{[t_l,t_k]} dx, \tag{4.44}$$

which reduces to

$$x^+(t_k) - x^-(t_k) = \int_{\{t_k\}} dx \tag{4.45}$$

for a singleton $\{t_k\}$. In fact, it is also possible to take the singular part of a function of bounded variation into account. We will assume, however, that all singular parts vanish, as fractal-like solutions are not of interest in the evolution problems we would like to describe. Such functions are called special functions of bounded variation (SBV), see [6].

With the differential inclusion (4.3)

$$\dot{x}(t) \in \mathcal{F}(t, x(t)),$$

in which $\mathcal{F}(t, x(t))$ is a set-valued mapping, we are able to describe a non-smooth absolutely continuous time-evolution $x(t)$. The solution $x(t)$ fulfills the differential inclusion almost everywhere, because $\dot{x}(t)$ does not exist on a Lebesgue negligible set D of non-smooth time-instances $t_i \in D$. Instead of using the density $\dot{x}(t)$, we can also write the differential inclusion using the differential measure

$$dx \in \mathcal{F}(t, x(t))dt, \tag{4.46}$$

which yields a measure differential inclusion. The solution $x(t)$ fulfills the measure differential inclusion (4.46) *for all* $t \in I$ because of the underlying integration process associated with measures. Moreover, writing the dynamics in terms of a measure differential inclusion allows us to study a larger class of functions $\boldsymbol{x}(t)$, as we can let $\mathrm{d}\boldsymbol{x}$ contain parts other than the Lebesgue integrable part. In order to describe a time-evolution of bounded variation but discontinuous at isolated time-instances, we let the differential measure $\mathrm{d}\boldsymbol{x}$ also have an atomic part such as in (4.43) and therefore extend the measure differential inclusion (4.46) with an atomic part as well

$$\mathrm{d}\boldsymbol{x} \in \mathcal{F}(t, \boldsymbol{x}(t))\mathrm{d}t + \mathcal{G}(t, \boldsymbol{x}(t))\mathrm{d}\eta. \tag{4.47}$$

Here, $\mathcal{G}(t, \boldsymbol{x}(t))$ is a single-valued or set-valued mapping, which is in general dependent on t, $\boldsymbol{x}^-(t)$ and $\boldsymbol{x}^+(t)$. Following [122], we simply write $\mathcal{G}(t, \boldsymbol{x}(t))$ although a dependence on t, $\boldsymbol{x}^-(t)$ and $\boldsymbol{x}^+(t)$ is meant. If $\mathcal{G}$ is only a function of $\boldsymbol{x}^+(t)$, then we write $\mathcal{G}(\boldsymbol{x}^+(t))$. Furthermore, if $\mathcal{G}(t, \boldsymbol{x}(t))$ is set-valued, then it is often assumed to take only conic values: $\lambda \mathcal{G}(t, \boldsymbol{x}(t)) \subset \mathcal{G}(t, \boldsymbol{x}(t))$ for all $\lambda \in [0, \infty)$ and $(t, \boldsymbol{x}) \in I \times X$. More conveniently, we write the measure differential inclusion (4.47) as

$$\mathrm{d}\boldsymbol{x} \in \mathrm{d}\boldsymbol{\Gamma}(t, \boldsymbol{x}(t)), \tag{4.48}$$

where $\mathrm{d}\boldsymbol{\Gamma}(t, \boldsymbol{x}(t))$ is a set-valued measure function defined as

$$\mathrm{d}\boldsymbol{\Gamma}(t, \boldsymbol{x}(t)) = \mathcal{F}(t, \boldsymbol{x}(t))\mathrm{d}t + \mathcal{G}(t, \boldsymbol{x}(t))\mathrm{d}\eta. \tag{4.49}$$

The measure differential inclusion (4.48) has to be understood in the sense of integration. A solution $\boldsymbol{x}(t)$ of (4.48) is a function of locally bounded variation which fulfills

$$\boldsymbol{x}^+(t) = \boldsymbol{x}^-(t_0) + \int_I \boldsymbol{f}(t, \boldsymbol{x})\mathrm{d}t + \boldsymbol{g}(t, \boldsymbol{x})\mathrm{d}\eta, \tag{4.50}$$

for every compact interval $I = [t_0, t]$, where the functions $\boldsymbol{f}(t, \boldsymbol{x})$ and $\boldsymbol{g}(t, \boldsymbol{x})$ have to obey

$$\boldsymbol{f}(t, \boldsymbol{x}) \in \mathcal{F}(t, \boldsymbol{x}), \qquad \boldsymbol{g}(t, \boldsymbol{x}) \in \mathcal{G}(t, \boldsymbol{x}). \tag{4.51}$$

Note that a function $\boldsymbol{x}(t)$ can be undefined on its discontinuity points and still satisfy (4.50) *for all* $t \geq 0$, i.e. the function $\boldsymbol{x}(t)$ does not have to be right-continuous. Substitution of (4.43) in the measure differential inclusion (4.48) gives

$$\dot{\boldsymbol{x}}(t)\mathrm{d}t + (\boldsymbol{x}^+(t) - \boldsymbol{x}^-(t))\mathrm{d}\eta \in \mathcal{F}(t, \boldsymbol{x}(t))\mathrm{d}t + \mathcal{G}(t, \boldsymbol{x}(t))\mathrm{d}\eta, \tag{4.52}$$

which we can separate in the Lebesgue integrable part and atomic part

$$\dot{\boldsymbol{x}}(t)\mathrm{d}t \in \mathcal{F}(t, \boldsymbol{x}(t))\mathrm{d}t, \qquad (\boldsymbol{x}^+(t) - \boldsymbol{x}^-(t))\mathrm{d}\eta \in \mathcal{G}(t, \boldsymbol{x}(t))\mathrm{d}\eta \tag{4.53}$$

from which we can retrieve (4.3) and the jump condition

$$\boldsymbol{x}^+(t) - \boldsymbol{x}^-(t) \in \mathcal{G}(t, \boldsymbol{x}(t)). \tag{4.54}$$

The latter clearly reveals that state jumps are induced by non-zero $\boldsymbol{g}(t, \boldsymbol{x}) \in \mathcal{G}(t, \boldsymbol{x})$. Moreover, by considering the limits $t \downarrow t_i$ and $t \uparrow t_i$ we obtain the differential inclusions for post- and pre-jump times

$$\dot{\boldsymbol{x}}^{+}(t) \in \mathcal{F}(t, \boldsymbol{x}^{+}(t)), \qquad \dot{\boldsymbol{x}}^{-}(t) \in \mathcal{F}(t, \boldsymbol{x}^{-}(t)), \qquad (4.55)$$

which we call the directional differential inclusions.

A special class of systems is described by set-valued measure functions $\mathrm{d}\boldsymbol{\varGamma}(t, \boldsymbol{x}(t))$ for which each density function is a conic subset of $\mathbb{R}^n$. In particular, the set-valued functions $\mathcal{F}(t, \boldsymbol{x}(t))$ and $\mathcal{G}(t, \boldsymbol{x}(t))$ are often equal to the same cone $K(t, \boldsymbol{x}(t))$, i.e. $\mathcal{F}(t, \boldsymbol{x}(t)) = \mathcal{G}(t, \boldsymbol{x}(t)) = K(t, \boldsymbol{x}(t))$. Following Moreau [122], we write in this case

$$\mathrm{d}\boldsymbol{x} \in K(t, \boldsymbol{x}(t)), \qquad (4.56)$$

and refrain from prescribing a measure in the right-hand side in advance. It is to be understood from (4.56) that, if $\mathrm{d}\boldsymbol{x}$ possesses a density function $\boldsymbol{f}'_{\mu}$ with respect to the non-negative differential measure $\mathrm{d}\mu$, then this density function belongs to the cone K, i.e. $\boldsymbol{f}'_{\mu}(t, \boldsymbol{x}) \in K(t, \boldsymbol{x})$. In particular, this applies for the Lebesgue measure as well as for the atomic measure. More explicitly, $\mathrm{d}\boldsymbol{x} \in K(t, \boldsymbol{x}(t))$ implies

$$\dot{\boldsymbol{x}}^{\pm}(t) \in K(t, \boldsymbol{x}^{\pm}(t)) \quad \text{and} \quad \boldsymbol{x}^{+}(t) - \boldsymbol{x}^{-}(t) \in K(t, \boldsymbol{x}(t)). \qquad (4.57)$$

In Lagrangian dynamics, the measure differential inclusion (4.48) typically describes the time-evolution of absolutely continuous generalised coordinates $\boldsymbol{q}(t)$ and generalised velocities $\boldsymbol{u}(t)$, which are of locally bounded variation. The set-valued measure function $\mathrm{d}\boldsymbol{\varGamma}(t, \boldsymbol{x}(t))$ typically contains indicator functions, which impose constraints on the system. Unilateral constraints $g(\boldsymbol{q}) \geq 0$ and bilateral constraints $g(\boldsymbol{q}) = 0$ restrict the generalised coordinates $\boldsymbol{q}(t)$ to an admissible set. Consider for instance a ball with mass m of which we measure its height above the ground with the coordinate q and its velocity with u. The ground imposes a unilateral constraint $q(t) \geq 0$ which restricts the coordinate q to the set $\mathbb{R}^{+}$. If the ball hits the ground, i.e. $q(t) = 0$, then a contact impulse can cause a velocity jump in $u(t)$. We consider the impact law $u^{+}(t) = -eu^{-}(t)$, $0 \leq e \leq 1$, where e is the restitution coefficient. The unilateral constraint $q(t) \geq 0$ does not only restrict the generalised coordinate, but if $q(t) = 0$ then also the post-impact velocity $u^{+}(t)$ is restricted to $u^{+}(t) \geq 0$. Moreover, if we initialise our model for the ball with admissible initial conditions

$$q(t_0) \geq 0, \qquad u^{+}(t_0) : \begin{cases} u^{+}(t_0) > 0 & q(t_0) = 0 \\ u^{+}(t_0) \text{ arbitrary} & q(t_0) > 0, \end{cases}$$

then the time-evolution $q(t)$ remains admissible for all $t > t_0$. In this sense, the system is consistent. However, we can also specify the left-limit $u^{-}(t_0)$ of the velocity as initial condition. In the latter case there is no restriction on $u^{-}(t_0)$, because a possible impulse will force $u^{+}(t_0)$ to be admissible, because $e \geq 0$. Hence, the system is not consistent for $e < 0$. From the above example it becomes clear that the states $\boldsymbol{x}(t)$ of the measure differential inclusion (4.48) are restricted to an admissible set $\mathcal{A}$. What $\mathcal{A}$ is depends on how the initial condition is understood.

Following [122], we imagine that the investigated process is already in progress before t_0. While the evolution for $t \geq t_0$ is the time-domain of prediction, the initial data are understood to convey the abridged information on the system history before t_0. We therefore interpret the initial condition as the left-limit, i.e. $x^-(t_0) = x_0$. Using this interpretation of the initial condition, we find that the admissible set in the bouncing ball example is $\mathcal{A} = \{(q, u) \in \mathbb{R} \times \mathbb{R} \mid q \geq 0\}$, which does not involve a restriction on the velocities. The condition $u^+(t) \geq 0$ for $q(t) = 0$ with $t \geq t_0$, being the restriction on the velocity, is fulfilled by consistency of the impact law ($e \geq 0$).

For the initial value problem, consisting of the measure differential inclusion (4.48) with initial condition $x^-(t_0) = x_0$, we define a solution as follows:

Definition 4.7 (Solution of a Measure Differential Inclusion). *A solution* $x(t) = \varphi(t, t_0, x_0)$ *of the measure differential inclusion (4.48) with the initial condition* $x^-(t_0) = x_0$ *is a function* $x : \mathbb{R} \to \mathbb{R}^n$, *being of locally bounded variation, which fulfills (4.48) for all* $t \geq t_0$ *and which is not defined at its discontinuity points.*

Remark that a solution $x(t)$ is not defined on the discontinuity points t_i, but still fulfills the measure differential inclusion for all $t \geq t_0$ in the sense of (4.50). From the physical point of view (or modelling point of view), there exists no value that we can meaningfully assign to $x(t)$ on its discontinuity points. It therefore makes sense to leave the function $x(t)$ undefined at its discontinuity points. For this reason, we will consider solutions of (4.48) which are not defined at their discontinuity points.

We often prefer to write $\varphi(t, t_0, x_0) := x(t)$ to explicitly state its dependence on the initial condition $x^-(t_0) = x_0$. The solution starting from a specific initial condition (t_0, x_0) is generally not unique, i.e. there exists a set $\mathcal{S}(\mathrm{d}\Gamma, t_0, x_0)$ of solution curves such that

$$\varphi(\cdot, t_0, x_0) \in \mathcal{S}(\mathrm{d}\Gamma, t_0, x_0). \tag{4.58}$$

If $\mathcal{S}(\mathrm{d}\Gamma, t_0, x_0)$ contains a single element, then the solution is unique. If the set of solutions is empty, $\mathcal{S}(\mathrm{d}\Gamma, t_0, x_0) = \emptyset$, then a solution does not exist.

A solution curve $\varphi(t, t_0, x_0)$ with $x_0 \in \mathcal{A}$ is called 'viable' if it remains within the set $\mathcal{A}$ for all future times $t \geq t_0$ for which it is defined. If all solution curves of a measure differential inclusion are viable, and at least one solution curve exists for each initial condition, then the system is called consistent.

Definition 4.8 (Consistency of a Measure Differential Inclusion). *A measure differential inclusion (4.48) is called consistent if*

$$x_0 \in \mathcal{A} \Rightarrow \varphi(t, t_0, x_0) \in \mathcal{A} \text{ for almost all } t \geq t_0$$

for each solution curve $\varphi(\cdot, t_0, x_0) \in \mathcal{S}(\mathrm{d}\Gamma, t_0, x_0) \neq \emptyset$.

The admissible set $\mathcal{A}$ of a consistent measure differential inclusion is there-fore a positively invariant set (see Definition 6.4) of the system. This puts a tangency restriction on the differential measure $d\boldsymbol{x}$ on the set $\mathcal{A}$. If the sys-tem (4.48) is consistent, and if the solution $\boldsymbol{x}(t)$ is absolutely continuous, then it must hold that

$$d\boldsymbol{x} = \dot{\boldsymbol{x}}dt \in K_{\mathcal{A}}(\boldsymbol{x}), \qquad (4.59)$$

where $K_{\mathcal{A}}(\boldsymbol{x})$ is the contingent cone of $\mathcal{A}$ at the point $\boldsymbol{x}$ (Definition 2.25). In the following chapters, we will assume that the considered systems are consistent.

A set-valued measure function $d\boldsymbol{\Gamma}(\boldsymbol{x}) = \boldsymbol{\mathcal{F}}(\boldsymbol{x})dt + \boldsymbol{\mathcal{G}}(\boldsymbol{x}^+)d\eta$ is called (max-imal) monotone if the set-valued density functions $\boldsymbol{\mathcal{F}}(\boldsymbol{x})$ and $\boldsymbol{\mathcal{G}}(\boldsymbol{x}^+)$ are (max-imal) monotone. A sufficient criterion for uniqueness of solutions can again be gained from a monotonicity property of the system (see Section 4.2 for the specific case of differential inclusions).

Theorem 4.9 (Uniqueness of solutions through monotonicity). *Con-sider a measure differential inclusion of the form*

$$d\boldsymbol{x} \in -d\boldsymbol{A}(\boldsymbol{x}^+).$$

Let the system be consistent and have existence of solutions for all initial conditions $\boldsymbol{x}_0 \in \mathcal{A}$. If $d\boldsymbol{A}(\boldsymbol{x}^+)$ is a monotone set-valued measure function, then the measure differential inclusion has a unique solution $\boldsymbol{x}(t)$ for all initial conditions $\boldsymbol{x}(0) = \boldsymbol{x}_0 \in \mathcal{A}$.

Proof: The proof is similar to the proof of Theorem 4.5. Let $\boldsymbol{x}_1(\cdot) \in \mathcal{S}(-d\boldsymbol{A}, 0, \boldsymbol{x}_{10})$ and $\boldsymbol{x}_2(\cdot) \in \mathcal{S}(-d\boldsymbol{A}, 0, \boldsymbol{x}_{20})$ be solutions of the measure dif-ferential inclusion with $\boldsymbol{x}_{10}, \boldsymbol{x}_{20} \in \mathcal{A}$. For the differential measure of $d\boldsymbol{x}$ we write

$$d\boldsymbol{x} = -\boldsymbol{a}_t dt - \boldsymbol{a}_\eta d\eta, \qquad (4.60)$$

where the single-valued densities obey the set-valued force laws

$$-\dot{\boldsymbol{x}} = \boldsymbol{a}_t \in \boldsymbol{A}_t(\boldsymbol{x}), \qquad -(\boldsymbol{x}^+ - \boldsymbol{x}^-) = \boldsymbol{a}_\eta \in \boldsymbol{A}_\eta(\boldsymbol{x}^+). \qquad (4.61)$$

The set-valued operators $\boldsymbol{A}_t$ and $\boldsymbol{A}_\eta$ are maximal monotone operators and are related to the set-valued measure function by

$$d\boldsymbol{A}(\boldsymbol{x}^+) = \boldsymbol{A}_t(\boldsymbol{x})dt + \boldsymbol{A}_\eta(\boldsymbol{x}^+)d\eta. \qquad (4.62)$$

Consider the positive definite function $V \in \mathbb{R}$

$$V(t) = \frac{1}{2}\|\boldsymbol{x}_2(t) - \boldsymbol{x}_1(t)\|^2. \qquad (4.63)$$

The differential measure dV of V has a density $\dot{V}$ with respect to the Lebesgue measure and a density $V^+ - V^-$ with respect to the atomic measure $d\eta$

$$dV = \dot{V}dt + (V^+ - V^-)d\eta. \tag{4.64}$$

The Lebesgue measurable part yields

$$\begin{aligned}
\dot{V} &= (\boldsymbol{x}_2 - \boldsymbol{x}_1)^{\mathrm{T}}(\dot{\boldsymbol{x}}_2 - \dot{\boldsymbol{x}}_1) \\
&= -(\boldsymbol{x}_2 - \boldsymbol{x}_1)^{\mathrm{T}}(\boldsymbol{a}_{t2} - \boldsymbol{a}_{t1}) \\
&\leq 0,
\end{aligned} \tag{4.65}$$

with $\boldsymbol{a}_{t1} \in \mathcal{A}_t(\boldsymbol{x}_1), \boldsymbol{a}_{t2} \in \mathcal{A}_t(\boldsymbol{x}_2)$, due to the monotonicity of $\mathcal{A}_t(\cdot)$. Evaluation of the atomic part gives

$$\begin{aligned}
V^+ - V^- &= \frac{1}{2}(\boldsymbol{x}_2^+ + \boldsymbol{x}_2^- - \boldsymbol{x}_1^+ - \boldsymbol{x}_1^-)^{\mathrm{T}}(\boldsymbol{x}_2^+ - \boldsymbol{x}_2^- - \boldsymbol{x}_1^+ + \boldsymbol{x}_1^-) \\
&= -\frac{1}{2}(2\boldsymbol{x}_2^+ - 2\boldsymbol{x}_1^+ + \boldsymbol{a}_{\eta2} - \boldsymbol{a}_{\eta1})^{\mathrm{T}}(\boldsymbol{a}_{\eta2} - \boldsymbol{a}_{\eta1}) \\
&= -(\boldsymbol{x}_2^+ - \boldsymbol{x}_1^+)^{\mathrm{T}}(\boldsymbol{a}_{\eta2} - \boldsymbol{a}_{\eta1}) - \frac{1}{2}\|\boldsymbol{a}_{\eta2} - \boldsymbol{a}_{\eta1}\|^2 \\
&\leq 0,
\end{aligned} \tag{4.66}$$

with $\boldsymbol{a}_{\eta1} \in \mathcal{A}_\eta(\boldsymbol{x}_1^+)$, $\boldsymbol{a}_{\eta2} \in \mathcal{A}_\eta(\boldsymbol{x}_2^+)$, in which we used the monotonicity of $\mathcal{A}_\eta(\cdot)$. Hence, the function V can not increase, meaning that the distance between $\boldsymbol{x}_1(t)$ and $\boldsymbol{x}_2(t)$ can not increase, i.e. $\|\boldsymbol{x}_1^+(t) - \boldsymbol{x}_2^+(t)\| \leq \|\boldsymbol{x}_{10} - \boldsymbol{x}_{20}\|$ for all $t \geq t_0$. Taking $\boldsymbol{x}_{10} = \boldsymbol{x}_{20}$ it follows that $\boldsymbol{x}_1(t) = \boldsymbol{x}_2(t)$ for almost all t. Using the fact that $\boldsymbol{x}_1(t)$ and $\boldsymbol{x}_2(t)$ are of locally bounded variation, it follows that $\boldsymbol{x}_1(t)$ and $\boldsymbol{x}_2(t)$ have the same discontinuity points. The functions $\boldsymbol{x}_1(t)$ and $\boldsymbol{x}_2(t)$ are not defined at their discontinuity points. Consequently, the trajectories $\boldsymbol{x}_1(t)$ and $\boldsymbol{x}_2(t)$ are identical, i.e. the system has uniqueness of solutions. □

4.4 Summary

Differential equations with a non-smooth continuous right-hand side have, under mild conditions (Lipschitz constant), a unique solution in forward and backward time. When dealing with differential equations with a discontinuous right-hand side, one has to think about an appropriate solution concept. Filippov's solution concept consists of filling in the graph with a convex set, which yields a differential inclusion. Differential inclusions can describe the (non-smooth) time-evolution of systems with an absolutely continuous state. The differential measure of the state, which usually consists of a Lebesgue measureable part, can be augmented with an atomic part. An inclusion based on this differential measure yields a measure differential inclusion, which can describe the time-evolution of systems with a discontinuous state (of locally bounded variation). A sufficient condition for uniqueness of solutions of (measure) differential inclusions follows from the monotonicity of the negative right-hand side. The framework of measure differential inclusions will be used in the following chapters to set up a Lyapunov stability theory for non-smooth systems and to describe the non-smooth dynamics of mechanical systems with frictional unilateral contacts.

5

Mechanical Systems with Set-valued Force Laws

Some theoretical background on non-smooth systems has been discussed in the previous chapter. Mechanical multibody systems form a special and important class of non-smooth systems, because they can be cast in an elegant structured form. The special structure of mechanical systems is due to the fact that the dynamics is described by the Lagrangian formalism, which links dynamics to variational calculus. Moreover, contact forces are incorporated in the equation of motion by using the Lagrange multiplier theorem. But most importantly, contact forces are (mostly) derived from (pseudo-)potentials or dissipation functions.

In this chapter we will discuss the mathematical formulation of Lagrangian mechanical systems with unilateral contact and friction modelled with set-valued force laws. It is important to note that (finite-dimensional) Lagrangian mechanical systems encompass rigid multibody systems as well as discretized continuous systems (e.g. through a Ritz approach or a finite-element discretization) with possible frictional unilateral contacts. First, we discuss how set-valued force laws can be derived from non-smooth potentials. Subsequently, we treat the contact laws for unilateral contact and various types of friction within the setting of non-smooth potential theory. This leads to a unified approach with which all set-valued forces can be formulated. Finally, we incorporate the set-valued forces as Lagrangian multipliers in the Newton-Euler equations. The notation in this chapter is kept as close as possible to the notation of Glocker [63].

5.1 Non-smooth Potential Theory

Mechanical constitutive laws are commonly expressed by derivatives of potentials. Both conservative and non-conservative forces can be derived from potentials. However, non-conservative potentials, also called dissipation functions, are non-integrable, i.e. the value of the integral is path-dependent.

The potential of a linear mechanical spring is given by $U(q) = \frac{1}{2}kq^2$, where k is the spring stiffness and q is the elongation of the spring. The force law of a linear spring can be expressed as

$$-\lambda = \frac{\partial U(q)}{\partial q} = kq, \tag{5.1}$$

where λ is the force exerted by the spring. Similarly, the potential (usually called dissipation function) for a linear viscous damper is given by $\Phi(u) = \frac{1}{2}cu^2$, where c is the damping constant and u is the time-derivative of the elongation of the damper. The force law of a linear damper can be expressed as

$$-\lambda = \frac{\partial \Phi(u)}{\partial u} = cu. \tag{5.2}$$

The potentials $U(q)$ and $\Phi(u)$ of a linear spring and damper are quadratic smooth functions of the displacement q or velocity u. Non-quadratic smooth potentials describe the energy or dissipation rate of nonlinear smooth force laws. However, many force laws are expressed by non-smooth potentials [61, 63].

Force laws that can be derived from a potential or dissipation function are generally of the form[1]

$$-\lambda \in \partial\pi(s), \tag{5.3}$$

in which λ is the force in the force element, s is the flow variable and $\pi(s)$ is the potential. The potential $\pi(s)$ is assumed to be differentiable in the generalised sense of Clarke, i.e. the generalised differential $\partial\pi(s)$ exists. The force law is stated in the form of an inclusion because the generalised differential $\partial\pi(s)$ is set-valued if $\pi(s)$ is non-smooth. Following [61, 63], our analysis of non-smooth potential functions will be restricted to functions $\pi : \mathbb{R} \to (-\infty, \infty]$ which can be decomposed into a differentiable function $\pi_D(s)$, an indicator function $\pi_S(s)$ and a convex potential $\pi_P(s)$ with polyhedral epigraph:

$$\pi(s) = \pi_D(s) + \pi_S(s) + \pi_P(s). \tag{5.4}$$

Without loss of generality we set $\pi(0) = 0$ and therefore consider force elements for which the energy or dissipation rate is zero for a vanishing flow variable s. The indicator function π_S is defined as $\pi_S(s) = \Psi_S(s)$ on some closed interval $S = [s_A, s_E]$ and the convex potential $\pi_P(s)$ can be written in the form

$$\pi_P(s) = \sum_{i=1}^{m} \pi_{Pi}(s), \qquad \pi_{Pi}(s) = \begin{cases} 0 & s \le s_i \\ a_i(s - s_i) & s > s_i \end{cases}, \tag{5.5}$$

with $a_i > 0$. The decomposition (5.4) is illustrated in Figure 5.1.

---

[1] The notation differs from [101] in which the force law is defined as $\lambda \in \partial\pi(s)$.

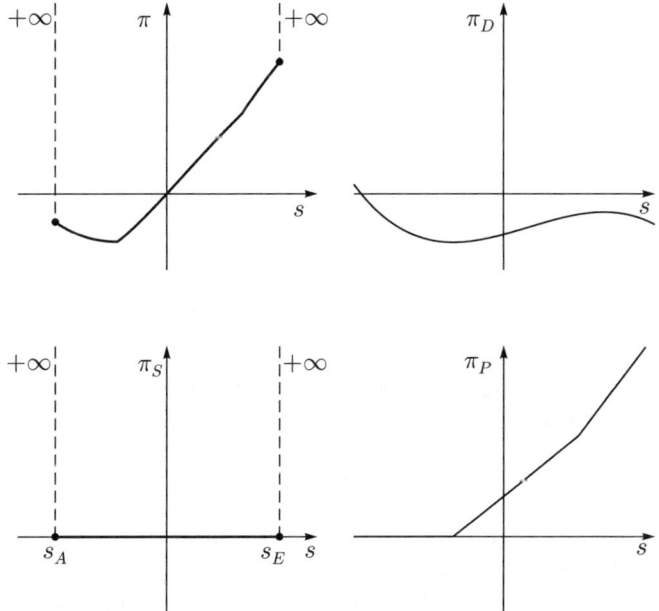

Fig. 5.1. The decomposition of a potential $\pi(s) = \pi_D(s) + \pi_S(s) + \pi_P(s)$.

The generalised differential ∂ of $\pi_D(s)$ equals the gradient ∇, because $\pi_D(s)$ is differentiable. The potentials $\pi_S(s)$ and $\pi_P(s)$ are convex and their generalised differential therefore equals the subdifferential. Hence, we obtain

$$\partial\pi(s) = \nabla\pi_D(s) + \partial\pi_S(s) + \partial\pi_P(s). \tag{5.6}$$

We can therefore also decompose the force λ in the same fashion:

$$\lambda = \lambda_D + \lambda_S + \lambda_P, \tag{5.7}$$

with the force laws

$$-\lambda_D = \nabla\pi_D(s), \qquad -\lambda_S \in \partial\pi_S(s), \qquad -\lambda_P \in \partial\pi_P(s). \tag{5.8}$$

Note that only the differentiable part $\pi_D(s)$ is allowed to be non-convex.

We now study the case that the potential $\pi(s)$ is convex, i.e. $\pi_D(s)$ is convex or vanishes. A convex potential $\pi(s)$ has a conjugate potential $\pi^*(-\lambda)$ (see Definition 2.28)

$$\pi^*(-\lambda) = \sup_{s}\{-s\lambda - \pi(s)\}. \tag{5.9}$$

The force law can now be also be stated in its conjugate form

$$-\lambda \in \partial\pi(s) \iff s \in \partial\pi^*(-\lambda), \quad s \in S, \ -\lambda \in C \tag{5.10}$$

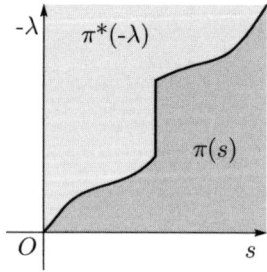

Fig. 5.2. Complementary behaviour of $\pi(s)$ and $\pi^*(-\lambda)$.

where C denotes the effective domain of $\pi^*(-\lambda)$. If the force law $-\lambda \in \partial\pi(s)$ is fulfilled, then it holds that

$$-\lambda s = \pi(s) + \pi^*(-\lambda), \tag{5.11}$$

from which we see (Figure 5.2) that the potential $\pi(s)$ and its conjugate $\pi^*(-\lambda)$ describe complementary areas in the first quadrant of the s–λ plane. The conjugate potential $\pi^*(-\lambda)$ is therefore known as 'complementary energy function' in structural mechanics.

We will study the convex potentials $\pi_S(s)$ and $\pi_P(s)$ in more detail. The subdifferential of the indicator function $\pi_S(s) = \Psi_S(s)$ is the normal cone to S, i.e.

$$-\lambda_S \in N_S(s). \tag{5.12}$$

Similarly, taking the conjugate of the potential $\pi_{Pi}(s)$

$$
\begin{aligned}
\pi_{Pi}^*(-\lambda_{Pi}) &= \sup_s\{-s\lambda_{Pi} - \pi_{Pi}(s)\} \\
&= \sup_s \left\{ \begin{array}{ll} -s\lambda_{Pi} & \text{if } s \le s_i \\ -s\lambda_{Pi} - a_i(s - s_i) & \text{if } s > s_i \end{array} \right\} \\
&= \left\{ \begin{array}{l} -s_i\lambda_{Pi} \text{ if } 0 \le -\lambda_{Pi} \le a_i \\ \infty \qquad \text{else} \end{array} \right. \\
&= \Psi_{A_i}(-\lambda_{Pi}) - s_i\lambda_{Pi}, \qquad A_i = [0, a_i],
\end{aligned}
\tag{5.13}
$$

and using

$$s \in \partial\pi_{Pi}^*(-\lambda_{Pi}) \implies s \in \partial\Psi_{A_i}(-\lambda_{Pi}) + s_i, \tag{5.14}$$

we deduce that it holds that

$$s - s_i \in N_{A_i}(-\lambda_{Pi}). \tag{5.15}$$

The latter exposition shows how the force laws associated with the potentials $\pi_S(s)$ and $\pi_P(s)$ can be written in a 'normal cone' formulation.

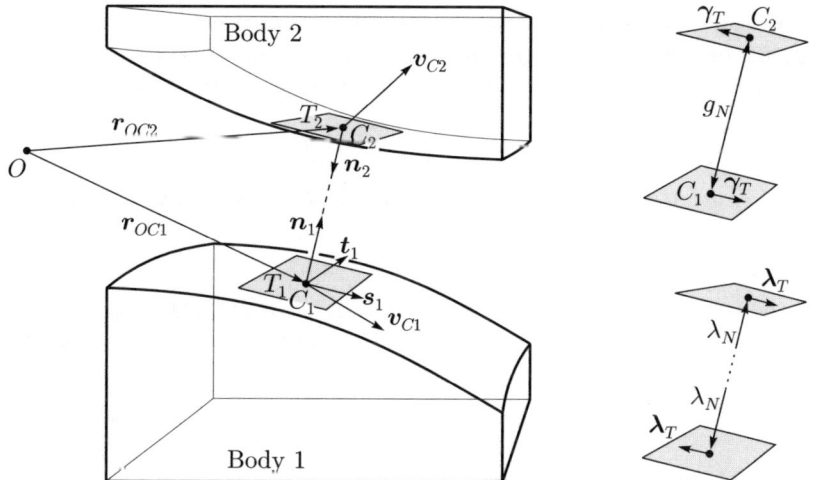

Fig. 5.3. Contact distance g_N and tangential velocity γ_T between two rigid bodies.

5.2 Contact Geometry

In the following section, we study a number of set-valued force laws that are common in the description of contact problems of mechanical systems. We therefore need to describe what we mean with contact points, their distance and their relative velocity. Consider two convex bodies, say body 1 and 2, which have a locally smooth surface in the region where they are close to each other, as shown in Figure 5.3. We denote the point on body 1, which is closest to body 2, by P_1. Similarly, the point P_2 is the point on body 2, which is closest to body 1. Note that the points P_i wander over the body while the bodies move in time. Let the body-fixed point C_1 on body 1 momentarily coincide with the point P_1, i.e. $r_{OP1} = r_{OC1}$. Similarly, let the body-fixed point C_2 on body 2 momentarily coincide with the point P_2. For each point C_j on the locally smooth surface of body j we can define an outward normal n_j and tangent plane T_j, $j = 1, 2$. Two points C_1 and C_2, momentarily coinciding with P_1 and P_2, on the surface of body 1 and 2 respectively are said to form a *contact pair* or *contact point* if their outward normals oppose each other:

$$n_1 = -n_2, \tag{5.16}$$

where n_j is the outward normal on body j at C_j. The tangent planes T_1 and T_2 are therefore necessarily parallel. The tangent plane T_j is spanned by the vectors t_j and s_j, such that (s_j, t_j, n_j) is a body-fixed orthonormal frame at the body-fixed point C_j. The points P_1 and P_2 are the closest extremal points of the two bodies under consideration and are candidates to contact. The contact distance

$$g_N = (r_{OP2} - r_{OP1})^T n_1 = (r_{OP1} - r_{OP2})^T n_2 \tag{5.17}$$

is the distance between the two bodies. The bodies are separated when $g_N > 0$, are in contact when $g_N = 0$, and penetrate each other when $g_N < 0$. The bodies are approaching each other with the normal contact velocity

$$\gamma_N = (v_{C2} - v_{C1})^T n_1, \tag{5.18}$$

where v_{Cj} are the absolute velocities of the body-fixed points C_j. The difference of velocities

$$v_{CjPj} = v_{Pj} - v_{Cj} \tag{5.19}$$

gives the velocity with which the point P_j wanders over body j. The points C_j move also relative to each other along the tangent plane T_1 with the tangential contact velocity

$$\gamma_T = \begin{bmatrix} \gamma_{T1} \\ \gamma_{T2} \end{bmatrix} \in T_1, \tag{5.20}$$

with the components

$$\gamma_{T1} = (v_{C2} - v_{C1})^T t_1, \quad \gamma_{T2} = (v_{C2} - v_{C1})^T s_1. \tag{5.21}$$

If the bodies are in contact, i.e. $g_N = 0$, then the tangential contact velocity γ_T becomes the relative velocity with which the bodies slide over each other expressed in the frame (t_1, s_1), and is a two-dimensional vector in the tangent plane $T = T_1 = T_2$.

5.3 Force Laws for Frictional Unilateral Contact

In this section we will discuss various set-valued force laws for the constitutive description of unilateral contact with friction. Bodies in contact naturally deform under the influence of contact forces yielding a contact area. Usually, the local deformations at the contact area, as well as the size of the contact area, are negligibly small compared with the global motion of the contacting bodies. We can therefore often assume that the bodies are deformable but neglect the local indentation and idealise the contact area to be a contact point [128]. A possible further modelling step, although not essential, is the assumption of rigidity of the contacting bodies, in which also the deformation of the bodies is neglected [139].

5.3.1 Signorini's Contact Law

Contact in normal direction between contacting bodies is usually described by Signorini's law which is the most elementary set-valued force law. Signorini's law describes the impenetrability of the contact, i.e. the contact distance is nonnegative ($g_N \geq 0$). A contact force λ_N along the line C_1–C_2 can be positive when contact is present ($g_N = 0$), but must vanish when the contact is open ($g_N > 0$). The normal contact force λ_N between the bodies is nonnegative

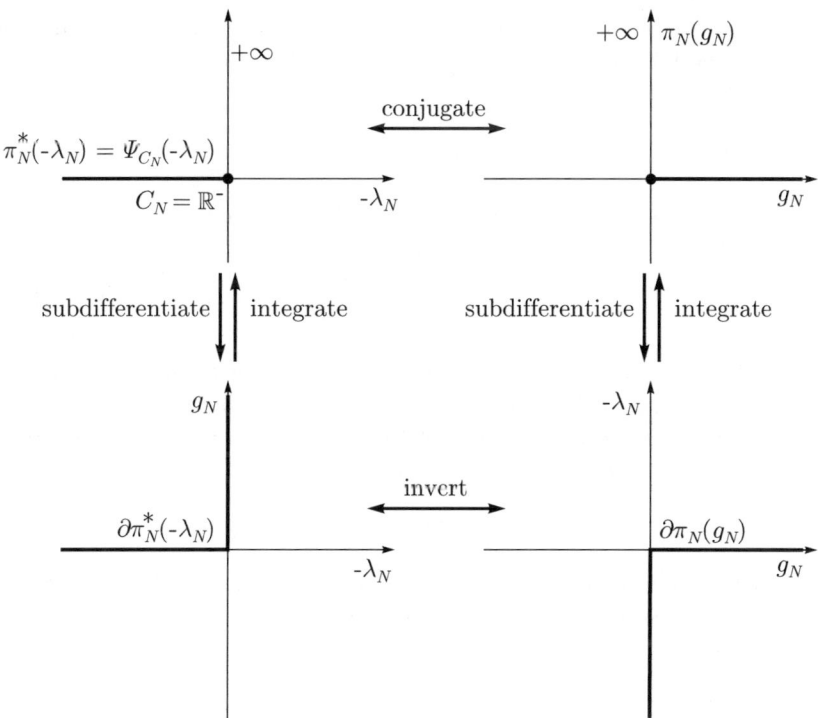

Fig. 5.4. Potential, conjugate potential and Signorini's force law.

because the bodies can only push on each other, i.e. the constraint is unilateral. Only two situations may occur:

$$g_N = 0 \wedge \lambda_N \geq 0 \quad \text{contact}$$
$$g_N > 0 \wedge \lambda_N = 0 \quad \text{no contact} \tag{5.22}$$

From (5.22) we see that the normal contact law shows a complementarity behaviour: the product of the contact force and normal contact distance is always zero:

$$g_N \lambda_N = 0. \tag{5.23}$$

Hence, Signorini's law does not lead to energy creation or dissipation. The relation between the normal contact force and the normal contact distance is therefore given by

$$g_N \geq 0, \quad \lambda_N \geq 0, \quad g_N \lambda_N = 0, \tag{5.24}$$

which is the inequality complementarity condition between g_N and λ_N, also known as the Signorini-Fichera condition. We now put the Signorini-Fichera condition (5.24) in the framework of non-smooth potential theory. Consider $s = g_N$ to be the flow variable with its dual force $\lambda = \lambda_N$. The Signorini-Fichera condition (5.24) appears to be a maximal monotone set-valued force law $-\lambda_N(g_N)$ (see the lower right figure in Figure 5.4). The maximal monotonicity of the force law implies that it can be derived from a convex potential $\pi_N(g_N)$. We therefore write

$$-\lambda_N \in \partial \pi_N(g_N) \tag{5.25}$$

and find that the associated energy potential is $\pi_N(g_N)$:

$$\pi_N(g_N) = \begin{cases} \infty, & g_N < 0, \\ 0, & g_N \geq 0. \end{cases} \tag{5.26}$$

We see that the potential $\pi_N(g_N)$ purely consists of an indicator function $\pi_N(g_N) = \pi_{NS}(g_N) = \Psi_{\mathbb{R}+}(g_N)$. The conjugate potential $\pi_N^*(-\lambda_N)$ can easily be shown to be equal to the indicator function $\Psi_{C_N}(-\lambda_N)$ on the convex set $C_N = \mathbb{R}^-$. We interpret the set

$$C_N = \{-\lambda_N \in \mathbb{R} | \lambda_N \geq 0\} \tag{5.27}$$

as the set of admissible (negative) normal contact forces. Equivalently, we state the force law in its conjugate form

$$g_N \in \partial \pi_N^*(-\lambda_N). \tag{5.28}$$

We now use the equivalence (2.27) between the normal cone and the subdifferential of the indicator function and obtain $N_{C_N}(-\lambda_N) = \partial \Psi_{C_N}(-\lambda_N) = \partial \pi_N^*(-\lambda_N)$ with which we cast the force law in a normal cone formulation

$$g_N \in N_{C_N}(-\lambda_N). \tag{5.29}$$

If the system has a number of contact points $i = 1, \ldots, n_C$, then the force law

$$g_{Ni} \in N_{C_{Ni}}(-\lambda_{Ni}), \qquad C_{Ni} = \{-\lambda_{Ni} \in \mathbb{R} | \lambda_{Ni} \geq 0\}, \tag{5.30}$$

holds for each contact point i. Note that all sets C_{Ni} are identical, $C_{Ni} = C_N = \mathbb{R}^-$. Using the short-hand notation

$$\mathbf{x} \geq \mathbf{0} \quad \Longleftrightarrow \quad x_i \geq 0 \;\; \forall i, \tag{5.31}$$

the normal contact laws (5.30) can be written in vector form as

$$\mathbf{g}_N \in N_{C_N}(-\boldsymbol{\lambda}_N), \qquad C_N = \{-\boldsymbol{\lambda}_N \in \mathbb{R}^n | \boldsymbol{\lambda}_N \geq \mathbf{0}\}, \tag{5.32}$$

where $\boldsymbol{\lambda}_N$ is the vector containing the normal contact forces λ_{Ni} and $\mathbf{g}_N$ is the vector of normal contact distances g_{Ni}.

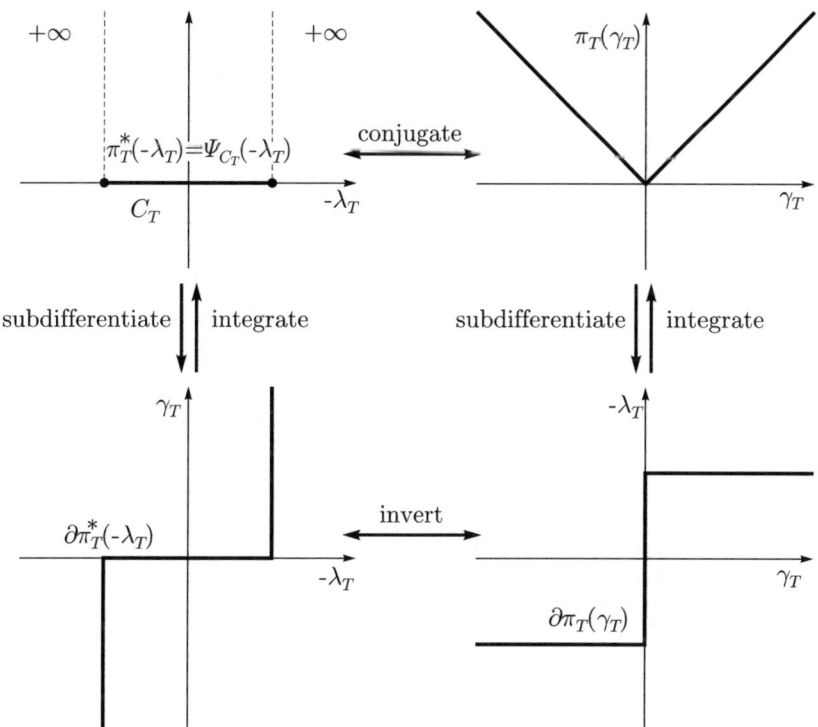

Fig. 5.5. Potential, conjugate potential and Coulomb's friction law.

5.3.2 Coulomb's Friction Law

Coulomb's friction law is another classical example of a force law that can be described by a non-smooth potential. Before discussing Coulomb's spatial friction law, we first study the planar case for which the relative sliding velocity γ_T is a scalar. If contact is present between the bodies, i.e. $g_N = 0$, then the friction between the bodies imposes a scalar force λ_T along the tangent line of the contact point. When the bodies are sliding over each other, the friction force λ_T equals to $\mu\lambda_N$ and acts in the direction of $-\gamma_T$

$$-\lambda_T = \mu\lambda_N \operatorname{sign}(\gamma_T), \quad \gamma_T \neq 0, \tag{5.33}$$

where μ is the friction coefficient and λ_N is the normal contact force. The classical Coulomb friction law considers the friction coefficient to be constant. If the relative tangential velocity vanishes, i.e. $\gamma_T = 0$, then the bodies purely roll over each other without slip. Pure rolling, or no slip for locally flat objects, is denoted by *stick*. If the bodies stick, then the friction force must lie in the interval $-\mu\lambda_N \leq \lambda_T \leq \mu\lambda_N$. For unidirectional friction, i.e. for planar contact problems, the following three cases are possible:

$$\gamma_T = 0 \Rightarrow |\lambda_T| \leq \mu\lambda_N \quad \text{sticking,}$$
$$\gamma_T < 0 \Rightarrow \lambda_T = +\mu\lambda_N \quad \text{negative sliding,} \qquad (5.34)$$
$$\gamma_T > 0 \Rightarrow \lambda_T = -\mu\lambda_N \quad \text{positive sliding.}$$

We can express the friction force by a (pseudo-)potential $\pi_T(\gamma_T)$, which we mechanically interpret as a dissipation function,

$$-\lambda_T \in \partial\pi_T(\gamma_T), \quad \pi_T(\gamma_T) = \mu\lambda_N|\gamma_T|, \qquad (5.35)$$

from which follows the maximal monotone set-valued force law

$$-\lambda_T \in \begin{cases} \mu\lambda_N, & \gamma_T > 0, \\ [-1,1]\mu\lambda_N, & \gamma_T = 0, \\ -\mu\lambda_N, & \gamma_T < 0, \end{cases} \qquad (5.36)$$

or

$$-\lambda_T \in \mu\lambda_N \operatorname{Sign}(\gamma_T). \qquad (5.37)$$

The admissible values of the (negative) tangential force $-\lambda_T$ form a convex set C_T which is bounded by the values of the normal force [139]:

$$C_T = \{-\lambda_T \mid -\mu\lambda_N \leq \lambda_T \leq +\mu\lambda_N\}. \qquad (5.38)$$

Coulomb's law can be expressed with the aid of the indicator function of C_T as

$$\gamma_T \in \partial\Psi_{C_T}(-\lambda_T) \Leftrightarrow \gamma_T \in N_{C_T}(-\lambda_T), \qquad (5.39)$$

where the indicator function Ψ_{C_T} is the conjugate potential of π_T, i.e. $\pi_T(\gamma_T) = \Psi^*_{C_T}(\gamma_T)$, see Figure 5.5. We call $\pi_T(\gamma_T)$ a pseudo-potential, because it is dependent on the normal contact force, which is in turn dependent on the motion of the system, and is therefore not a true potential. In the following, the word 'pseudo' will be omitted for the sake of brevity.

The classical Coulomb's friction law for spatial contact formulates a two-dimensional friction force

$$\boldsymbol{\lambda}_T = \begin{bmatrix} \lambda_{T1} \\ \lambda_{T2} \end{bmatrix}, \qquad (5.40)$$

with the components

$$\lambda_{T1} = \boldsymbol{\lambda}^\mathrm{T}\boldsymbol{t}_1, \quad \lambda_{T2} = \boldsymbol{\lambda}^\mathrm{T}\boldsymbol{s}_1, \qquad (5.41)$$

where $\boldsymbol{\lambda} = \lambda_N\boldsymbol{n}_1 + \lambda_{T1}\boldsymbol{t}_1 + \lambda_{T2}\boldsymbol{s}_1$ is the total contact force exerted by body 1 on body 2. The set of admissible friction forces is a convex set $C_T \subset \mathbb{R}^2$ which is a disk for isotropic Coulomb friction:

$$C_T = \{-\boldsymbol{\lambda}_T \mid \|\boldsymbol{\lambda}_T\| \leq \mu\lambda_N\}. \qquad (5.42)$$

Using the set C_T, the spatial Coulomb friction law can be formulated as

$$\gamma_T \in \partial\Psi_{C_T}(-\boldsymbol{\lambda}_T) \Leftrightarrow -\boldsymbol{\lambda}_T \in \partial\Psi^*_{C_T}(\gamma_T) \Leftrightarrow \gamma_T \in N_{C_T}(-\boldsymbol{\lambda}_T), \qquad (5.43)$$

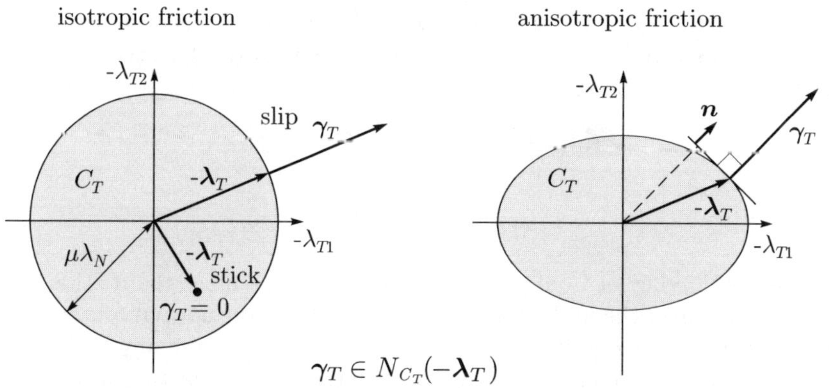

isotropic friction   anisotropic friction

$$\gamma_T \in N_{C_T}(-\boldsymbol{\lambda}_T)$$

Fig. 5.6. Graphical interpretation of Coulomb's friction law.

in which $\gamma_T \in \mathbb{R}^2$ is the relative sliding velocity. Similarly, an elliptic choice of C_T would result in an orthotropic Coulomb friction law (see Figure 5.6). Note that the dual variables $-\boldsymbol{\lambda}_T$ and γ_T are no longer aligned if C_T is non-circular. For a fixed vector γ_T, the vector $-\boldsymbol{\lambda}_T$ is that vector in C_T which causes the inner product $-\gamma_T^{\mathrm{T}}\boldsymbol{\lambda}_T$ to be maximal. In other words: a vector γ_T defines a unique tangent plane on the boundary of C_T such that $\boldsymbol{n} = \gamma_T/\|\gamma_T\|$ is the normal to the tangent plane. The projection of $-\boldsymbol{\lambda}_T$ on the normal $\boldsymbol{n}$ to this tangent plane is maximal. The force law (5.43) corresponds therefore to a principle of maximal dissipation $-\gamma_T^{\mathrm{T}}\boldsymbol{\lambda}_T$.

Coulomb's friction law in its classical setting considers the friction coefficient to be constant. In order to model the Stribeck effect, which can lead to the occurrence of friction induced vibrations [101, 102], the friction coefficient is usually considered to be dependent on the relative sliding velocity γ_T, i.e.

$$-\lambda_{\mathrm{Stribeck}} \in \mu(\gamma_T)\lambda_N \operatorname{Sign}(\gamma_T), \qquad (5.44)$$

where $\mu : \mathbb{R} \to \mathbb{R}^+$ is a continuous function. Such a type of law is associated with a non-convex potential $\pi_{\mathrm{Stribeck}}(-\gamma_T)$ such that $-\lambda_{\mathrm{Stribeck}} = \partial\pi_{\mathrm{Stribeck}}(\gamma_T)$. The friction force $\lambda_{\mathrm{Stribeck}}$ can be decomposed in a force λ_T, which obeys the standard Coulomb set-valued friction law, and a force $\lambda_{\mathrm{smooth}}$ being a smooth function of γ_T:

$$
\begin{aligned}
-\lambda_{\mathrm{Stribeck}} &\in \mu(\gamma_T)\lambda_N \operatorname{Sign}(\gamma_T) \\
&= \mu(0)\lambda_N \operatorname{Sign}(\gamma_T) + (\mu(\gamma_T) - \mu(0))\lambda_N \operatorname{Sign}(\gamma_T) \\
&= \mu(0)\lambda_N \operatorname{Sign}(\gamma_T) + (\mu(\gamma_T) - \mu(0))\lambda_N \operatorname{sign}(\gamma_T) \\
&= -\lambda_T - \lambda_{\mathrm{smooth}}
\end{aligned}
\qquad (5.45)
$$

with

$$-\lambda_T \in \mu(0)\lambda_N \operatorname{Sign}(\gamma_T), \qquad -\lambda_{\mathrm{smooth}} = \underbrace{(\mu(\gamma_T) - \mu(0))\lambda_N}_{\mu_{\mathrm{smooth}}} \operatorname{sign}(\gamma_T), \quad (5.46)$$

and $\mu_{\text{smooth}}(0) = 0$. The potential $\pi_{\text{Stribeck}}(\gamma_T)$ can be decomposed accordingly in a smooth potential $\pi_{\text{smooth}}(\gamma_T)$ and the convex potential $\pi_T(\gamma_T)$ of the classical Coulomb friction law:

$$\pi_{\text{Stribeck}}(\gamma_T) = \pi_{\text{smooth}}(\gamma_T) + \pi_T(\gamma_T), \tag{5.47}$$

with

$$-\lambda_{\text{smooth}} = \nabla \pi_{\text{smooth}}(\gamma_T), \quad -\lambda_T \in \partial \pi_T(\gamma_T). \tag{5.48}$$

The friction law of Coulomb, as defined above, assumes the friction forces to be a function of the unilateral normal forces. Both the normal contact forces and the friction forces have to be determined. However, in many applications the situation is less complicated as the normal force is constant or at least a given function of time. A known normal contact force allows for a simplified contact law. The tangential friction forces are assumed to obey either one of the following friction laws:

1. <u>Non-associated Coulomb's law</u> for which the normal force is dependent on the generalised coordinates q and/or generalised velocities u and therefore not known in advance. The set of (negative) admissible contact forces is given by

$$C_T(\lambda_N) = \{-\lambda_T \mid \|\lambda_T\| \le \mu \lambda_N\}, \tag{5.49}$$

which is dependent on the normal contact forces λ_N and friction coefficient μ. The dissipation function is a pseudo-potential.

2. <u>Associated Coulomb's law</u> for which the normal force is known in advance. The set of (negative) admissible contact forces is given by

$$C_T = \{-\lambda_T \mid \|\lambda_T\| \le \mu F_N\}, \tag{5.50}$$

which is dependent on the known normal forces F_N and friction coefficient μ. The dissipation function is a true potential.

5.3.3 Coulomb-Contensou Friction Law

Bodies in contact do not only slide over each other, but also pivot with respect to each other. Their relative pivoting velocity

$$\omega_N = (\Omega_2 - \Omega_1)^{\mathrm{T}} n_1 \tag{5.51}$$

depends on the angular velocity vectors Ω_j of each body j. In the following, we refer to ω_N as the normal spin. A combined friction law, which takes into account sliding friction as well a pivoting (or drilling) friction, can be formulated using a three-dimensional set of admissible (generalised) friction forces and is called the spatial Coulomb-Contensou friction law [99].

The set-valued force law for spatial Coulomb friction describes the friction force λ_T of a single contact point with two components $\lambda_T = \begin{bmatrix} \lambda_{T1} & \lambda_{T2} \end{bmatrix}^{\mathrm{T}}$. The

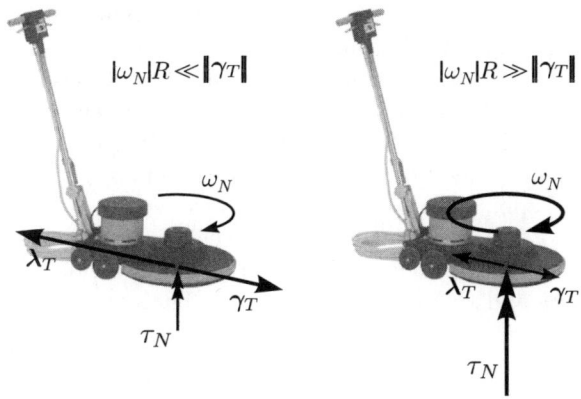

Fig. 5.7. The Contsensou-effect explained on an electric polishing machine.

friction force $\boldsymbol{\lambda}_T$ lies within a disk $\|\boldsymbol{\lambda}_T\| \leq \mu\lambda_N$ on the tangent plane of the contact point. The friction disk constitutes a friction cone in the $(\lambda_{T1}, \lambda_{T2}, \lambda_N)$ space. The pair of contacting bodies are assumed to be rigid and impenetrable within the framework of rigid multibody dynamics and the contact is therefore idealised to be a point. The contact point can not transmit a friction torque and the influence of normal spin and pivoting friction on the sliding friction $\boldsymbol{\lambda}_T$ is therefore usually neglected. In reality, the stiff (but still deformable) bodies deform and touch each other on a small contact surface, being more or less circular. The small deformations of stiff bodies are negligible compared to the geometry of the bodies and the global rigid body motion they undergo, but lead to contacting areas which can influence the dynamics of the system. A contact *surface* can not only transmit a sliding force $\boldsymbol{\lambda}_T$ tangent to the contact surface, but also transmit a friction torque τ_N normal to the contact surface. The effective radius of the contact surface is influenced by the normal contact force λ_N, the elasticity of the contacting bodies, the surface roughness and the pressure distribution. The friction torque τ_N is, in the absence of sliding and tangential contact force, more or less proportional to the normal contact force, the friction coefficient and the effective radius, i.e $\tau_N \propto \mu\lambda_N R_{\text{eff}}$. The influence of the friction torque is in most applications neglected because the effective radius is very small in practice. If, however, an object is spinning fastly, then the influence of normal spin and pivoting friction on the dynamics becomes large and can no longer be neglected.

Contensou [39] realised that pivoting friction and especially normal spin are important for the dynamics of fastly spinning tops. The pivoting friction is obviously necessary to describe the gradual loss of energy of the top which brings it back to rest. More of interest is the influence of the spinning velocity on the sliding friction force. A fastly spinning top experiences very little resistance in sliding direction and easily wanders over the floor. The same phenomenon occurs in an electric polishing machine with turning brushes used to clean floors [110]. The machine (Figure 5.7) is hard to move when the brushes are not rotating (Coulomb friction), but the machine can easily be pushed over the floor with rotating brushes. This is called the Contensou effect. Note that the opposite holds for the torque which the motor delivers. If the machine is not moved then the motor has to deliver a high torque and is experiencing pure Coulomb friction in a rotational sense. If the sliding speed is much higher than the normal spin, then the torque is lowered.

Consider a contact surface, which may be elliptic or even non-convex, with a normal pressure distribution σ. The relative sliding velocity of the contact is denoted by γ_T and the normal spin by ω_N. The normal spin ω_N is scaled to a velocity

$$\gamma_P = \omega_N R, \tag{5.52}$$

where R is some characteristic length of the contact surface (e.g. the radius if the contact surface is circular). The vector

$$\gamma_F = \begin{bmatrix} \gamma_{T1} \\ \gamma_{T2} \\ \gamma_P \end{bmatrix} \tag{5.53}$$

expresses the generalised sliding velocity. Coulomb's spatial friction law is assumed to hold on an arbitrary surface element dA with sliding velocity w. The sliding velocity w is a function of the generalised sliding velocity γ_F. The magnitude of the friction force on dA is according to Coulomb's law $d\lambda_T = -\mu\sigma dA \frac{w}{\|w\|}$ for $w \neq 0$. The velocity potential (or dissipation function) for dA is $d\pi_F = -w^T d\lambda_T = \mu\sigma\|w\|dA$. The potential for the total contact surface is obtained by integrating over the contact surface

$$\pi_F(\gamma_F; \sigma) = \iint_A \mu\sigma\|w(\gamma_F)\|dA. \tag{5.54}$$

It can be proven that the potential $\pi_F(\gamma_F; \sigma)$ is a support function on a convex set $B_F(\sigma)$ which we call the friction ball:

$$\pi_F(\gamma_F; \sigma) = \Psi^*_{B_F(\sigma)}(\gamma_F). \tag{5.55}$$

We can therefore express the combined Coulomb-Contensou friction law as

$$\gamma_F \in N_{B_F(\sigma)}(-\lambda_F), \tag{5.56}$$

where $N_{B_F(\sigma)}$ is the normal cone on the friction ball $B_F(\sigma)$. The generalised force vector $-\lambda_F$ is the dual variable to γ_F and can be interpreted as

$$\lambda_F = \begin{bmatrix} \lambda_{T1} \\ \lambda_{T2} \\ \lambda_P \end{bmatrix}, \tag{5.57}$$

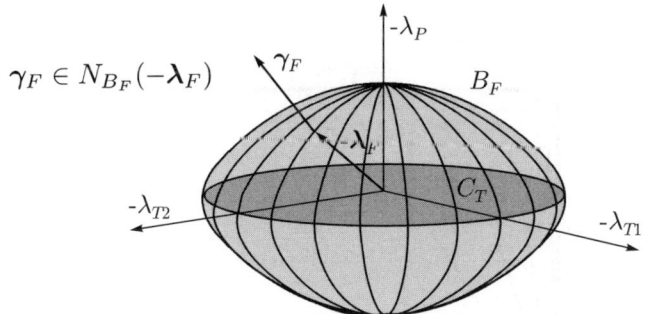

$$\gamma_F \in N_{B_F}(-\boldsymbol{\lambda}_F)$$

Fig. 5.8. Coulomb-Contensou friction law with the friction ball B_F.

in which $\lambda_P = \tau_N/R$ is the scaled pivoting torque. In [99], the potential π_F and the friction ball B_F have been derived for a circular contact with a uniform or parabolic pressure distribution σ. The friction ball B_F (see Figure 5.8) is in these cases squashed but axisymmetric around the $-\lambda_P$ axis. The intersection of B_F with the $(-\lambda_{T1}, -\lambda_{T2})$ plane is formed by the friction disk C_T.

Denote the magnitude of the sliding velocity by $v = \|\boldsymbol{\gamma}_T\|$. The friction characteristic $\boldsymbol{\lambda}_T(v)$ for a fixed value of γ_P can be derived from (5.56) and is shown in Figure 5.9a for different fixed values of γ_P. A uniform pressure distribution and a circular contact surface have been assumed. The curves for $\gamma_P > 0$ are all single-valued functions of v. Apparently, a superimposed normal spin ω_N on a sliding velocity v causes a smoothing effect of the friction characteristic $\boldsymbol{\lambda}_T(v)$. The friction characteristic for $\gamma_P = 0$ is (for $\lambda_P = 0$) the classical set-valued friction characteristic of Coulomb. Similarly, the dependence of λ_P on γ_P, for different fixed values of v, is shown in Figure 5.9b. The same smoothing effect occurs in the $\lambda_P(\gamma_P)$ relationship due to a superimposed velocity $v > 0$. Again, a set-valued relationship is obtained for $v = 0$, which corresponds (for $\boldsymbol{\lambda}_T = 0$) to the classical set-valued friction law of a purely spinning contact. The boundaries $\|\boldsymbol{\lambda}_T\| = \mu\lambda_N$ and $|\lambda_P| = \frac{2}{3}\mu\lambda_N$ of the sets for $v = \gamma_P = 0$ are the extreme values of the friction ball $B_F(\lambda_N)$ along its principal axes under the assumption of a uniform pressure distribution and a circular contact surface. For the classical set-valued friction characteristic of Coulomb (i.e. a purely translational contact $\lambda_P = 0$, $\gamma_P = 0$), the magnitude friction force $\|\boldsymbol{\lambda}_T\|$ rises up to the value $\mu\lambda_N$ when the contact changes from a sticking state to sliding. This is in general *not* the case for Coulomb-Contensou friction as soon as we have some spin. A sticking contact obeying Coulomb-Contensou's friction law will start slipping for $\|\boldsymbol{\lambda}_T\| < \mu\lambda_N$ if $\lambda_P \neq 0$. The slip-criterion for Coulomb-Contensou friction is given by $-\boldsymbol{\lambda}_F \in \text{bdry}\, B_F(\lambda_N)$.

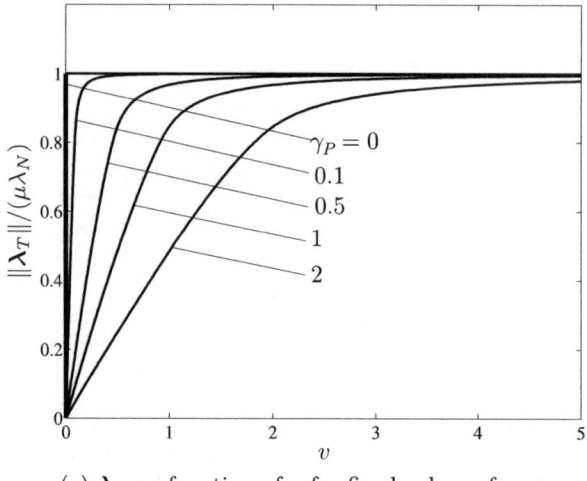

(a) $\boldsymbol{\lambda}_T$ as function of v for fixed values of γ_P.

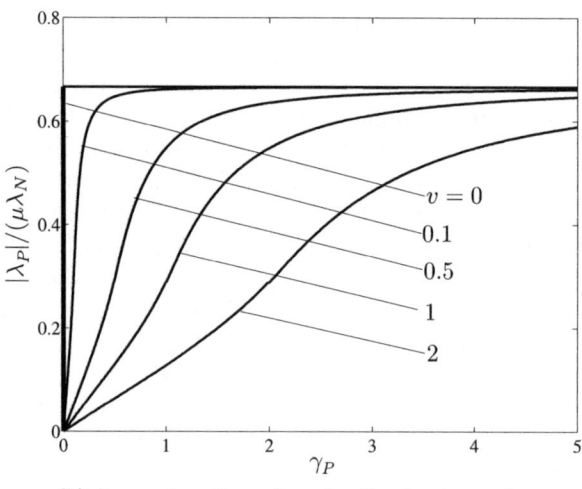

(b) $\boldsymbol{\lambda}_P$ as function of γ_P for fixed values of v.

Fig. 5.9. Friction curves for a uniform pressure distribution and circular contact surface.

5.3.4 Rolling Friction

A resistance against sliding and pivoting of bodies in contact is modelled by the Coulomb-Contensou friction model of the previous section. However, the bodies may also experience a resistance against rolling over each other. At this point we have to ask ourselves what we exactly mean when we say that bodies 'roll' over each other. We may call 'rolling' the movement of the contact point P_i over the surface of body i. A resistance against such a type of movement will be called contour friction [95]. Usually, the term rolling is associated with resistance against a difference in angular velocity of the contacting bodies, i.e. $\Omega_2 - \Omega_1$, tangential to the tangent plane of contact (see for instance [83]). This will be called classical rolling friction. Contour friction and classical rolling friction may be identical to each other or be essentially different, depending on the type of system. For instance, if a planar wheel rolling over a flat floor is considered, then the two types of rolling friction yield the same kind of dissipation mechanism, because the velocity of the contact point over the contour of the wheel is directly related to the angular velocity of the wheel. However, the two types of rolling friction are essentially different if we consider two wheels in contact. If we let the two wheels move over each other with no relative angular velocity, i.e. $\Omega_2 = \Omega_1$, then the contact point still moves over the contour of both wheels. The fact that the classical rolling friction model yields no dissipation for this case shows that this rolling friction model is cumbersome. The classical rolling friction model becomes also questionable when applied to two wheels in contact with different diameters. The main problem is that it is not clear which dissipation mechanism the classical rolling friction model tries to model. A mathematical formulation for contour and classical rolling friction will be discussed in the following.

Contour friction [95] is defined as the resistance against the contour velocity on body j

$$\gamma_{C,j} = \begin{bmatrix} \gamma_{C1,j} \\ \gamma_{C2,j} \end{bmatrix} \in \mathbb{R}^2, \tag{5.58}$$

which has the components

$$\gamma_{C1,j} = v_{D_jP_j}^{\mathrm{T}} t_j, \quad \gamma_{C2,j} = v_{D_jP_j}^{\mathrm{T}} s_j, \tag{5.59}$$

where $v_{D_jP_j}$ is the velocity with which contact point P_j moves with respect to an arbitrary body-fixed point D_j on body j. The velocity $v_{D_jP_j}$ is necessarily a vector in the tangent plane T_j of the contact point P_j, because $v_{D_jP_j}^{\mathrm{T}} n_j = 0$. Each body $j \in \{1,2\}$ has its own contour velocity $\gamma_{C,j}$ for which we can set up a contour friction law. In the following, we will suppress the subscript $_{,j}$. In order to set up the contact law, we choose a potential $\pi_C(\gamma_C)$ which acts as dissipation function and which leads to the force law

$$-\lambda_C \in \partial \pi_C(\gamma_C), \tag{5.60}$$

where λ_C is the contour friction force, being the dual variable to γ_C. If we choose the $\pi_C(\gamma_C)$ to be a support function on a convex set C_C

$$\pi_C(\gamma_C) = \Psi_{C_C}^*(\gamma_C), \tag{5.61}$$

then we obtain a Coulomb-like friction law which can be expressed as a normal cone inclusion

$$\gamma_C \in N_{C_C}(-\lambda_C). \tag{5.62}$$

The set C_C of admissible contour friction forces can be chosen to be dependent on the normal contact force λ_N

$$C_C = \{-\boldsymbol{\lambda}_C \mid \|\boldsymbol{\lambda}_C\| \leq \mu_C \lambda_N\}, \tag{5.63}$$

where μ_C is the contour friction coefficient.

Although the classical rolling friction model is cumbersome (see the above discussion), we will still briefly discuss how it can be mathematically described. Classical rolling friction [83] is similar to pivoting friction, which describes dissipation due to a normal spin ω_N. Classical rolling friction defines a resistance against the tangential spin of the two contacting bodies

$$\boldsymbol{\omega}_T = \begin{bmatrix} \omega_{T1} \\ \omega_{T2} \end{bmatrix}, \tag{5.64}$$

which has the components

$$\omega_{T1} = (\boldsymbol{\Omega}_2 - \boldsymbol{\Omega}_1)^{\mathrm{T}} \boldsymbol{t}_1, \quad \omega_{T2} = (\boldsymbol{\Omega}_2 - \boldsymbol{\Omega}_1)^{\mathrm{T}} \boldsymbol{s}_1. \tag{5.65}$$

The tangential spin $\boldsymbol{\omega}_T$ is a vector in the tangent plane of the contact point, which can be normalised to a rolling velocity

$$\boldsymbol{\gamma}_R = \boldsymbol{\omega}_T L, \tag{5.66}$$

where L is some characteristic length of body 1 or 2. Just as before, we choose a potential $\pi_R(\boldsymbol{\gamma}_R) = \partial \Psi_{C_R}^*(\boldsymbol{\gamma}_R)$ in the form of a support function which acts as dissipation function and which leads to the force law

$$-\boldsymbol{\lambda}_R \in \partial \pi_R(\boldsymbol{\gamma}_R) \quad \Longleftrightarrow \quad \boldsymbol{\gamma}_R \in N_{C_R}(-\boldsymbol{\lambda}_R), \tag{5.67}$$

where $\boldsymbol{\lambda}_R$ is the classical rolling friction force, being the dual variable to $\boldsymbol{\gamma}_R$ and C_R is the admissible set of $-\boldsymbol{\lambda}_R$. The classical rolling friction force $\boldsymbol{\lambda}_R$ can be interpreted as a normalised rolling friction moment $\boldsymbol{\tau}_T$:

$$\boldsymbol{\tau}_T = \boldsymbol{\lambda}_R L, \tag{5.68}$$

with components τ_{T1} and τ_{T2} dual to ω_{T1} and ω_{T1}. The total torque transmitted by the contact is therefore the vector

$$\boldsymbol{\tau} = \tau_N \boldsymbol{n}_1 + \tau_{T1} \boldsymbol{t}_1 + \tau_{T2} \boldsymbol{s}_1. \tag{5.69}$$

Coulomb friction, Coulomb-Contensou and two types of rolling friction have been discussed in the previous sections. If dry friction laws are considered, then all these friction laws can be formulated as a normal cone inclusion

$$\boldsymbol{\gamma}_D \in N_{C_D}(-\boldsymbol{\lambda}_D), \qquad \boldsymbol{\gamma}_D, \boldsymbol{\lambda}_D \in \mathbb{R}^p, \quad C_D \subset \mathbb{R}^p. \tag{5.70}$$

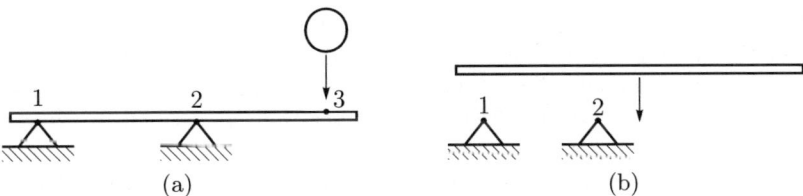

Fig. 5.10. Multi-contact collision.

The dimension p is 1 for planar Coulomb friction and 2 for spatial Coulomb friction. Coulomb friction and pivoting friction were considered to be coupled to each other which has been described by the Coulomb-Contensou friction law for which $p = 3$. Contour and classical rolling friction have dimension $p = 2$. However, it might also be possible to set up a dissipation function for combined sliding–pivoting–rolling, which would result in $p = 5$. In the following, we will just write (5.70) to denote any frictional dissipation law, being derived from a velocity pseudo-potential $\pi_{D(\lambda_N)}(\gamma_D) = \Psi^*_{D(\lambda_N)}(\gamma_D)$. This notation allows us to describe the mechanical system in a general way and to prove the stability results for arbitrary friction laws.

5.3.5 Impact Laws

Signorini's law and Coulomb's friction law are set-valued force laws for non-impulsive forces. In order to describe impact, we need to introduce impact laws for the contact impulses. We will consider a Newton-type of restitution law,

$$\gamma_N^+ = -e_N \gamma_N^-, \qquad g_N = 0, \qquad 0 \le e_N \le 1, \tag{5.71}$$

which relates the post-impact velocity γ_N^+ of a contact point to the pre-impact velocity γ_N^- by Newton's coefficient of restitution e_N. The case $e_N = 1$ corresponds to a completely elastic contact, whereas $e_N = 0$ corresponds to a completely inelastic contact. The impact, which causes the sudden change in relative velocity, is accompanied by a normal contact impulse $\Lambda_N > 0$. Following [62], suppose that, for any reason, the contact does not participate in the impact, i.e. that the value of the normal contact impulse Λ_N is zero, although the contact is closed. This happens normally for multi-contact situations. Consider for example a rod, depicted in Figure 5.10a, which is supported on two supports, somewhat offset to the right. A ball is hitting the rod on its right end. Three contact points are closed when the ball hits the rod: 1. the rod–left support contact, 2. the rod right-support contact, and 3. the ball–rod contact. Contact 2 and 3 will transmit a contact impulse which ensures the the impenetrability of these contacts. Contact 1, however, contact 1 will not transmit a contact impulse. Such a contact will be called superfluous. The pre-impact velocity of contact 1 is zero and its post-impact velocity is positive due to the transmitted impacts at contact points 2 and 3. Another example shows

Figure 5.10b. A horizontal rod is fall on two supports, both located on the left side of the rod's center of mass. Again, contact 1 is superfluous. This time, however, its pre-impact relative velocity is negative. For superfluous contacts we allow post-impact relative velocities higher than prescribed by Newton's impact law in the case of a non-vanishing impulse, $\gamma_N^+ > -e_N \gamma_N^-$, in order to express that the contact is superfluous and could be removed without changing the contact-impact process. Summarising, two cases can occur at a closed contact

1. The contact is actively participating in the impact process, i.e. $\Lambda_N > 0$ and $\gamma_N^+ = -e_N \gamma_N^-$,
2. The contact is superfluous, i.e. $\Lambda_N = 0$ and $\gamma_N^+ \geq -e_N \gamma_N^-$.

We can combine these two cases in an impact law formulated as an inequality complementarity condition on velocity–impulse level:

$$\Lambda_N \geq 0, \quad \xi_N \geq 0, \quad \Lambda_N \xi_N = 0, \tag{5.72}$$

with $\xi_N = \gamma_N^+ + e_N \gamma_N^-$ (see [62]). Similarly to Signorini's law on velocity level we can write the impact law in normal direction as

$$\xi_N \in N_{C_N}(-\Lambda_N), \quad g_N = 0, \quad C_N = \mathbb{R}^-, \tag{5.73}$$

or by using the support function

$$-\Lambda_N \in \partial \Psi_{C_N}^*(\xi_N), \quad g_N = 0. \tag{5.74}$$

A normal contact impulse Λ_N at a frictional contact leads to a frictional contact impulse $\boldsymbol{\Lambda}_D \in \mathbb{R}^p$ with $-\boldsymbol{\Lambda}_D \in C_D(\Lambda_N)$. We therefore have to specify a frictional impact law as well, taking into account a possible frictional restitution. Restitution in tangential direction occurs for instance in the motion of the Super Ball, being a very elastic ball used on play grounds. The frictional impact law can be formulated in a similar way as has been done for the normal impact law:

$$-\boldsymbol{\Lambda}_D \in \partial \Psi_{C_D(\Lambda_N)}^*(\boldsymbol{\xi}_D), \quad g_N = 0, \tag{5.75}$$

with $\boldsymbol{\xi}_D = \boldsymbol{\gamma}_D^+ + e_D \boldsymbol{\gamma}_D^-$ and $|e_D| \leq 1$.

Let us analyse the frictional impact law (5.75) for planar Coulomb friction, i.e.

$$-\Lambda_T \in \partial \Psi_{C_T(\Lambda_N)}^*(\xi_T), \quad g_N = 0, \tag{5.76}$$

with $\xi_T = \gamma_T^+ + e_T \gamma_T^-$ and $C_T(\Lambda_N) = \{\Lambda_T \mid -\mu \Lambda_N \leq \Lambda_T \leq \mu \Lambda_N\}$. This friction law is equivalent to

$$-\Lambda_T \in \mu \Lambda_N \operatorname{Sign}(\gamma_T^+ + e_T \gamma_T^-), \tag{5.77}$$

which is similar to (5.37). We distinguish three cases:

- impact with stick

$$|\Lambda_T| \leq \mu \Lambda_N, \quad \gamma_T^+ = -e_T \gamma_T^-,$$

- impact with forward slip

$$\Lambda_T = -\mu\Lambda_N, \quad \gamma_T^+ > -e_T\gamma_T^-,$$

- impact with backward slip

$$\Lambda_T = \mu\Lambda_N, \quad \gamma_T^+ < -e_T\gamma_T^-.$$

In this section we have presented a generalisation of Newton's impact law. Such a generalisation is also possible for Poisson-type of impact laws [139], which distinguish between a compression and an expansion phase. Poisson's impact law relates the impulse in the expansion phase to the impulse in the compression phase through a restitution coefficient [62]. A generalisation of Poisson's impact law involves two complementarity inequalities, one for each phase.

A restitution coefficient must be regarded as an impact process parameter, not as a material constant. Consider for instance a ball which is falling on a flexible bar. The dynamic response will greatly depend on the location of impact. If the ball falls on a node of the first bending eigenmode of the bar, then this will cause another response than if the ball hits an antinode. Hence, the impact process is governed by wave effects and the dissipation caused by wave effects. In a rigid-body approach, we are not interested in the wave effects themselves, but we only want to capture the macroscopic motion of the bodies. The wave and damping effects are therefore represented by the restitution coefficients of the impact law.

The class of impact laws (5.74) and (5.75) does not parameterise the whole set of possible post-impact velocities that are kinematically admissible (see [4]). The impact laws presented in this section are local, i.e. the post-impact velocity of a contact point is related to the pre-impact velocity of the same contact point. Non-local effects, which are due to wave effects, play for instance a role in Newton's cradle [4]. Non-local effects can be taken into account by specifying restitution coefficients between different contact points.

5.4 Measure Newton-Euler Equations

In this section, we set up the measure Newton-Euler equations which form together with the set-valued force laws for the contact forces a measure differential inclusion which describes the temporal dynamics of mechanical systems with discontinuities in the velocity. Subsequently, we study the equilibrium set of the measure differential inclusion. For simplicity, we will restrict the study to time-autonomous systems, although the measure Newton-Euler equations can also be set up for the non-autonomous case.

We assume that the time-autonomous mechanical systems under consideration exhibit only bilateral holonomic frictionless constraints and unilateral constraints in which dry friction can be present. This assumption is not essential. We can very well model bilateral constraints with friction as we will do in Section 7.3.2, but to simplify the notation we will start with bilateral frictionless constraints and frictional unilateral constraints. Furthermore, we assume that a set of independent generalised coordinates, $q \in \mathbb{R}^n$, for which these bilateral constraints are eliminated from the formulation of the dynamics of the system, is known. The generalised coordinates $q(t)$ are assumed to be absolutely continuous functions of time t. Also, we assume the generalised velocities, $u(t) = \dot{q}(t)$ for almost all t, to be functions of locally bounded variation. At each time-instance it is therefore possible to define a left limit u^- and a right limit u^+ of the velocity. The generalised accelerations $\dot{u}$ are therefore not for all t defined. The set of discontinuity points $\{t_j\}$ for which $\dot{u}$ is not defined is assumed to be Lebesgue negligible. We formulate the dynamics of the system using a Lagrangian approach, resulting in[2]

$$\left(\frac{d}{dt}(T_{,u}) - T_{,q} + U_{,q}\right)^{\mathrm{T}} = f^{\mathrm{nc}} + \sum_{i=1}^{m}(w_{Ni}(q)\lambda_{Ni} + W_{Di}(q)\lambda_{Di}). \quad (5.78)$$

The scalar T represents the kinetic energy and U denotes the potential energy. The column-vector f^{nc} in (5.78) represents all smooth generalised non-conservative forces. The vector $w_{Ni} \in \mathbb{R}^n$ and matrices $W_{Di} \in \mathbb{R}^{n \times p}$ define the force directions of the normal contact forces λ_{Ni} and frictional contact forces $\lambda_{Di} \in \mathbb{R}^p$.

Alternatively we write the dynamics as

$$M(q)\dot{u} - h(q, u) = W_N(q)\lambda_N + W_D(q)\lambda_D, \quad (5.79)$$

which is a differential equation for the non-impulsive part of the motion. Herein, $M(q) = M^{\mathrm{T}}(q) > 0$ is the mass-matrix, and we assume that the kinetic energy for an autonomous system can be written as $T = \frac{1}{2}u^{\mathrm{T}}M(q)u$. The state-dependent column-vector $h(q, u)$ in (5.79) contains all differentiable forces (both conservative and non-conservative), such as spring forces, gravitational forces, smooth damper forces and gyroscopic terms. The contact forces are assembled in vectors λ_N and λ_D and their corresponding force directions in matrices W_N and W_D:

$$\lambda_N = \{\lambda_{Ni}\}, \quad \lambda_D = \{\lambda_{Di}\}, \quad W_N = \{w_{Ni}\}, \quad W_D = \{W_{Di}\}. \quad (5.80)$$

Due to the fact that the kinetic energy can be described by

$$T = \frac{1}{2}u^{\mathrm{T}}M(q)u = \frac{1}{2}\sum_{r,s}M_{rs}u^r u^s, \quad (5.81)$$

with $M(q) = M^{\mathrm{T}}(q)$, we can write in tensorial language

---

[2] Note that the subscript $,x$ indicates a partial derivative operation $\partial/\partial x$.

$$\frac{\partial T}{\partial q^k} = \frac{1}{2} \sum_{r,s} \left(\frac{\partial M_{rs}}{\partial q^k} \right) u^r u^s, \quad \frac{\partial T}{\partial u^k} = \sum_r M_{kr} u^r,$$

$$\frac{\mathrm{d}}{\mathrm{d}t} \left(\frac{\partial T}{\partial u^k} \right) = \sum_r M_{kr} \dot{u}^r + \sum_{r,s} \left(\frac{\partial M_{kr}}{\partial q^s} \right) u^r u^s$$

$$= \sum_r M_{kr} \dot{u}^r + 2\frac{\partial T}{\partial q^k} + \sum_{r,s} \left(\frac{\partial M_{kr}}{\partial q^s} - \frac{\partial M_{rs}}{\partial q^k} \right) u^r u^s \tag{5.82}$$

$$\frac{\mathrm{d}}{\mathrm{d}t} \left(T_{,u} \right) = \dot{u}^{\mathrm{T}} M(q) + 2T_{,q} - (f^{\mathrm{gyr}})^{\mathrm{T}} \qquad \text{for almost all } t,$$

with the gyroscopic forces [129]

$$f^{\mathrm{gyr}} = \{ f_k^{\mathrm{gyr}} \}, \qquad f_k^{\mathrm{gyr}} = -\sum_{r,s} \left(\frac{\partial M_{kr}}{\partial q^s} - \frac{\partial M_{rs}}{\partial q^k} \right) u^r u^s. \tag{5.83}$$

The gyroscopic forces f_{gyr} have zero power [129]

$$u^{\mathrm{T}} f^{\mathrm{gyr}} = \sum_k u^k f_k^{\mathrm{gyr}} = -\sum_{k,r,s} \left(\frac{\partial M_{kr}}{\partial q^s} - \frac{\partial M_{rs}}{\partial q^k} \right) u^r u^s u^k = 0. \tag{5.84}$$

In the same way as before, we can write the differential measure of $T_{,u}$ as

$$\mathrm{d} \left(T_{,u} \right) = \mathrm{d}u^{\mathrm{T}} M(q) + 2T_{,q} \mathrm{d}t - (f^{\mathrm{gyr}})^{\mathrm{T}} \mathrm{d}t \qquad \forall t. \tag{5.85}$$

Comparison with (5.79) and (5.78) yields

$$h = f^{\mathrm{nc}} + f^{\mathrm{gyr}} - (T_{,q} + U_{,q})^{\mathrm{T}}, \tag{5.86}$$

or in index notation

$$\begin{aligned} h_k &= f_k^{\mathrm{nc}} - \frac{\partial U}{\partial q^k} - \frac{\partial T}{\partial q^k} + f_k^{\mathrm{gyr}} \\ &= f_k^{\mathrm{nc}} - \frac{\partial U}{\partial q^k} - \frac{\partial T}{\partial q^k} - \sum_{r,s} \left(\frac{\partial M_{kr}}{\partial q^s} - \frac{\partial M_{rs}}{\partial q^k} \right) u^r u^s \\ &= f_k^{\mathrm{nc}} - \frac{\partial U}{\partial q^k} - \frac{1}{2} \sum_{r,s} \left(2\frac{\partial M_{kr}}{\partial q^s} - \frac{\partial M_{rs}}{\partial q^k} \right) u^r u^s \\ &= f_k^{\mathrm{nc}} - \frac{\partial U}{\partial q^k} - \frac{1}{2} \sum_{r,s} \left(\frac{\partial M_{kr}}{\partial q^s} + \frac{\partial M_{ks}}{\partial q^r} - \frac{\partial M_{rs}}{\partial q^k} \right) u^r u^s \\ &= f_k^{\mathrm{nc}} - \frac{\partial U}{\partial q^k} - \sum_{r,s} \Gamma_{k,rs} u^r u^s \end{aligned} \tag{5.87}$$

in which we recognise the holonomic Christoffel symbols of the first kind [129]

$$\Gamma_{k,rs} = \Gamma_{k,sr} := \frac{1}{2} \left(\frac{\partial M_{kr}}{\partial q^s} + \frac{\partial M_{ks}}{\partial q^r} - \frac{\partial M_{rs}}{\partial q^k} \right). \tag{5.88}$$

We introduce the following index sets:

$$I_G = \{1, \ldots, m\} \qquad \text{the set of all contacts,}$$
$$I_N = \{i \in I_G \mid g_{Ni}(\boldsymbol{q}) = 0\} \text{ the set of all closed contacts,} \qquad (5.89)$$

and set up the force laws and impact laws of each contact as has been elaborated in the previous sections. The normal contact distances $g_{Ni}(\boldsymbol{q})$ depend on the generalised coordinates $\boldsymbol{q}$ and are gathered in a vector $\boldsymbol{g}_N(\boldsymbol{q})$.

During a non-impulsive part of the motion, the normal contact force $-\lambda_{Ni} \in C_N$ and friction force $-\boldsymbol{\lambda}_{Di} \in C_{Di} \subset \mathbb{R}^p$ of each closed contact $i \in I_N$, are assumed to be associated with a non-smooth potential, being the support function of a convex set, i.e.

$$-\lambda_{Ni} \in \partial \Psi^*_{C_N}(\gamma_{Ni}), \qquad -\boldsymbol{\lambda}_{Di} \in \partial \Psi^*_{C_{Di}}(\boldsymbol{\gamma}_{Di}), \qquad (5.90)$$

where $C_N = \mathbb{R}^-$ and the set C_{Di} can be dependent on the normal contact force λ_{Ni}. The normal and frictional relative velocities are gathered in columns $\boldsymbol{\gamma}_N = \{\gamma_{Ni}\}$ and $\boldsymbol{\gamma}_D = \{\boldsymbol{\gamma}_{Di}\}$, for $i = 1, \ldots, m$. We assume that these contact velocities are related to the generalised velocities through:

$$\boldsymbol{\gamma}_N(\boldsymbol{q}, \boldsymbol{u}) = \boldsymbol{W}_N^T(\boldsymbol{q})\boldsymbol{u}, \qquad \boldsymbol{\gamma}_D(\boldsymbol{q}, \boldsymbol{u}) = \boldsymbol{W}_D^T(\boldsymbol{q})\boldsymbol{u}. \qquad (5.91)$$

This assumption is very important as it excludes rheonomic contacts[3]. Note also that $\boldsymbol{W}_X^T(\boldsymbol{q}) = \frac{\partial \boldsymbol{\gamma}_X}{\partial \boldsymbol{u}}$ for $X = N, D$, which also holds for rheonomic contacts.

Equation (5.79) together with the set-valued force laws (5.90) form a differential inclusion

$$\boldsymbol{M}(\boldsymbol{q})\dot{\boldsymbol{u}} - \boldsymbol{h}(\boldsymbol{q}, \boldsymbol{u}) \in \sum_{i \in I_N} \left(-\boldsymbol{w}_{Ni}(\boldsymbol{q})\partial \Psi^*_{C_N}(\gamma_{Ni}) - \boldsymbol{W}_{Di}(\boldsymbol{q})\partial \Psi^*_{C_{Di}}(\boldsymbol{\gamma}_{Di}) \right),$$
$$(5.92)$$

for almost all t. Differential inclusions of this type are called Filippov systems, which obey the solution concept given by Definition 4.2. The differential inclusion (5.92) only holds for impact free motion.

Subsequently, we define for each contact point the constitutive impact laws

$$-\Lambda_{Ni} \in \partial \Psi^*_{C_N}(\xi_{Ni}), \qquad -\boldsymbol{\Lambda}_{Di} \in \partial \Psi^*_{C_{Di}(\Lambda_{Ni})}(\boldsymbol{\xi}_{Di}), \qquad i \in I_N, \qquad (5.93)$$

with

$$\xi_{Ni} = \gamma_{Ni}^+ + e_{Ni}\gamma_{Ni}^-, \qquad \boldsymbol{\xi}_{Di} = \boldsymbol{\gamma}_{Di}^+ + e_{Di}\boldsymbol{\gamma}_{Di}^-, \qquad (5.94)$$

in which $0 \le e_{Ni} \le 1$ and $|e_{Di}| \le 1$ are the normal and frictional restitution coefficients respectively. The impact laws (5.93) can be generalised by replacing the restitution coefficient e_{Di} by a matrix.

---

[3] A rheonomic contact is characterised by a contact distance $g(\boldsymbol{q}, t)$ being explicitly dependent on time, or a contact velocity $\boldsymbol{\gamma}(\boldsymbol{q}, \boldsymbol{u}, t)$ being an affine function in $\boldsymbol{u}$, i.e. $\boldsymbol{\gamma} = \boldsymbol{W}^T(\boldsymbol{q}, t)\boldsymbol{u} + \boldsymbol{w}(\boldsymbol{q}, t)$. A contact which is not rheonomic is called scleronomic.

The force laws for non-impulsive motion can be put in the same form as (5.93)

$$-\boldsymbol{\lambda}_{Ni} \in \partial\Psi^*_{C_N}(\xi_{Ni}), \qquad -\boldsymbol{\lambda}_{Di} \in \partial\Psi^*_{C_{Di}(\lambda_{Ni})}(\boldsymbol{\xi}_{Di}). \qquad (5.95)$$

because $\boldsymbol{u}^! = \boldsymbol{u}^-$ holds (for non-impulsive motion) and because of the positive homogeneity of the support function (see Section 2.5):

$$\Psi^*_C(\xi) = \Psi^*_C(\gamma + e\gamma) = \Psi^*_C((1+e)\gamma) = (1+e)\Psi^*_C(\gamma)$$

$$\Rightarrow \partial\Psi^*_C(\xi) = \partial\Psi^*_C(\gamma).$$

We now replace the differential inclusion (5.92), which holds for almost all t, by an equality of measures

$$\boldsymbol{M}(\boldsymbol{q})\mathrm{d}\boldsymbol{u} - \boldsymbol{h}(\boldsymbol{q},\boldsymbol{u})\mathrm{d}t = \boldsymbol{W}_N(\boldsymbol{q})\mathrm{d}\boldsymbol{P}_N + \boldsymbol{W}_D(\boldsymbol{q})\mathrm{d}\boldsymbol{P}_D \quad \forall t, \qquad (5.96)$$

which holds for all time-instances t. The differential measure of the contact percussions $\mathrm{d}\boldsymbol{P}_N$ and $\mathrm{d}\boldsymbol{P}_D$ contains a Lebesgue measurable part $\boldsymbol{\lambda}\mathrm{d}t$ and an atomic part $\boldsymbol{\Lambda}\mathrm{d}\eta$

$$\mathrm{d}\boldsymbol{P}_N = \boldsymbol{\lambda}_N\mathrm{d}t + \boldsymbol{\Lambda}_N\mathrm{d}\eta, \qquad \mathrm{d}\boldsymbol{P}_D = \boldsymbol{\lambda}_D\mathrm{d}t + \boldsymbol{\Lambda}_D\mathrm{d}\eta, \qquad (5.97)$$

which can be expressed as inclusions:

$$-\mathrm{d}\boldsymbol{P}_{Ni} \in \partial\Psi^*_{C_N}(\xi_{Ni})\mathrm{d}t + \partial\Psi^*_{C_N}(\xi_{Ni})\mathrm{d}\eta,$$
$$-\mathrm{d}\boldsymbol{P}_{Di} \in \partial\Psi^*_{C_{Di}(\lambda_{Ni})}(\boldsymbol{\xi}_{Di})\mathrm{d}t + \partial\Psi^*_{C_{Di}(\Lambda_{Ni})}(\boldsymbol{\xi}_{Di})\mathrm{d}\eta. \qquad (5.98)$$

The set $\partial\Psi^*_{C_N}(\xi_{Ni})$ is a cone and the measure constitutive law for normal contact (5.98) and positive measures $\mathrm{d}t$ and $\mathrm{d}\eta$ can therefore be written as (4.56)

$$-\mathrm{d}\boldsymbol{P}_{Ni} \in \partial\Psi^*_{C_N}(\xi_{Ni}), \qquad (5.99)$$

which means that the density functions $-\lambda_{Ni}$ and $-\Lambda_{Ni}$ both belong to this cone (see Section 4.3). The fundamental idea behind the equality of measures (5.96) is that it describes the differential equation for impact-free motion and the impact equations by a single equation. Indeed, if we integrate the equation of measures over an arbitrary impact-free time-interval I for which $\boldsymbol{\Lambda}_N = \boldsymbol{\Lambda}_D = \boldsymbol{0}$, then we obtain the integral equation

$$\int_I \boldsymbol{M}(\boldsymbol{q})\mathrm{d}\boldsymbol{u} = \int_I \boldsymbol{h}(\boldsymbol{q},\boldsymbol{u})\mathrm{d}t + \int_I \boldsymbol{W}_N(\boldsymbol{q})\boldsymbol{\lambda}_N\mathrm{d}t + \int_I \boldsymbol{W}_D(\boldsymbol{q})\boldsymbol{\lambda}_D\mathrm{d}t.$$

Hence, the differential measure $\mathrm{d}\boldsymbol{u}$ can only have a density with respect to the Lebesgue measure $\mathrm{d}t$, i.e. $\mathrm{d}\boldsymbol{u} = \dot{\boldsymbol{u}}\mathrm{d}t$, from which we retrieve the differential equation of motion (5.79)

$$\boldsymbol{M}(\boldsymbol{q})\dot{\boldsymbol{u}} = \boldsymbol{h}(\boldsymbol{q},\boldsymbol{u}) + \boldsymbol{W}_N(\boldsymbol{q})\boldsymbol{\lambda}_N + \boldsymbol{W}_D(\boldsymbol{q})\boldsymbol{\lambda}_D.$$

Similarly, if we consider an integration interval consisting of a singleton $\{t_i\}$ for which the impulses $\boldsymbol{\Lambda}_N$ and/or $\boldsymbol{\Lambda}_D$ are non-zero, then we obtain the integral equation

$$\int_{\{t_i\}} \boldsymbol{M}(\boldsymbol{q})\mathrm{d}\boldsymbol{u} = \int_{\{t_i\}} \boldsymbol{W}_N(\boldsymbol{q})\boldsymbol{\Lambda}_N\mathrm{d}\eta + \int_{\{t_i\}} \boldsymbol{W}_D(\boldsymbol{q})\boldsymbol{\Lambda}_D\mathrm{d}\eta,$$

where we already used that the integral of the Lebesgue measure $\mathrm{d}t$ over the singleton $\{t_i\}$ vanishes. Hence, the differential measure $\mathrm{d}\boldsymbol{u}$ must have a density with respect to the atomic measure $\mathrm{d}\eta$, i.e. $\mathrm{d}\boldsymbol{u} = (\boldsymbol{u}^+ - \boldsymbol{u}^-)\mathrm{d}\eta$, from which we retrieve the impact equation which descibes the velocity jump at $t = t_i$:

$$\boldsymbol{M}(\boldsymbol{q})(\boldsymbol{u}^+ - \boldsymbol{u}^-) = \boldsymbol{W}_N(\boldsymbol{q})\boldsymbol{\Lambda}_N + \boldsymbol{W}_D(\boldsymbol{q})\boldsymbol{\Lambda}_D. \qquad (5.100)$$

As an abbreviation of the equality of measures (5.96) we write

$$\boldsymbol{M}(\boldsymbol{q})\mathrm{d}\boldsymbol{u} - \boldsymbol{h}(\boldsymbol{q},\boldsymbol{u})\mathrm{d}t = \boldsymbol{W}(\boldsymbol{q})\mathrm{d}\boldsymbol{P} \quad \forall t, \qquad (5.101)$$

using

$$\mathrm{d}\boldsymbol{P} = \begin{bmatrix} \mathrm{d}\boldsymbol{P}_N \\ \mathrm{d}\boldsymbol{P}_D \end{bmatrix}, \quad \boldsymbol{W} = \begin{bmatrix} \boldsymbol{W}_N & \boldsymbol{W}_D \end{bmatrix}, \quad \boldsymbol{\gamma} = \begin{bmatrix} \boldsymbol{\gamma}_N \\ \boldsymbol{\gamma}_D \end{bmatrix}. \qquad (5.102)$$

Furthermore we introduce the variables $\boldsymbol{\xi}$ and $\boldsymbol{\delta}$

$$\boldsymbol{\xi} = \boldsymbol{\gamma}^+ + \boldsymbol{E}\boldsymbol{\gamma}^-, \qquad \boldsymbol{\delta} = \boldsymbol{\gamma}^+ - \boldsymbol{\gamma}^-, \qquad (5.103)$$

with $\boldsymbol{E} = \mathrm{diag}(\{e_{Ni}, e_{Di}\})$ from which we deduce

$$\begin{aligned} \boldsymbol{\gamma}^+ &= (\boldsymbol{I} + \boldsymbol{E})^{-1}(\boldsymbol{\xi} + \boldsymbol{E}\boldsymbol{\delta}), \\ \boldsymbol{\gamma}^- &= (\boldsymbol{I} + \boldsymbol{E})^{-1}(\boldsymbol{\xi} - \boldsymbol{\delta}). \end{aligned} \qquad (5.104)$$

The equality of measures (5.101) together with the set-valued force laws (5.98) form a measure differential inclusion which describes the time-evolution of a mechanical system with discontinuities in the generalised velocities.

Here, we introduced the restitution coefficient matrix $\boldsymbol{E}$, which has the diagonal elements e_{ii}. In Chapter 7, we will make use of the dissipation index matrix $\boldsymbol{\Delta}$ defined by

$$\boldsymbol{\Delta} := (\boldsymbol{I} - \boldsymbol{E})(\boldsymbol{I} + \boldsymbol{E})^{-1}, \qquad (5.105)$$

which is a diagonal matrix with dissipation indices $\Delta_{ii} = \frac{1-e_{ii}}{1+e_{ii}}$. If $|e_{ii}| < 1$, then it holds that $\boldsymbol{\Delta} > 0$. Note that $\Delta_{\max} = \frac{1-e_{\min}}{1+e_{\min}}$ and $\Delta_{\min} = \frac{1-e_{\max}}{1+e_{\max}}$. Moreover, if all dissipation indices Δ_{ii} are equal, then the global dissipation index δ of Moreau [123] is related to the dissipation index matrix by $\boldsymbol{\Delta} = \delta\boldsymbol{I}$.

The measure differential inclusion described by (5.101) and (5.98) may exhibit equilibrium sets, i.e. simply connected sets of equilibrium points. Note that $\boldsymbol{u} = \boldsymbol{0}$ implies $\boldsymbol{\gamma}_D = \boldsymbol{0}$, see (5.91). This means that every equilibrium point implies sticking in all closed contact points. We will reserve the word *equilibrium point* $\boldsymbol{x}^*$ for an equilibrium of a measure differential inclusion in

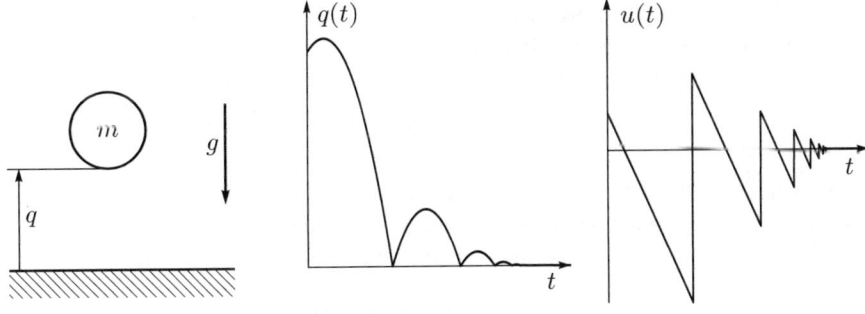

Fig. 5.11. Bouncing ball example.

first-order form and use the term *equilibrium position* q^* to denote an equilibrium configuration of a Lagrangian mechanical system. Hence, if the measure differential inclusion (6.1) in first-order form represents the Lagrangian mechanical system (5.101) then it holds that

$$x^* = \begin{bmatrix} q^* \\ 0 \end{bmatrix}. \tag{5.106}$$

Every equilibrium position q^* has to obey the equilibrium inclusion:

$$h(q^*,0) - \sum_{i \in I_N} \left(w_{Ni}(q^*) \partial \Psi^*_{C_N}(0) + W_{Di}(q^*) \partial \Psi^*_{C_{Di}}(0) \right) \ni 0 \tag{5.107}$$

or using $C = \partial \Psi^*_C(0)$

$$h(q^*,0) - \sum_{i \in I_N} \left(w_{Ni}(q^*) \mathbb{R}^- + W_{Di}(q^*) C_{Di} \right) \ni 0. \tag{5.108}$$

An equilibrium set $\mathcal{E}$, being a simply connected set of equilibrium points (Definition 6.8), therefore obeys

$$\mathcal{E} \subset \left\{ (q,u) \,\middle|\, (u = 0) \wedge h(q,0) + \sum_{i \in I_N} \left(w_{Ni}(q) \mathbb{R}^+ - W_{Di}(q) C_{Di} \right) \ni 0 \right\} \tag{5.109}$$

and is positively invariant (see Definition 6.4 in Chapter 6) if we assume uniqueness of the solutions in forward time. It should be noted that due to the fact that nonlinear mechanical systems, without dry friction, can exhibit multiple equilibrium points, the system with dry friction may exhibit multiple equilibrium sets. A formal definition of an equilibrium set in a measure differential inclusion will be given in Definition 6.8.

In order to illustrate the modelling techniques introduced in the chapter, we study the dynamics of the bouncing ball example shown in Figure 5.11. A rigid ball with height q and mass m is falling on a rigid floor. Only vertical motion is considered. The restitution coefficient is e_N. The position $q(t)$ is an absolutely continuous function in time with differential measure

$$dq = udt, \qquad (5.110)$$

where u is the velocity of the ball. The velocity u is assumed to be of locally bounded variation with vanishing singular part and its differential measure therefore contains a Lebesque measurable part and an atomic part

$$du = \dot{u}dt + (u^+ - u^-)d\eta, \qquad (5.111)$$

where $\dot{u}$ is the acceleration and $u^+ - u^-$ is a jump in the velocity. The equation of motion in terms an equality of measures (5.101) gives for this system

$$mdu + mgdt = dP_N, \qquad q \geq 0, \qquad (5.112)$$

where the contact effort dP_N obeys the set-valued force law

$$\begin{aligned} dP_N &= 0 \text{ if } q > 0, \\ -dP_N &\in \partial\Psi^*_{C_N}(\xi_N) \text{ if } q = 0, \end{aligned} \qquad (5.113)$$

with $C_N = \mathbb{R}^-$ and $\xi_N = u^+ + e_N u^-$. Simplifying the term $\Psi^*_{C_N}(\xi_N)$ with $\Psi^*_{\mathbb{R}^-} = \Psi_{\mathbb{R}^+}$ and $\partial\Psi_C = N_C$ gives

$$\begin{aligned} dP_N &= 0 \text{ if } q > 0, \\ -dP_N &\in N_{\mathcal{K}}(\xi_N) \text{ if } q = 0, \end{aligned} \qquad (5.114)$$

where $\mathcal{K} = \{q \mid q \geq 0\} = \mathbb{R}^+$ is the admissible set of positions q. We can further abbreviate the set-valued force law to (see [63])

$$-dP_N \in N_{K_{\mathcal{K}}(q)}(\xi_N), \qquad (5.115)$$

where $K_{\mathcal{K}}$ is the contingent cone to $\mathcal{K}$. Hence, the system is completely described by the measure differential inclusion

$$\begin{aligned} dq &= udt, \\ mdu + mgdt &= dP_N, \\ -dP_N &\in N_{K_{\mathcal{K}}(q)}(\xi_N), \quad \xi_N = u^+ + e_N u^- \quad \mathcal{K} = \{q \mid q \geq 0\}. \end{aligned} \qquad (5.116)$$

5.5 Summary

The description of Lagrangian mechanical systems with frictional unilateral contact in the form of measure differential inclusions has been discussed in this chapter. Set-valued force laws, which are used to describe constitutive behaviour, can be derived from non-smooth potentials. The non-smooth potential is decomposed in a differentiable function, an indicator function and a convex potential with polyhedral epigraph. The smooth part of the set-valued force is derived from the differentiable function while the non-smooth set-valued part stems from the subdifferential of the convex remainder. Set-valued force laws for sliding and pivoting friction, rolling friction as well as impact have been set up as inclusions to normal cones on convex sets of admissible contact forces. The contact forces can be incorporated in the Newton-Euler equations as Lagrangian multipliers. The Newton-Euler equations in terms of differential measures, together with set-valued force laws for the contact forces, form a measure differential inclusion in second-order form. The stability of equilibrium sets of measure differential inclusions, and in particular of those stemming from Lagrangian mechanical systems, will be discussed in Chapters 6 and 7.

6

Lyapunov Stability Theory for Measure Differential Inclusions

Lyapunov stability theory has originally been developed for smooth ordinary differential equations. In this chapter, we generalise the stability theory of Lyapunov to measure differential inclusions of the form

$$\mathrm{d}\boldsymbol{x} \in \mathrm{d}\boldsymbol{\Gamma}(\boldsymbol{x}), \tag{6.1}$$

being time-autonomous, or non-autonomous

$$\mathrm{d}\boldsymbol{x} \in \mathrm{d}\boldsymbol{\Gamma}(t, \boldsymbol{x}). \tag{6.2}$$

We assume that the measure differential inclusion is consistent (see Definition 4.8) and that it has at least one solution $\boldsymbol{\varphi}(t, t_0, \boldsymbol{x}_0)$ for all admissible initial conditions $\boldsymbol{x}_0 \in \mathcal{A}$, i.e. it holds that $\mathcal{S}(\cdot, t_0, \boldsymbol{x}_0) \neq \emptyset \, \forall t_0, \, \forall \boldsymbol{x}_0 \in \mathcal{A}(t_0)$. In other words: existence is assumed for all admissible conditions. First, some mathematical prerequisites are presented in Section 6.1. Subsequently, the invariance of sets and limit sets of the autonomous measure differential inclusion (6.1) are studied in Section 6.2. Definitions of stability properties of equilibria and positively invariant sets of autonomous measure differential inclusions are given in Section 6.3. In Sections 6.2 and 6.3, we restrict ourselves, for the sake of simplicity, to the case of autonomous measure differential inclusions, because the stability results in Sections 6.5, 6.6 and 6.7 are formulated for autonomous measure differential inclusions. Moreover, in Chapter 7, we apply these results to autonomous mechanical systems.

In Section 6.4, however, we propose stability definitions of solutions of non-autonomous measure differential inclusions, from which stability definitions of, for example, equilibria of non-autonomous measure differential inclusions can be directly deducted as non-autonomous variants of the definitions in Section 6.3. Moreover, these stability notions for solutions of non-autonomous measure differential inclusions are needed in defining the notion of convergence for such systems in Chapter 8.

In Section 6.5, we give generalisations of the direct method of Lyapunov. Subsequently, a generalisation is given of LaSalle's invariance principle in

Section 6.6 and of Chetaev's instability theorem in Section 6.7 for measure differential inclusions (6.1). Although in the latter sections only the autonomous case is considered, the results can be readily extended to the non-autonomous case.

6.1 Mathematical Preliminaries

Classically, the direct method of Lyapunov considers a Lyapunov function $V(\boldsymbol{x})$, being a (locally) positive definite function, and concludes stability properties by analysing the time-derivative $\dot{V}(t) = \nabla V^{\mathrm{T}} \dot{\boldsymbol{x}}(t)$ along solution curves of the system. In this chapter, we will generalise the direct method of Lyapunov to measure differential inclusions. The inequality constraints on the state $\boldsymbol{x}(t)$, which are implicitly defined by the measure differential inclusion, need to be taken into account in the Lyapunov function. For this reason, we will consider a Lyapunov function $V : \mathbb{R}^n \to \mathbb{R} \cup \{\infty\}$ which can take infinite values on its domain. Such a function is called an extended function. Furthermore, we will need to generalise the notion of the derivative of the Lyapunov function V. By considering a lower-semi-continuous Lyapunov function $V(\boldsymbol{x}(t)$, we are able to define a subderivative of V if $\boldsymbol{x}(t)$ is absolutely continuous in time.

An extended lower semi-continuous function $V(\boldsymbol{x})$, where $V : \mathbb{R}^n \to \mathbb{R} \cup \{\infty\}$, is called a

- **radially unbounded function** if

$$V(\boldsymbol{x}) \to \infty \quad \text{as} \quad \|\boldsymbol{x}\| \to \infty, \tag{6.3}$$

- **locally positive definite function** (LPDF) if there exists an $h > 0$ such that

$$V(\boldsymbol{0}) = 0 \quad \text{and} \quad V(\boldsymbol{x}) > 0 \, \forall \boldsymbol{x} \in B_h \backslash \{\boldsymbol{0}\}, \tag{6.4}$$

- **positive definite function** (PDF) if $V(\boldsymbol{x})$ is radially unbounded and

$$V(\boldsymbol{0}) = 0 \quad \text{and} \quad V(\boldsymbol{x}) > 0 \, \forall \boldsymbol{x} \neq \boldsymbol{0}. \tag{6.5}$$

Definition 6.1 (Class K function and class KR function). *A function* $\alpha : \mathbb{R}^+ \to \mathbb{R}^+$, *with* $\alpha(0) = 0$, *which is continuous and strictly increasing is called a class K function. If in addition* $\alpha(x) \to \infty$ *as* $x \to \infty$, *then function* α *is called a class KR function.*

Hence, $V(\boldsymbol{x})$ is PDF if and only if it is bounded from below by a class KR function $\alpha(\|\boldsymbol{x}\|)$ [152], i.e.

$$V(\boldsymbol{x}) \geq \alpha(\|\boldsymbol{x}\|) \quad \forall \boldsymbol{x} \in \mathbb{R}^n. \tag{6.6}$$

In the above definition, a PDF function is by definition radially unbounded. Other authors, e.g. [85], do not immediately associate positive definiteness with radially unboundedness and have to state radially unboundedness explicitly when required.

The function $V(\boldsymbol{x})$ is called quadratic if it is of the form $V(\boldsymbol{x}) = \boldsymbol{x}^{\mathrm{T}} \boldsymbol{B} \boldsymbol{x}$ with $\boldsymbol{B} \in \mathbb{R}^{n \times n}$. It is always possible to express a quadratic form with a symmetric matrix $\boldsymbol{A} = \frac{1}{2}(\boldsymbol{B} + \boldsymbol{B}^{\mathrm{T}})$ as

$$V(\boldsymbol{x}) = \boldsymbol{x}^{\mathrm{T}} \boldsymbol{A} \boldsymbol{x}, \quad \boldsymbol{A} = \boldsymbol{A}^{\mathrm{T}} \in \mathbb{R}^{n \times n}. \tag{6.7}$$

If the symmetric matrix $\boldsymbol{A} \in \mathbb{R}^{n \times n}$ is positive definite (PD), i.e.

$$\boldsymbol{x}^{\mathrm{T}} \boldsymbol{A} \boldsymbol{x} > 0 \quad \forall \boldsymbol{x} \neq \boldsymbol{0}, \tag{6.8}$$

then it holds that

$$\lambda_{\min}(\boldsymbol{A}) \|\boldsymbol{x}\|^2 \leq \boldsymbol{x}^{\mathrm{T}} \boldsymbol{A} \boldsymbol{x} \leq \lambda_{\max}(\boldsymbol{A}) \|\boldsymbol{x}\|^2, \quad \forall \boldsymbol{x} \tag{6.9}$$

where $\lambda_{\min}(\boldsymbol{A}), \lambda_{\max}(\boldsymbol{A}) > 0$ are the minimal and maximal eigenvalues of $\boldsymbol{A}$ and the quadratic function (6.7) is therefore radially unbounded. By definition it holds that $V(\boldsymbol{x}) = \boldsymbol{x}^{\mathrm{T}} \boldsymbol{A} \boldsymbol{x} > 0$ for all $\boldsymbol{x} \neq \boldsymbol{0}$. A quadratic function V with positive definite matrix $\boldsymbol{A}$ is therefore PDF. For a quadratic positive definite function V it holds that

$$\boldsymbol{x}^{\mathrm{T}} \nabla V(\boldsymbol{x}) = 2\boldsymbol{x}^{\mathrm{T}} \boldsymbol{A} \boldsymbol{x} > 0 \quad \forall \boldsymbol{x} \neq \boldsymbol{0}, \tag{6.10}$$

which means that the vector $\boldsymbol{x}$ and the gradient always form an acute angle.

For an extended lower semi-continuous function $V : \mathbb{R}^n \to \mathbb{R} \cup \{\infty\}$ and $c \geq 0$ we define the level set Ω_c as

$$\Omega_c = \{\boldsymbol{x} \in \mathbb{R}^n \mid V(\boldsymbol{x}) \leq c\}, \tag{6.11}$$

with the corresponding level surface $\mathcal{L}_c$

$$\mathcal{L}_c = \mathrm{bdry}\, \Omega_c = \{\boldsymbol{x} \in \mathbb{R}^n \mid V(\boldsymbol{x}) = c\}. \tag{6.12}$$

The level sets of quadratic positive definite functions are ellipsoids centred at the origin (see Figure 6.1). Level sets of (extended) lower semi-continuous PDF functions are closed and bounded.

Positive definite functions have important properties.

Proposition 6.2. *If an extended lower semi-continuous function V is a PDF, then there exists a $c^* > 0$ such that each level set $\Omega_c = \{\boldsymbol{x} \in \mathbb{R}^n \mid V(\boldsymbol{x}) \leq c\}$ with $c \leq c^*$ is simply connected.*

Proof: *Reductio ad absurdum: if Ω_c is not simply connected then there exists a local minimum of $V(\boldsymbol{x})$ at $\boldsymbol{x}_{\min} \neq \boldsymbol{0}$ with $V(\boldsymbol{x}_{\min}) \leq c$. If $V(\boldsymbol{x})$ is PDF then $V(\boldsymbol{x}) > 0$ for all $\boldsymbol{x} \neq \boldsymbol{0}$, and the origin is therefore the global minimum and the only minimum of $V(\boldsymbol{x})$ in some neighbourhood B_ε of the origin. For this B_ε, there exists a c^* such that $\Omega_c \subset B_\varepsilon$ for all $c \leq c^*$. Hence, if $\Omega_c \subset B_\varepsilon$,*

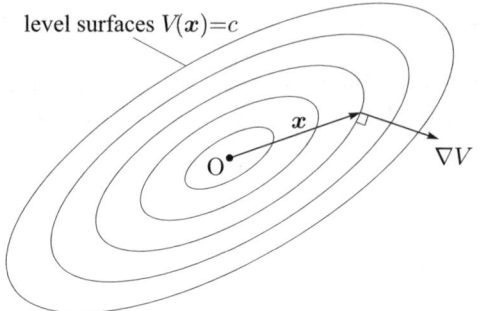

Fig. 6.1. Level surfaces of a quadratic positive definite function.

then the origin is the only minimum in Ω_c, which implies that Ω_c is simply connected if $c \leq c^*$.   □

Quadratic PDF's have a unique minimum at $\boldsymbol{x} = \boldsymbol{0}$ and each level set is therefore simply connected.

Let the state $\boldsymbol{x}(t)$ of a system be governed by the autonomous ordinary differential equation $\dot{\boldsymbol{x}}(t) = \boldsymbol{f}(\boldsymbol{x}(t))$. A function $V(\boldsymbol{x}(t))$ depends therefore implicitly on time t. If $V(\boldsymbol{x})$ is continuous in $\boldsymbol{x}$ and $\boldsymbol{x}(t)$ is continuous in t, then $V(\boldsymbol{x}(t))$ varies continuously in time. A smooth function V decreases if it holds that

$$\dot{V}(\boldsymbol{x}(t)) = \frac{\partial V}{\partial \boldsymbol{x}} \frac{\mathrm{d}\boldsymbol{x}}{\mathrm{d}t} = \nabla V(\boldsymbol{x}(t))^{\mathrm{T}} \boldsymbol{f}(\boldsymbol{x}(t)) < 0 \qquad (6.13)$$

along trajectories $\boldsymbol{x}(t)$ of the system.

If the state $\boldsymbol{x}(t)$ of a system is governed by a measure differential inclusion (6.1), then $V(\boldsymbol{x}(t))$ is in general not continuous in t because $\boldsymbol{x}(t)$ might have discontinuities in t (even if V is continuous in $\boldsymbol{x}$). Moreover, if V is discontinuous in $\boldsymbol{x}$, then $V(\boldsymbol{x}(t))$ is in general discontinuous in t. When studying measure differential inclusions, we relax the continuity with respect to time of functions $V(\boldsymbol{x}(t))$ to locally bounded variation in time.

Proposition 6.3 (Lyapunov functions of locally bounded variation).
Let $V : X \to \mathbb{R}$ be a function which is Lipschitz continuous on the closed domain $D \subset X$ with Lipschitz constant K. If it holds that $\boldsymbol{x} \in \mathrm{lbv}(I, D)$, and consequently $\boldsymbol{x}(t) \in D \,\forall t \in I$, then it holds that the function $\tilde{V}(t) = V \circ \boldsymbol{x} = V(\boldsymbol{x}(t))$ is of locally bounded variation on I.

Proof: Because $V : X \to \mathbb{R}$ is Lipschitz continuous on the closed domain $D \subset X$, it holds that

$$\exists K < \infty, \quad \|V(\boldsymbol{x}) - V(\boldsymbol{y})\| \leq K \|\boldsymbol{x} - \boldsymbol{y}\| \quad \forall \boldsymbol{x}, \boldsymbol{y} \in D. \qquad (6.14)$$

The variation of $\tilde{V}$ on a compact interval $[a, b] \subset I$ gives

$$\text{var}(\tilde{V}, [a,b]) = \sup \sum_{i=1}^{n} \|\tilde{V}(t_i) - \tilde{V}(t_{i-1})\|$$

$$= \sup \sum_{i=1}^{n} \|V(\boldsymbol{x}(t_i)) - V(\boldsymbol{x}(t_{i-1}))\|$$

$$\leq K \sup \sum_{i=1}^{n} \|\boldsymbol{x}(t_i) - \boldsymbol{x}(t_{i-1})\| \qquad (6.15)$$

$$\leq K \, \text{var}(\boldsymbol{x}, [a,b])$$

$$< \infty,$$

since $\boldsymbol{x}(t)$ is of locally bounded variation. This proves that $\tilde{V}(t) = V \circ \boldsymbol{x} = V(\boldsymbol{x}(t))$ is of locally bounded variation on I.   □

In particular, if $V(\boldsymbol{x}) = v(\boldsymbol{x}) + \varPsi_D(\boldsymbol{x})$, where $v : X \to \mathbb{R}$ is a Lipschitz continuous function on X and $\boldsymbol{x} \in \text{lbv}(I, D)$, then it follows that $\tilde{V} = V \circ \boldsymbol{x} \in \text{lbv}(I, \mathbb{R})$.

We consider the differential measure of V to be composed of an absolutely continuous part and an atomic part:

$$dV := d(V \circ \boldsymbol{x}) = \dot{V}(\boldsymbol{x}(t))dt + (V^+(\boldsymbol{x}(t)) - V^-(\boldsymbol{x}(t)))d\eta. \qquad (6.16)$$

The function $V(\boldsymbol{x})$ will be assumed to be an extended lower semi-continuous function $V : \mathbb{R}^n \to \mathbb{R} \cup \{\infty\}$. For a number of special cases we can express the differential measure dV in more detail:

1. If $V(\boldsymbol{x})$ is a locally continuous function in $\boldsymbol{x}$, then it holds that $V^+(\boldsymbol{x}(t)) = V(\boldsymbol{x}^+(t))$ and $V^-(\boldsymbol{x}(t)) = V(\boldsymbol{x}^-(t))$.
2. If $\boldsymbol{x}(t)$ is an absolutely continuous function in time and V is an extended lower semi-continuous function, then V admits a subderivative $dV(\boldsymbol{x})(\boldsymbol{v})$ at $\boldsymbol{x}$ in the direction $\boldsymbol{v}$ (see Section 2.6). Using the chain rule we express the differential measure of V in the differential measure $d\boldsymbol{x} = \dot{\boldsymbol{x}}dt$ as

$$dV = dV(\boldsymbol{x})(d\boldsymbol{x}), \qquad d\boldsymbol{x} \in d\boldsymbol{\varGamma}(\boldsymbol{x}) = \boldsymbol{\mathcal{F}}(\boldsymbol{x})dt. \qquad (6.17)$$

Note that dV denotes the differential measure of V and dV is used to denote the subderivative of V.

3. If $\boldsymbol{x}(t)$ is of locally bounded variation and if V is a quadratic form $V(\boldsymbol{x}) = \frac{1}{2}\boldsymbol{x}^T\boldsymbol{A}\boldsymbol{x}$, with $\boldsymbol{A}^T = \boldsymbol{A}$, then it follows from (3.64) that

$$dV = (\boldsymbol{x}^+ + \boldsymbol{x}^-)^T\boldsymbol{A}d\boldsymbol{x}. \qquad (6.18)$$

Substitution of $d\boldsymbol{x} = \dot{\boldsymbol{x}}dt + (\boldsymbol{x}^+ - \boldsymbol{x}^-)d\eta$ gives

$$dV = (\boldsymbol{x}^+ + \boldsymbol{x}^-)^T\boldsymbol{A}(\dot{\boldsymbol{x}}dt + (\boldsymbol{x}^+ - \boldsymbol{x}^-)d\eta)$$

$$= \boldsymbol{x}^T\boldsymbol{A}\dot{\boldsymbol{x}}dt + (\boldsymbol{x}^+ + \boldsymbol{x}^-)^T\boldsymbol{A}(\boldsymbol{x}^+ - \boldsymbol{x}^-)d\eta$$

$$= \boldsymbol{x}^T\boldsymbol{A}\dot{\boldsymbol{x}}dt + (\boldsymbol{x}^{+T}\boldsymbol{A}\boldsymbol{x}^+ - \boldsymbol{x}^{-T}\boldsymbol{A}\boldsymbol{x}^-)d\eta \qquad (6.19)$$

$$= (\nabla V(\boldsymbol{x}))^T \dot{\boldsymbol{x}}dt + (V(\boldsymbol{x}^+) - V(\boldsymbol{x}^-))d\eta.$$

In many practical applications (e.g. Lagrangian mechanical systems with unilateral constraints), the system is such that some of the states are absolutely
continuous (e.g. generalised positions) while the other states are of locally
bounded variation (e.g. generalised velocities). If the function V consists of
a lower semi-continuous function, dependent on the absolutely continuous
states, and a quadratic function, dependent on the other states, then the
differential measure of V can be obtained using the above special cases.

6.2 Invariant Sets and Limit Sets

The state-space of a dynamical system may contain equilibrium points and
sets, periodic solutions and other types of attractors, repellors or saddles such
as quasi-periodic solutions and chaotic attractors. For smooth dynamical systems, these kinds of motions have one thing in common: each of these motions
forms a point or set in the state-space from which the solution curves remain
within the set. We denote these sets as invariant sets. In this section, we consider in particular positively invariant sets of autonomous measure differential
inclusions of the form (6.1).

Definition 6.4 (Positively Invariant Set). *A set $\mathcal{M} \subset \mathbb{R}^n$ is said to be
positively invariant with respect to the measure differential inclusion (6.1) if
for every initial condition $x_0 \in \mathcal{M}$ it holds that each forward solution curve
$\varphi(\cdot, t_0, x_0) \in \mathcal{S}(\mathrm{d}\boldsymbol{\Gamma}, t_0, x_0) \neq \emptyset$ satisfies*

$$\varphi(t, t_0, x_0) \in \mathcal{M} \quad \text{for almost all } t > t_0.$$

The solution $\varphi(t, t_0, x_0)$ is not required to stay in $\mathcal{M}$ on its discontinuity
points t_i, because $\varphi(t_i, t_0, x_0)$ is not defined. However, due to the fact that
$\varphi(t, t_0, x_0)$ is of locally bounded variation, it still holds on a discontinuity
point $t_i > t_0$ that

$$\varphi(t_i, t_0, x_0)^+ \in \mathcal{M}, \quad \varphi(t_i, t_0, x_0)^- \in \mathcal{M},$$

when $\mathcal{M}$ is a positively invariant set of (6.1). In other words, the solution
before and after the discontinuity are in $\mathcal{M}$. Moreover, every positively invariant set $\mathcal{M}$ of (6.1) is a subset of the admissible set $\mathcal{A}$, which is the largest
positively invariant set of the system. The solution curve $\varphi(\cdot, t_0, x_0)$ is not
necessarily unique. Positive invariance of a set $\mathcal{M}$ means that *each* solution
curve of (6.1) starting from (t_0, x_0) remains within $\mathcal{M}$ for all $x_0 \in \mathcal{M}$ and for
almost all succeeding times $t > t_0$. This is sometimes referred to as strongly
invariant, in contrast to weakly invariant which only demands that at least
one solution curve remains within $\mathcal{M}$ [37, 156]. A negatively invariant set $\mathcal{M}$
consists of initial conditions from which all backward solution curves remain
within $\mathcal{M}$, i.e. $\varphi(t, t_0, x_0) \in \mathcal{M}$ for almost all $t < t_0$. A set which is both
positively and negatively invariant is called an invariant set.

Positive invariance of a level set Ω_c (6.11) of a function $V(\boldsymbol{x})$ with respect to a measure differential inclusion can be proven using the following proposition.

Proposition 6.5. *Let $\Omega_c = \{\boldsymbol{x} \in \mathbb{R}^n \mid V(\boldsymbol{x}) \leq c\}$ be a bounded set. Consider the differential measure dV (6.16) and the measure differential inclusion (6.1). If $dV \leq 0$ for all $\boldsymbol{x} \in \Omega_c$, then the set Ω_c is positively invariant with respect to (6.1).*

Proof: The function V can not increase in Ω_c because $dV \leq 0$ for all $\boldsymbol{x} \in \Omega_c$. For $\boldsymbol{x}_0 \in \Omega_c$ it holds that

$$V(\boldsymbol{x}^+(t)) = V(\boldsymbol{x}_0) + \int_{[t_0,t]} dV$$
$$\leq V(\boldsymbol{x}_0)$$
$$\leq c$$

for all solution curves $\boldsymbol{x}(t) = \varphi(\cdot, t_0, \boldsymbol{x}_0) \in \mathcal{S}(d\boldsymbol{\Gamma}, t_0, \boldsymbol{x}_0)$. Consequently, all solution curves $\boldsymbol{x}(t)$ remain in Ω_c for almost all t (except at time-instances for which $\boldsymbol{x}(t)$ is not defined). $\qquad\square$

Note that if V is a PDF, and therefore by definition radially unbounded, then Ω_c is bounded for all bounded values c. Moreover, the level set Ω_c is in general not simply connected and may consist of connected components Ω_{ci}, $i = 1, 2, \ldots$. For example, consider the level set Ω_c in Figure 6.2 which consists of Ω_{c1}, Ω_{c2} and Ω_{c3}. We assume $dV \leq 0$ in Ω_c. If a solution $x(t)$ of (6.1) starts in Ω_{c2}, then the function $V(x)$ will not increase along the solution curve, forcing the solution to stay within the level set Ω_c. Hence, the level set is positively invariant. However, the solution $x(t)$ might jump during the time-evolution to another connected component of Ω_c, e.g. to Ω_{c3} as is depicted in Figure 6.2. Consequently, the connected components of a positively invariant level set Ω_c are therefore generally *not* positively invariant.

The most elementary invariant set in a smooth dynamical system is an invariant equilibrium point.

Definition 6.6 (Equilibrium Point). *A point $\boldsymbol{x}^*$ is called an equilibrium point of (6.1) if there exists a forward solution curve $\varphi(\cdot, t_0, \boldsymbol{x}^*) \in \mathcal{S}(d\boldsymbol{\Gamma}, t_0, \boldsymbol{x}^*)$ such that*

$$\varphi(t, t_0, \boldsymbol{x}^*) = \boldsymbol{x}^*, \quad \forall t > t_0.$$

Corollary 6.7. *It holds that $\boldsymbol{0} \in d\boldsymbol{\Gamma}(\boldsymbol{x}^*)$ if and only if $\boldsymbol{x}^*$ is an equilibrium point of (6.1).*

Proof: If $\boldsymbol{0} \in d\boldsymbol{\Gamma}(\boldsymbol{x}^*)$, then there exists a forward solution curve $\varphi(t, t_0, \boldsymbol{x}^*) = \boldsymbol{x}^*$ for all $t > t_0$, i.e. $\boldsymbol{x}^*$ is an equilibrium point. Moreover, because an equilibrium point $\boldsymbol{x}^*$ is a constant solution of (6.1), it must hold that $d\boldsymbol{x} = \boldsymbol{0}$ along a constant solution curve $\varphi(t, t_0, \boldsymbol{x}^*) = \boldsymbol{x}^*$, i.e. $\boldsymbol{0} \in d\boldsymbol{\Gamma}(\boldsymbol{x}^*)$. $\qquad\square$

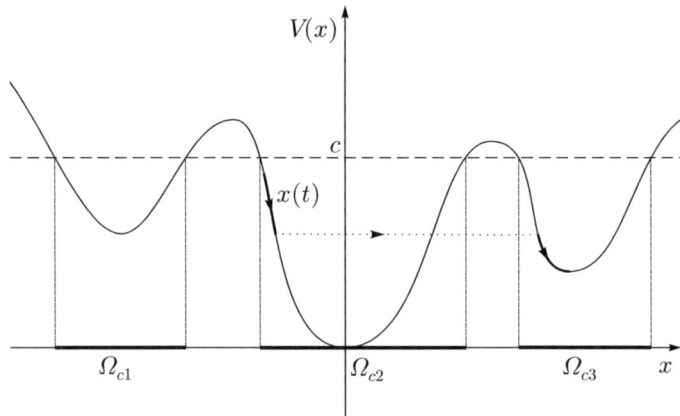

Fig. 6.2. Non-invariance of connected components Ω_{ci} of an invariant level set Ω_c.

Non-uniqueness of solutions in non-smooth dynamical systems may destroy the invariance of an equilibrium point. If the forward solution from an equilibrium point is non-unique, then there exists a solution curve which does not remain on the equilibrium point. The equilibrium point is in this case not a positively invariant set. Non-smooth dynamical systems may also exhibit simply connected sets of equilibrium points which may or may not be positively invariant.

Definition 6.8 (Equilibrium Set). *A simply connected set $\mathcal{E}$ is called an equilibrium set of (6.1) if every point $\boldsymbol{x}^* \in \mathcal{E}$ is an equilibrium point and if there exists no equilibrium point $\boldsymbol{x}^* \notin \mathcal{E}$ infinitely close to $\mathcal{E}$.*

Periodic solutions in smooth dynamical systems form invariant closed orbits[1] in the state-space. Periodic solutions in non-smooth systems still have a periodicity property

$$\varphi(t, t_0, \boldsymbol{x}_0) = \varphi(t + kT, t_0, \boldsymbol{x}_0), \qquad k = 1, 2, 3, \ldots, \quad \forall t > t_0, \ t \neq t_i, \quad (6.20)$$

but the solution $\varphi(t, t_0, \boldsymbol{x}_0)$ is not defined on the discontinuity points $\{t_i\}$. The image of the periodic solution is in general therefore *not* a closed orbit. We therefore will use the term *periodic orbit* for the set $\mathcal{O}$ defined by

$$\mathcal{O} = \mathrm{cl}\{\boldsymbol{x} \in \mathbb{R}^n \mid \boldsymbol{x} = \varphi(t, t_0, \boldsymbol{x}_0), t_0 \leq t \leq t_0 + T\}, \qquad (6.21)$$

which is the closure of the image of the periodic solution $\varphi(t, t_0, \boldsymbol{x}_0)$. The set $\mathcal{O}$ is closed, in the sense that it contains its boundary, but $\mathcal{O}$ is not necessarily a closed orbit (Figure 6.3). If the periodic solution $\varphi(t, t_0, \boldsymbol{x}_0)$ is positively invariant, then its corresponding periodic orbit is a closed positively invariant set.

---

[1] The term *orbit* shows the influence of astronomy on the terminology in stability theory.

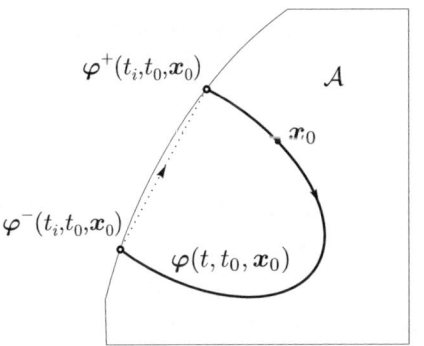
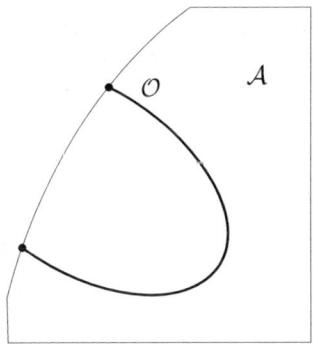

Fig. 6.3. Periodic solution $\varphi(t, t_0, \boldsymbol{x}_0)$ and periodic orbit $\mathcal{O}$ within the admissible set $\mathcal{A}$.

Closely related to the invariance of a set is the behaviour of solution curves when time approaches infinity. After Birkhoff [20], we introduce the terms positive limit point and set (see also [70, 85, 89]).

Definition 6.9 (Positive Limit Point). *A point* $\boldsymbol{p} \in \mathbb{R}^n$ *is a positive limit point of a solution curve* $\varphi(\cdot, t_0, \boldsymbol{x}_0) \in \mathcal{S}(\mathrm{d}\boldsymbol{\Gamma}, t_0, \boldsymbol{x}_0)$ *of (6.1) if there exists a sequence* $\{t_j\}$ *with* $t_j \to +\infty$ *as* $j \to \infty$, *such that* $\varphi(t_j, t_0, \boldsymbol{x}_0) \to \boldsymbol{p}$ *for* $j \to \infty$.

Definition 6.10 (Positive Limit Set). *A positive limit set* $\mathcal{L}^+$ *of a solution curve* $\varphi(\cdot, t_0, \boldsymbol{x}_0) \in \mathcal{S}(\mathrm{d}\boldsymbol{\Gamma}, t_0, \boldsymbol{x}_0)$ *is the set of all its positive limit points.*

A positive limit set $\mathcal{L}^+$ is a set which is approached for $t \to \infty$. Similarly, a negative limit set $\mathcal{L}^-$ is a set which is approached for $t \to -\infty$. For smooth dynamical systems it holds, under the assumption of bounded state behaviour, that every positive limit set is an invariant set. This assertion is in general not true for measure differential inclusions (6.1). Consider for instance the scalar measure differential inclusion

$$\mathrm{d}x = -x(t)\mathrm{d}t + a(x(t))\mathrm{d}\eta, \quad \text{with } a(x(t)) = \begin{cases} 1, & x^-(t) = 0, \\ 0, & x^-(t) \neq 0. \end{cases} \quad (6.22)$$

The system is described by the differential equation $\dot{x}(t) = -x(t)$, with the solution $x(t) = x_0 e^{-(t-t_0)}$ for $x_0 \neq 0$. If $x^-(t) = 0$, then an impulse occurs which brings the state to $x^+(t) = 1$. A solution with $x_0 \neq 0$ will asymptotically approach the origin $p = 0$, but will never reach it. The point $p = 0$ is therefore a positive limit point, but is not a positively invariant point because $\varphi(t, t_0, p) \neq p$ for $t > t_0$. The non-invariance of the limit point p is caused by a discontinuous dependence of the solution on the initial condition. Continuous dependence with respect to the initial condition is defined in [50] for differential inclusions. Here, we generalise the definition to measure differential inclusions.

Definition 6.11 (Continuous Dependence on the Initial Condition).
Consider an autonomous measure differential inclusion $\mathrm{d}\boldsymbol{x} \in \mathrm{d}\boldsymbol{\Gamma}(\boldsymbol{x}(t))$ *with the admissible set* $\mathcal{A}$. *The system has a continuous dependence on the initial condition if for any finite time-interval* $[t_0, t_1]$ *and any solution* $\boldsymbol{\varphi}(t, t_0, \boldsymbol{x}_{01}) \in \mathcal{S}(\boldsymbol{\Gamma}, t_0, \boldsymbol{x}_{01})$ *with* $\boldsymbol{x}_{01} \in \mathcal{A}$ *it holds that for each* ε *there exists a* $\delta(\varepsilon)$ *such that*

$$\|\boldsymbol{x}_{02} - \boldsymbol{x}_{01}\| < \delta \Rightarrow \|\boldsymbol{\varphi}(t, t_0, \boldsymbol{x}_{02}) - \boldsymbol{\varphi}(t, t_0, \boldsymbol{x}_{01})\| < \varepsilon \text{ for almost all } t \in [t_0, t_1],$$

for all $\boldsymbol{\varphi}(t, t_0, \boldsymbol{x}_{02}) \in \mathcal{S}(\boldsymbol{\Gamma}, t_0, \boldsymbol{x}_{02})$ *with* $\boldsymbol{x}_{02} \in \mathcal{A}$.

Continuous dependence on the initial condition implies that if two initial points are infinitely close to each other, then their solutions curves remain infinitely close. Non-uniqueness of solutions destroys this property. Hence, if the measure differential inclusion has a continuous dependence on the initial condition, then the solution is unique, i.e. $\mathcal{S}(\mathrm{d}\boldsymbol{\Gamma}, t_0, \boldsymbol{x}_0) = \{\boldsymbol{\varphi}(\cdot, t_0, \boldsymbol{x}_0)\}$. The converse is not true: a system can have uniqueness of solutions but have a discontinuous dependence of solutions on the initial conditions. The following proposition (taken from [33, 34]) states a sufficient condition for a positive limit set to be invariant:

Proposition 6.12 (Positive Invariance of a Positive Limit Set [33]).
If (6.1) is consistent and the solutions have a continuous dependence on the initial condition, then each positive limit set $\mathcal{L}^+$ *is positively invariant.*

Proof: Let $\boldsymbol{p} \in \mathcal{L}^+$ be a positive limit point of a solution curve $\boldsymbol{\varphi}(\cdot, t_0, \boldsymbol{x}_0)$. Then there exists a sequence $\{t_j\}$ with $t_j \to +\infty$ as $j \to \infty$, such that $\boldsymbol{\varphi}(t_j, t_0, \boldsymbol{x}_0) \to \boldsymbol{p}$ for $j \to \infty$. Uniqueness of solutions in forward time is guaranteed because of the continuous dependence on initial conditions, and allows us to speak of *the* solution curve. Moreover, if there is a continuous dependence on the initial condition, then the solution curve starting from $\boldsymbol{p}$ is identical to the solution curve starting from $\boldsymbol{\varphi}(t_j, t_0, \boldsymbol{x}_0)$, i.e.

$$\boldsymbol{\varphi}(t, t_0, \boldsymbol{p}) = \lim_{j \to \infty} \boldsymbol{\varphi}(t, t_0, \boldsymbol{\varphi}(t_j, t_0, \boldsymbol{x}_0)),$$

for almost all $t \geq t_0$. The concatenation of solution curves $\boldsymbol{\varphi}(t, t_0, \boldsymbol{\varphi}(t_j, t_0, \boldsymbol{x}_0))$ is equivalent to $\boldsymbol{\varphi}(t + t_j - t_0, t_0, \boldsymbol{x}_0)$ (where we again used uniqueness of solutions in forward time as well as the fact that (6.1) is time-autonomous), which gives after substitution in the above equation

$$\boldsymbol{\varphi}(t, t_0, \boldsymbol{p}) = \lim_{j \to \infty} \boldsymbol{\varphi}(t + t_j - t_0, t_0, \boldsymbol{x}_0),$$

for almost all $t \geq t_0$. Consequently, for almost all $t \in [t_0, \infty)$ there exists a sequence $\{t_j\}$ such that the solution curve $\boldsymbol{\varphi}(t + t_j - t_0, t_0, \boldsymbol{x}_0)$ converges to $\boldsymbol{\varphi}(t, t_0, \boldsymbol{p})$. This means that $\boldsymbol{\varphi}(t, t_0, \boldsymbol{p})$ is a limit point of the solution curve $\boldsymbol{\varphi}(\cdot, t_0, \boldsymbol{x}_0)$ for almost all $t \in [t_0, \infty)$. Using the definition of $\mathcal{L}^+$, it holds for each $\boldsymbol{p} \in \mathcal{L}^+$ that

$$\boldsymbol{\varphi}(t, t_0, \boldsymbol{p}) \in \mathcal{L}^+, \quad \text{for almost all } t \geq t_0,$$

which shows the positive invariance of the positive limit set $\mathcal{L}^+$. □

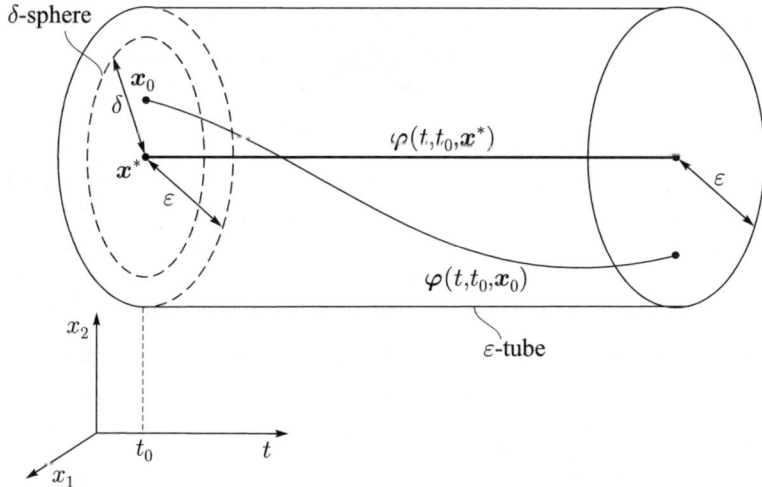

Fig. 6.4. The concept of Lyapunov stability of an equilibrium point.

6.3 Definitions of Stability Properties for Autonomous Systems

In this section we will give definitions of stability properties of equilibria and positively invariant sets of autonomous measure differential inclusions of the form (6.1) which are autonomous in time. Here, we restrict ourselves to the case of autonomous measure differential inclusions, because the stability results in Sections 6.5, 6.6 and 6.7 are formulated for autonomous measure differential inclusions.

The stability definition of Lyapunov defines an equilibrium point to be stable if all neighbouring solutions remain in the neighbourhood of the equilibrium point.

Definition 6.13 (Lyapunov Stability of an Equilibrium Point).
An equilibrium point $\boldsymbol{x}^*$ of the consistent time-autonomous measure differential inclusion (6.1) is Lyapunov stable if for each $\varepsilon > 0$ there exists a $\delta(\varepsilon) > 0$ such that for any $\boldsymbol{x}_0 \in A$ with

$$\|\boldsymbol{x}_0 - \boldsymbol{x}^*\| < \delta$$

each solution curve $\boldsymbol{\varphi}(\cdot, t_0, \boldsymbol{x}_0) \in \mathcal{S}(\mathrm{d}\boldsymbol{\Gamma}, t_0, \boldsymbol{x}_0) \neq \emptyset$ satisfies

$$\|\boldsymbol{\varphi}(t, t_0, \boldsymbol{x}_0) - \boldsymbol{x}^*\| < \varepsilon \quad \textit{for almost all } t \geq t_0.$$

The definition of Lyapunov involves an ε-δ argumentation. We consider a tube with radius ε around the equilibrium point $\boldsymbol{x}^*$. The stability concept demands that neighbouring solutions of $\boldsymbol{\varphi}(t, t_0, \boldsymbol{x}^*) = \boldsymbol{x}^*$ have to remain in the ε-tube, when the admissible initial condition $\boldsymbol{x}_0$ is chosen close enough to

x^* (in a δ-sphere around x^*). An initial condition is admissible if $x_0 \in \mathcal{A}$ and each solution curve $\varphi(t, t_0, x_0)$ remains within the admissible set for almost all $t > t_0$, because the system is assumed to be consistent. For every ε-tube we have to seek a sphere with radius δ centred at x^* for which holds that all solutions with admissible initial conditions in this δ-sphere remain in the ε-tube (see Figure 6.4). If we can indeed find such a δ for every ε, then the equilibrium point is said to be Lyapunov stable. Obviously, the δ we choose depends on ε, which we express by $\delta(\varepsilon)$ and it obviously holds that $\delta(\varepsilon) \leq \varepsilon$. An essential point in the definition is the fact that we have to find such a δ-sphere for arbitrary small values of ε, which leads to the following proposition.

Proposition 6.14. *A stable equilibrium point is positively invariant.*

Proof: Using the definition of stability and taking the limit $\varepsilon \downarrow 0$ gives $\|\varphi(t, t_0, x_0) - x^*\| < \varepsilon \downarrow 0$ with $\|x_0 - x^*\| < \delta \leq \varepsilon \downarrow 0$. We deduce that $\varphi(t, t_0, x^*) = x^*$ for almost all $t \geq t_0$. The equilibrium point x^* is therefore positively invariant. $\square$

Stability therefore implies positive invariance. The neighbouring solutions which start in the admissible part of the δ-sphere are required to remain in the ε-tube, that is to say they stay in the neighbourhood. The neighbouring solutions around a stable equilibrium point do not necessarily converge towards the equilibrium point. This additional property is called attractivity.

Definition 6.15 (Local Attractivity of an Equilibrium Point). *An equilibrium point x^* of the measure differential inclusion (6.1) is locally attractive if there exists a $\delta > 0$ such that for any $x_0 \in \mathcal{A}$ with*

$$\|x_0 - x^*\| < \delta$$

each solution curve $\varphi(\cdot, t_0, x_0) \in \mathcal{S}(\mathrm{d}\Gamma, t_0, x_0)$ converges to x^* in positive time:

$$\lim_{t \to \infty} \|\varphi(t, t_0, x_0) - x^*\| = 0.$$

An attractive equilibrium point has a domain of attraction. The admissible part of the δ-sphere in the above definition, i.e. $B_\delta \cap \mathcal{A}$, forms a conservative estimate for the domain of attraction. If the domain of attraction extends to the whole admissible set $\mathcal{A}$, then the equilibrium point is said to be globally attractive.

Definition 6.16 (Global Attractivity of an Equilibrium Point). *An equilibrium point x^* of the measure differential inclusion (6.1) is globally attractive if each solution curve $\varphi(\cdot, t_0, x_0) \in \mathcal{S}(\mathrm{d}\Gamma, t_0, x_0)$ converges to x^*:*

$$\lim_{t \to \infty} \|\varphi(t, t_0, x_0) - x^*\| = 0 \quad \forall x_0 \in \mathcal{A}.$$

An equilibrium point of a smooth dynamical system which is both stable and locally/globally attractive is called in literature locally/globally 'asymptotically stable'. The words 'local' or 'global' refer to the asymptotic behaviour of the solution curves in the neighbourhood of the equilibrium point, i.e. to the attractivity property and not to the stability (note that Lyapunov stability is in essence a local requirement, see Definition 6.13). Solution curves of a smooth dynamical system can never meet each other because of the uniqueness of solutions in forward and backward time. The attractivity in smooth systems is therefore always *asymptotic*[2] in the sense that neighbouring solution curves approach but never reach the equilibrium point when $t \to \infty$. The attractivity of an equilibrium point of a non-smooth system is not necessarily asymptotic as it might be reached in a finite time. For instance, if we consider the differential inclusion (4.9), $\dot{x} \in 1 - 2\,\mathrm{Sign}(x)$ with $x(0) = x_0$, then we immediately see that the equilibrium position $x = 0$ is reached when $t = x_0$ if $x_0 > 0$ or $t = -\frac{1}{3}x_0$ if $x_0 < 0$. We therefore refrain from the terminology 'asymptotic stability' to denote an equilibrium point which is both attractive and stable. Instead, we will use the terminology *attractive stability*. If the solution curves converge asymptotically to the equilibrium point, then we speak of *asymptotic attractivity*. If the solution curves converge to the equilibrium point in a finite time, then we speak of *symptotic attractivity*.

Definition 6.17 (Symptotic Attractivity of an Equilibrium Point).
An equilibrium point x^* of the measure differential inclusion (6.1) is symptotically attractive if there exists a $\delta > 0$ such that for any bounded $x_0 \in \mathcal{A}$ with

$$\|x_0 - x^*\| < \delta$$

each solution curve $\varphi(\cdot, t_0, x_0) \in \mathcal{S}(\mathrm{d}\Gamma, t_0, x_0)$ reaches x^* in a finite time, i.e.

$$\varphi(T + t_0, t_0, x_0) = x^*,$$

with $T < \infty$.

Symptotic attractive stability is sometimes called finite-time stability in literature [19].

The notion of stability and attractivity of an equilibrium point can be extended to the stability and attractivity of a set (e.g. an equilibrium set, periodic orbit or chaotic attractor). A stable set is necessarily positively invariant. Let $\mathcal{M}$ be a closed positively invariant set of (6.1). Define the open ε-neighbourhood of $\mathcal{M}$ by

$$U_\varepsilon(\mathcal{M}) = \{x \in \mathbb{R}^n \mid \mathrm{dist}_{\mathcal{M}}(x) < \varepsilon\}, \qquad (6.23)$$

where $\mathrm{dist}_{\mathcal{M}}(x)$ is the minimal distance from x to a point in $\mathcal{M}$ (see Definition 2.32).

---

[2] The word *asymptote* stems from the Greek words a (not) and *sympiptein* (to meet) and therefore means 'not meeting'. Proclus Diadochus (411-485 A.D.) writes in his *Commentary on Euclid's Elements* about asymptotic lines as well as symptotic lines (those that do meet).

Definition 6.18 (Stability of a Closed Positively Invariant Set). *A closed positively invariant set $\mathcal{M}$ of (6.1) is stable if for each $\varepsilon > 0$ there exists a $\delta(\varepsilon) > 0$ such that for any $\boldsymbol{x}_0 \in \mathcal{A}$ with*

$$\boldsymbol{x}_0 \in U_\delta(\mathcal{M})$$

each solution curve $\boldsymbol{\varphi}(\cdot, t_0, \boldsymbol{x}_0) \in \mathcal{S}(\mathrm{d}\boldsymbol{\Gamma}, t_0, \boldsymbol{x}_0)$ satisfies

$$\boldsymbol{\varphi}(t, t_0, \boldsymbol{x}_0) \in U_\varepsilon(\mathcal{M}) \quad \text{for almost all } t \geq t_0.$$

Definition 6.19 (Local/Global Attractivity of a Closed Positively Invariant Set). *A closed positively invariant set $\mathcal{M}$ of (6.1) is locally attractive if there exists a $\delta > 0$ such that for any $\boldsymbol{x}_0 \in \mathcal{A}$ with*

$$\boldsymbol{x}_0 \in U_\delta(\mathcal{M})$$

each solution curve $\boldsymbol{\varphi}(\cdot, t_0, \boldsymbol{x}_0) \in \mathcal{S}(\mathrm{d}\boldsymbol{\Gamma}, t_0, \boldsymbol{x}_0)$ converges to $\mathcal{M}$:

$$\lim_{t \to \infty} \mathrm{dist}_{\mathcal{M}}(\boldsymbol{\varphi}(t, t_0, \boldsymbol{x}_0)) = 0.$$

If convergence occurs for all $\boldsymbol{x}_0 \in \mathcal{A}$, then $\mathcal{M}$ is called globally attractive.

If a positively invariant equilibrium set $\mathcal{E}$ is stable in the sense of Definition 6.18, then each equilibrium point $\boldsymbol{x}^* \in \mathcal{E}$ is Lyapunov stable in the sense of Definition 6.13. However, if a positively invariant orbit $\mathcal{O}$ is stable in the sense of Definition 6.18, then neighbouring solutions of the periodic solution $\boldsymbol{\varphi}(t, t_0, \boldsymbol{x}_0)$ stay in the neighbourhood of $\mathcal{O}$, but do not necessarily stay close to $\boldsymbol{\varphi}(t, t_0, \boldsymbol{x}_0)$ because they generally encircle the orbit $\mathcal{O}$ in a time different from the period time of the periodic solution. A periodic solution with a 'stable' behaviour, i.e. it is associated with a stable closed orbit, is therefore called orbitally stable or Poincaré stable.

6.4 Definitions of Stability Properties of Solutions Non-autonomous Systems

The previous section dealt with the definition of stability properties of positively invariant sets of time-autonomous systems. In this section we will take a more general perspective and define stability properties for (time-varying) solutions of non-autonomous systems. The definitions for (uniform) stability and attractivity of solutions of differential equations have been well-defined, see [38, 132, 176]. Here, we generalise these definitions to differential inclusions and measure differential inclusions.

6.4.1 Differential Inclusions

Consider the differential inclusion

$$\dot{\boldsymbol{x}} \in \mathcal{F}(t, \boldsymbol{x}), \quad \boldsymbol{x} \in \mathbb{R}^n, \quad t \in \mathbb{R}, \tag{6.24}$$

where the function $\mathcal{F}(t, \boldsymbol{x})$ is a set-valued function. A solution $\boldsymbol{x}(t)$ of (6.24) is an absolutely continuous function, defined for all t (at least locally), which fulfills (6.24) for almost all t. The set of forward solutions of (6.24) with $\boldsymbol{x}(t_0) = \boldsymbol{x}_0$ is denoted by $\mathcal{S}(\mathcal{F}, t_0, \boldsymbol{x}_0)$.

Definition 6.20 (Stability). *A solution $\bar{\boldsymbol{x}}(t)$ of system (6.24), with $\bar{\boldsymbol{x}}(t_0) = \bar{\boldsymbol{x}}_0$ and which is defined on $t \in (t_*, +\infty)$, is said to be*

- stable *if for any $t_0 \in (t_*, +\infty)$ and $\varepsilon > 0$ there exists a $\delta = \delta(\varepsilon, t_0) > 0$ such that $\|\boldsymbol{x}_0 - \bar{\boldsymbol{x}}(t_0)\| < \delta$ implies that each forward solution $\boldsymbol{x}(t) \in \mathcal{S}(\mathcal{F}, t_0, \boldsymbol{x}_0)$ satisfies $\|\boldsymbol{x}(t) - \bar{\boldsymbol{x}}(t)\| < \varepsilon$ for all $t \geq t_0$.*
- uniformly stable *if it is stable and the number δ in the definition of stability is independent of t_0.*
- attractively stable *if it is stable and for any $t_0 \in (t_*, +\infty)$ there exists $\bar{\delta} = \bar{\delta}(t_0) > 0$ such that $\|\boldsymbol{x}_0 - \bar{\boldsymbol{x}}_0\| < \bar{\delta}$ implies that each forward solution $\boldsymbol{x}(t) \in \mathcal{S}(\mathcal{F}, t_0, \boldsymbol{x}_0)$ satisfies $\lim_{t \to +\infty} \|\boldsymbol{x}(t) - \bar{\boldsymbol{x}}(t)\| = 0$ for all $t \geq t_0$.*
- uniformly attractively stable *if it is uniformly stable and there exists $\bar{\delta} > 0$ (independent of t_0) such that for any $\varepsilon > 0$ there exists $T = T(\varepsilon) > 0$ such that $\|\boldsymbol{x}_0 - \bar{\boldsymbol{x}}_0\| < \bar{\delta}$ for $t_0 \in (t_*, +\infty)$ implies that each forward solution $\boldsymbol{x}(t) \in \mathcal{S}(\mathcal{F}, t_0, \boldsymbol{x}_0)$ satisfies $\|\boldsymbol{x}(t) - \bar{\boldsymbol{x}}(t)\| < \varepsilon$ for all $t \geq t_0 + T$.*

Note that if $\bar{\boldsymbol{x}}(t)$ is a stable solution, then it must be the unique forward solution from $(t_0, \bar{\boldsymbol{x}}_0)$, i.e. $\mathcal{S}(\mathcal{F}, t_0, \bar{\boldsymbol{x}}_0) = \{\bar{\boldsymbol{x}}(\cdot)\}$. Definitions of global attractive stability and global uniform attractive stability of a solution can be given in a similar way.

6.4.2 Measure Differential Inclusions

Consider the measure differential inclusion (6.2). As has been discussed in Section 4.3, a solution of (6.2) is a function of locally bounded variation that fulfills (6.2) in a measure sense *for all* t. A solution $\boldsymbol{x}(t)$ of (6.2) is defined for *almost all* t, i.e. not for a Lebesgue negligible set of time-instance for which the solution $\boldsymbol{x}(t)$ jumps. A measure differential inclusion usually describes a physical process. Set-valued force-laws restrict the state $\boldsymbol{x}$ in (6.2) to some admissible set $\mathcal{A}(t)$. For instance, contact laws and restitution laws prohibit penetration of a unilateral contact in a mechanical system and therefore restrict the position to some admissible set.

Definition 6.21 (Stability). *A solution $\bar{\boldsymbol{x}}(t)$ of system (6.2), with $\bar{\boldsymbol{x}}(t_0) = \bar{\boldsymbol{x}}_0 \in \mathcal{A}(t_0)$ and which is defined on $t \in (t_*, +\infty)$, is said to be*

- stable *if for any $t_0 \in (t_*, +\infty)$ and $\varepsilon > 0$ there exists a $\delta = \delta(\varepsilon, t_0) > 0$ such that $\|x_0 - \bar{x}(t_0)\| < \delta$ with $x_0 \in A(t_0)$ implies that each forward solution $x(t) \in S(\mathrm{d}\Gamma, t_0, x_0)$ satisfies $\|x(t) - \bar{x}(t)\| < \varepsilon$ for almost all $t \geq t_0$.*
- uniformly stable *if it is stable and the number δ in the definition of stability is independent of t_0.*
- attractively stable *if it is stable and for any $t_0 \in (t_*, +\infty)$ there exists $\bar{\delta} = \bar{\delta}(t_0) > 0$ such that $\|x_0 - \bar{x}_0\| < \bar{\delta}$ with $x_0 \in A(t_0)$ implies that each forward solution $x(t) \in S(\mathrm{d}\Gamma, t_0, x_0)$ satisfies $\lim_{t \to +\infty} \|x(t) - \bar{x}(t)\| = 0$ for almost all $t \geq t_0$.*
- uniformly attractively stable *if it is uniformly stable and there exists $\bar{\delta} > 0$ (independent of t_0) such that for any $\varepsilon > 0$ there exists $T = T(\varepsilon) > 0$ such that $\|x_0 - \bar{x}_0| < \bar{\delta}$ with $x_0 \in A(t_0)$ for $t_0 \in (t_*, +\infty)$ implies that each forward solution $x(t) \in S(\mathrm{d}\Gamma, t_0, x_0)$ satisfies $\|x(t) - \bar{x}(t)| < \varepsilon$ for almost all $t \geq t_0 + T$.*

So far, stability properties have been defined for equilibrium points, positively invariant sets and, in the current section, of any (time-varying) solution of the system under study. Let us now discuss a stability property on system level, called *incremental stability*, which implies the stability of all solutions of a system (with respect to each other) [7]. Incremental stability reflect a kind of contraction property of the system. We will define incremental stability for non-autonomous measure differential inclusions.

Definition 6.22 (Incremental Stability of a Measure Differential Inclusion). *The measure differential inclusion (6.2) is*

- incrementally stable *if for all t_0, any $x_1, x_2 \in A(t_0)$ and all corresponding solution curves $\varphi(\cdot, t_0, x_i) \in S(\mathrm{d}\Gamma, t_0, x_i)$ ($i = 1, 2$), it holds that for any $\varepsilon > 0$ there exists a $\delta = \delta(\varepsilon, t_0)$ such that $\|x_1 - x_2\| < \delta$ implies that $\|\varphi(t, t_0, x_1) - \varphi(t, t_0, x_2)\| < \varepsilon$, for almost all $t \geq t_0$.*
- *In addition, the system (6.2) is called* attractively incrementally stable *if it is incrementally stable and*

$$\lim_{t \to \infty} \|\varphi(t, t_0, x_1) - \varphi(t, t_0, x_2)\| = 0, \qquad \forall t_0, \forall x_1, x_2 \in A(t_0).$$

If the system is attractively incrementally stable, then all solutions converge to one another. In Chapter 8, we will discuss the property of convergence. Many Lyapunov characterisations of convergence (based on quadratic Lyapunov functions) also imply incremental stability.

Remark: Note that the definition of stability as given in Definition 6.21, which is based on comparing different solutions on the same time instant, may, in a certain sense, be rather restrictive for solutions with jumps. Namely, it excludes certain situations that we may intuitively consider to be stable. Let us, for example, consider a particular solution $\bar{x}(t)$ (where x is scalar), of which we study the stability properties, see Figure 6.5. If all other solutions, such as

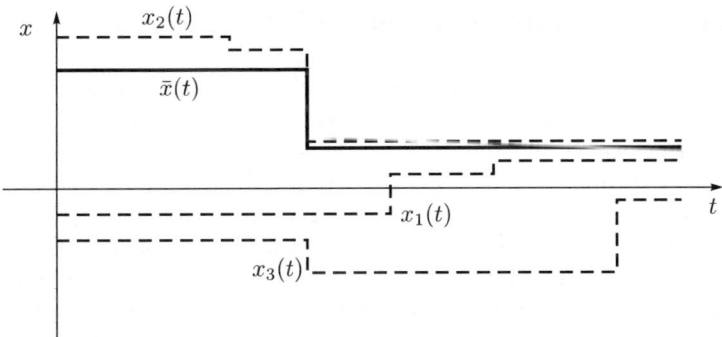

Fig. 6.5. Solution $\bar{x}(t)$ is a stable solution in the sense of Definition 6.21.

e.g. $x_1(t)$, $x_2(t)$ and $x_3(t)$, behave as depicted in Figure 6.5, then we would call $\bar{x}(t)$ stable (or even attractively stable) in the sense of Definition 6.21. More specifically, on all time instances $\bar{t}_i$, t_i^k corresponding to jump times of $\bar{x}(t)$ and $x_k(t)$, $k = 1, 2, 3, \ldots$, respectively, the distance between $\bar{x}(t)$ and $x_k(t)$ does not increase.

Next, we consider the situation as depicted in Figure 6.6. Clearly, $\bar{x}(t)$ is not stable in the sense of Definition 6.21 if solutions $x_k(t)$ starting arbitrarily close to $\bar{x}(t)$ at $t = t_0$ do not jump at the same times as $\bar{x}(t)$. However, if all other solutions $x_k(t)$ behave qualitatively similar to $x_1(t)$ in Figure 6.6, then one would be inclined to call $\bar{x}(t)$ stable (even attractively stable), because $\bar{x}(t)$ and $x_1(t)$ remain close and converge to each other (both) in a graphical sense. Therefore, we believe that in order to include such situations, new stability definitions for time-varying solutions with jumps should be developed, possibly based on certain graph-closeness properties of solutions with respect to each other, rather than comparing different solutions on the exact same time instant.

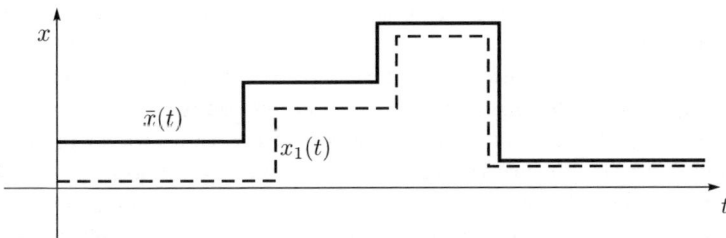

Fig. 6.6. Solution $\bar{x}(t)$ is not a stable solution in the sense of Definition 6.21.

6.5 Basic Lyapunov Theorems of Autonomous Systems

In this section we present generalised versions of basic Lyapunov stability results, that can be used to prove the (attractive) stability of equilibrium points and positively invariant sets of time-autonomous measure differential inclusions. Section 6.5.1 gives an introduction to Lyapunov-type theorems and presents several theorems that can be used to prove the (attractive) stability of equilibrium points of measure differential inclusions. These theorems are generalised in Section 6.5.2 to prove the (attractive) stability of equilibrium sets. The stability properties of systems with a monotonicity property are discussed in Section 6.5.3. A sufficient condition is given for incremental stability of an autonomous measure differential inclusion.

6.5.1 Lyapunov Stability of Equilibrium Points

The fundamental idea behind the original Lyapunov stability result is the stability theorem of Lagrange-Dirichlet. Consider the mathematical pendulum equation

$$ml^2\ddot{\theta} + mgl\sin\theta = 0, \tag{6.25}$$

in which m is the mass, l is the length of the pendulum, g is the gravitational acceleration and θ designates the angle of the pendulum relative to its downward hanging equilibrium position. The total energy E of the system consists of the kinetic energy $T(\dot{\theta}) = \frac{1}{2}ml^2\dot{\theta}^2$ and the potential energy $U(\theta) = mgl(1 - \cos\theta)$:

$$E := E(\theta, \dot{\theta}) = T(\dot{\theta}) + U(\theta). \tag{6.26}$$

The kinetic energy function is a positive definite function. The potential energy $U(\theta)$ is a locally positive definite function, which implies local positive definiteness of the total energy E. The downward hanging equilibrium position $(\theta, \dot{\theta}) = (0, 0)$ is therefore a local minimum of the total energy. Let $(\theta(t), \dot{\theta}(t))$ be the time-evolution of the system starting from the initial state $(\theta_0, \dot{\theta}_0)$ at $t = t_0$. Because the system is conservative, the total energy is constant. The time-evolution therefore has to remain on the level surface $E(\theta(t), \dot{\theta}(t)) = E(\theta_0, \dot{\theta}_0)$ for all $t > t_0$. In the neighbourhood of the origin, the level surfaces of E are concentric ellipses and the trajectories in the neighbourhood of the origin are therefore closed orbits. Neighbouring trajectories of the origin therefore stay in this neighbourhood. Consequently, the equilibrium at the origin is stable in the sense of Definition 6.13. Of course, if we add damping to the system then the system is no longer conservative and the energy can decrease, i.e. $\dot{E} \leq 0$ along trajectories of the system. Hence, the trajectories remain within the level set $E(\theta(t), \dot{\theta}(t)) \leq E(\theta_0, \dot{\theta}_0)$ for all $t > t_0$, which leads again to stability of the origin. Summarising, we can prove stability of an equilibrium with the Lagrange-Dirichlet theorem if the equilibrium is a local minimum of the total mechanical energy and if the total mechanical energy is not increasing along trajectories of the system.

Lyapunov showed that certain other functions can be used to prove the stability of an equilibrium point. This leads to what is called the *direct method* of Lyapunov or *method of Lyapunov functions*. The theorems of Lyapunov can therefore be seen as a generalisation of the stability theorem of Lagrange-Dirichlet. These functions, which we call Lyapunov functions, are characterised by (local) positive definiteness and by the property that they do not increase, or even strictly decrease, along trajectories of the system. Classically, smooth Lyapunov functions are used to study stability properties of time-autonomous ordinary differential equations (4.1). A Lyapunov function $V(\boldsymbol{x}(t))$ of a system (4.1) decreases if $\dot{V}(\boldsymbol{x}(t)) = (\nabla V(\boldsymbol{x}(t)))^{\mathrm{T}} \boldsymbol{f}(\boldsymbol{x}(t)) < 0$ along trajectories $\boldsymbol{x}(t)$ of the system (see (6.13)). When studying measure differential inclusions, we choose Lyapunov functions $V(\boldsymbol{x})$ which are bounded for $\boldsymbol{x} \in \mathcal{A}$ and unbounded for $\boldsymbol{x} \notin \mathcal{A}$, where $\mathcal{A}$ is the admissible set of the measure differential inclusion (see Section 4.3). A Lyapunov function $V(\boldsymbol{x})$ can therefore be decomposed into

$$V(\boldsymbol{x}) = v(\boldsymbol{x}) + \Psi_{\mathcal{A}}(\boldsymbol{x}), \tag{6.27}$$

where $v(\boldsymbol{x})$ is a locally bounded function and $\Psi_{\mathcal{A}}(\boldsymbol{x})$ is the indicator function (Definition 2.27) on the admissible set $\mathcal{A}$. We will assume that V is of the form (6.27) and that $v : \mathbb{R}^n \to \mathbb{R}$ is a smooth continuous function. Lyapunov functions of the form (6.27) are therefore extended lower semi-continuous functions. Moreover, the gradient ∇v as well as the subderivative of V exist. Note that if $v(\boldsymbol{x})$ is PDF, then also $V(\boldsymbol{x})$ is PDF, because $V(\boldsymbol{x}) \geq v(\boldsymbol{x})$.

If $\boldsymbol{x}(t)$ is absolutely continuous, then we can express the differential measure $\mathrm{d}V$ using the subderivative as in (6.17)

$$\begin{aligned} \mathrm{d}V &= \mathrm{d}V(\boldsymbol{x})(\mathrm{d}\boldsymbol{x}) \\ &= (\nabla v(\boldsymbol{x}))^{\mathrm{T}} \mathrm{d}\boldsymbol{x} + \mathrm{d}\Psi_{\mathcal{A}}(\boldsymbol{x})(\mathrm{d}\boldsymbol{x}), \end{aligned} \tag{6.28}$$

with $\mathrm{d}\boldsymbol{x} = \dot{\boldsymbol{x}}\mathrm{d}t \in \mathrm{d}\boldsymbol{\Gamma}(\boldsymbol{x}(t))$. The subderivative of the indicator function is equal to the indicator function of the associated contingent cone (see (2.54))

$$\mathrm{d}\Psi_{\mathcal{A}}(\boldsymbol{x})(\mathrm{d}\boldsymbol{x}) = \Psi_{K_{\mathcal{A}}(\boldsymbol{x})}(\mathrm{d}\boldsymbol{x}). \tag{6.29}$$

Moreover, the system is assumed to be consistent (Definition 4.8) and the differential measure $\mathrm{d}\boldsymbol{x} = \dot{\boldsymbol{x}}\mathrm{d}t$ lies therefore in the contingent cone of the admissible set $\mathcal{A}$, i.e. $\mathrm{d}\boldsymbol{x} \in K_{\mathcal{A}}(\boldsymbol{x})$. Hence, the subderivative of the indicator function vanishes, $\mathrm{d}\Psi_{\mathcal{A}}(\boldsymbol{x})(\mathrm{d}\boldsymbol{x}) = 0$, and the differential measure of V yields

$$\mathrm{d}V = (\nabla v(\boldsymbol{x}))^{\mathrm{T}} \mathrm{d}\boldsymbol{x} \tag{6.30}$$

where $\boldsymbol{x}(t)$ is an absolutely continuous trajectory.

We now present generalised versions of basic theorems of Lyapunov, that can be used to prove the (attractive) stability of equilibrium points of measure differential inclusions.

Theorem 6.23 (Lyapunov Stability of an Equilibrium Point). *Let* $x^* = 0 \in \mathcal{A}$ *be an equilibrium point of the consistent measure differential inclusion (6.1), which admits solutions* $\varphi(\cdot, t_0, x_0) \in \mathcal{S}(\mathrm{d}\Gamma, t_0, x_0)$ *for all* $x_0 \in \mathcal{A}$. *Let* $V : \mathbb{R}^n \rightarrow \mathbb{R} \cup \{\infty\}$ *be an extended lower semi-continuous function, which is PDF and of the form (6.27).*

a. *If* $\mathrm{d}V(x(t)) \leq 0 \ \forall x(t) \in B_h \cap \mathcal{A}$ *for some* $h > 0$, *then the equilibrium point is stable.*

b. *If there exists a function* β *of class* K *and a constant* $h > 0$ *such that* $\mathrm{d}V(x(t)) \leq -\beta(\|x\|)\mathrm{d}t \ \forall x(t) \in B_h \cap \mathcal{A}$, *then the equilibrium point is locally attractively stable.*

c. *If there exists a function* β *of class* K *such that* $\mathrm{d}V(x(t)) \leq -\beta(\|x\|)\mathrm{d}t$ $\forall x(t) \in \mathcal{A}$, *then the equilibrium point is globally attractively stable.*

Proof:

Theorem 6.23a: The function $V(x)$ is a PDF, so there exists a class KR function $\alpha(\cdot)$ (see Definition 6.1) such that

$$V(x) \geq \alpha(\|x\|) \quad \forall x \in \mathbb{R}^n. \tag{6.31}$$

Take $\varepsilon > 0$ and define c as $c = \alpha(\min(\varepsilon, h))$. The level set $\Omega_c = \{x \in \mathbb{R}^n \mid V(x) \leq c\}$ is a subset of B_ε as well as B_h (the proof is illustrated in Figure 6.7). Necessarily, it holds that $\Omega_c \subset \mathcal{A}$, because V is unbounded outside $\mathcal{A}$. Because $V(x)$ is a PDF, there exists a $\delta > 0$ such that B_δ is the largest ball which lies in $\Omega_c \cup \mathcal{A}^c$ (where $\mathcal{A}^c$ denotes the complement of $\mathcal{A}$), i.e.

$$\sup_{x \in B_\delta \cap \mathcal{A}} V(x) \leq c. \tag{6.32}$$

Hence, it holds that $(B_\delta \cap \mathcal{A}) \subset \Omega_c \subset B_\varepsilon$. Using $\mathrm{d}V(x(t)) \leq 0$ for all $x(t) \in B_h \cap \mathcal{A}$, the fact that $\Omega_c \subset B_h \cap \mathcal{A}$, and Proposition 6.5, it follows that Ω_c is positively invariant. Since $\varepsilon > 0$ can be arbitrarily chosen in the latter exposition, we can find for each $\varepsilon > 0$ a $\delta(\varepsilon)$ such that for each $x_0 \in \mathcal{A}$ it holds that

$$\|x_0\| < \delta(\varepsilon) \quad \Rightarrow \|\varphi(t, t_0, x_0)\| < \varepsilon \quad \forall t > t_0,$$

with $\varphi(\cdot, t_0, x_0) \in \mathcal{S}(\mathrm{d}\Gamma, t_0, x_0)$. This proves stability of the equilibrium point $x^* = 0$ in the sense of Definition 6.13.

Theorem 6.23b: The conditions of b are stronger than those of a and the equilibrium point $x^* = 0$ is therefore stable. Let $c^* = \alpha(h)$ and B_{δ^*} be the largest ball which lies in $\Omega_{c^*} \cup \mathcal{A}^c$. Note that V is bounded from below. Because $\mathrm{d}V(x(t)) < 0 \ \forall x(t) \in (B_h \cap \mathcal{A})\backslash\{0\}$ for some $h > 0$, it holds that if $x_0 \in \mathcal{A}$ then

$$\|x_0\| < \delta^* \Rightarrow \lim_{t \to \infty} V(\varphi(t, t_0, x_0)) = a \geq 0,$$

for each solution curve $\varphi(\cdot, t_0, x_0) \in \mathcal{S}(\mathrm{d}\Gamma, t_0, x_0)$, where a is some nonnegative constant. We now have to show that $a = 0$ for which we use a contradiction argument as in [85]. Suppose that $a > 0$ and choose d such that $B_d \subset \Omega_a$.

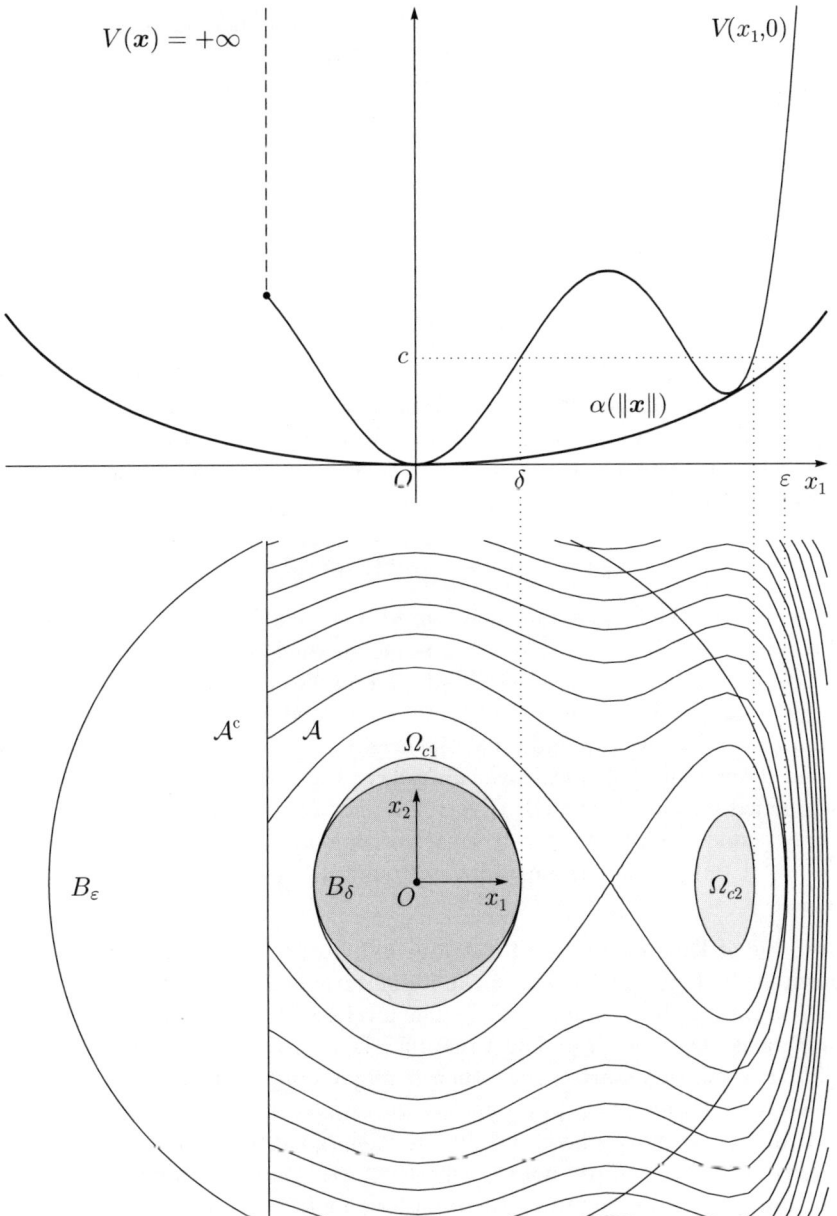

Fig. 6.7. Definition of the sets in the proof of Theorem 6.23 with $\varepsilon < h$, for the case that $x \in \mathbb{R}^2$.

The limit $V(\boldsymbol{x}(t)) \to a$ for $t \to \infty$ implies that $\boldsymbol{x}(t)$ lies outside the ball B_d for all $t \geq t_0$. Moreover, because $-dV$ is LPDF in the sense that there exists a function β of class K such that $dV(\boldsymbol{x}(t)) \leq -\beta(\|\boldsymbol{x}\|)dt \; \forall \boldsymbol{x}(t) \in B_h \cap \mathcal{A}$ we have

$$V(\boldsymbol{x}(t)) = V(\boldsymbol{x}(t_0)) + \int_{[t_0,t]} dV$$

$$\leq V(\boldsymbol{x}(t_0)) - \int_{t_0}^t \beta(\|\boldsymbol{x}(t)\|)dt$$

$$\leq V(\boldsymbol{x}(t_0)) - \beta(d)(t - t_0).$$

Since the right-hand side will eventually become negative, the inequality contradicts the assumption that $a > 0$. Hence, it holds that $V(\boldsymbol{x}(t)) \to a = 0$ for $t \to \infty$. The function $V(\boldsymbol{x})$ is PDF and $V(\boldsymbol{x}) = 0$ therefore implies $\boldsymbol{x} = \boldsymbol{0}$. Each solution curve $\varphi(\cdot, t_0, \boldsymbol{x}_0) \in \mathcal{S}(d\boldsymbol{\varGamma}, t_0, \boldsymbol{x}_0)$, which starts in the admissible part of the ball B_{δ^*}, is therefore attracted to the origin, i.e. for $\boldsymbol{x}_0 \in \mathcal{A}$ it holds that

$$\|\boldsymbol{x}_0\| < \delta^* \Rightarrow \lim_{t \to \infty} \varphi(t, t_0, \boldsymbol{x}_0) = \boldsymbol{0}.$$

Consequently, the equilibrium point $\boldsymbol{x}^* = \boldsymbol{0}$ is locally attractively stable.

Theorem 6.23c: The conditions of c are global versions of the conditions of b. The condition $dV(\boldsymbol{x}) \leq -\beta(\|\boldsymbol{x}\|)dt \; \forall \boldsymbol{x} \in \mathcal{A}$ implies that $\nabla v(\boldsymbol{x}) \neq \boldsymbol{0}$ for $\boldsymbol{x} \neq \boldsymbol{0}$. The level sets of V are therefore simply connected. Hence, using the positive definiteness of V, we conclude that the origin $\boldsymbol{x} = \boldsymbol{0}$ is the unique minimum of V and it therefore holds that we can take $h = \infty$. Moreover, $V(\boldsymbol{x})$ is radially unbounded because it is PDF. Every level set Ω_c of V is therefore bounded, simply connected and positively invariant. Following the proof of b but with $h = \infty$, it follows that the equilibrium point $\boldsymbol{x}^* = \boldsymbol{0}$ is globally attractively stable. $\qquad \Box$

The proof of Theorem 6.23a is illustrated in Figure 6.7. The function $V(x_1, x_2)$, shown in the figure, is PDF in x_1 and quadratic in x_2. To the chosen value of $\varepsilon < h$ corresponds a value $c = \alpha(\varepsilon)$. The level set Ω_c consists of the connected components Ω_{c1} and Ω_{c2}. The value of δ is such that B_δ is the largest ball in Ω_c. If a solution starts in B_δ, then it might jump to Ω_{c2}, but it can never escape from Ω_c and therefore remains within B_ε.

Lyapunov stability theorems give sufficient conditions for stability and attractive stability. They do not say whether these conditions are also necessary. When applying the Lyapunov theorems, we look for a suitable function that can serve as Lyapunov function. Such functions are referred to as Lyapunov function candidates. When it finally turns out that a Lyapunov function candidate can indeed be used to prove stability, then we say it is a Lyapunov function.

Theorem 6.23c is a generalisation of the Barbashin-Krasovskii theorem [15]. Radially unboundedness is essential in this theorem to ensure that each level set of the Lyapunov candidate function is bounded, so that solution curves can not escape to infinity. The set B_{δ^*} is a conservative estimate for the region of attraction in the proof of Theorem 6.23b. A less conservative estimate is the level set Ω_{c^*}.

Classically, the Lyapunov function $V(\boldsymbol{x})$ is only required to be LPDF in Theorem 6.23a and b [85, 152]. Indeed, *local* positive definiteness is enough when the time-evolution $\boldsymbol{x}(t)$ of the system is continuous. This is not the case for systems with a discontinuous state $\boldsymbol{x}(t)$. The level sets of $V(\boldsymbol{x})$ are generally not simply connected and a discontinuous solution may jump from a connected component $\Omega_{c1} \ni \boldsymbol{0}$ to another connected component Ω_{c2} (see Figure 6.2). If $V(\boldsymbol{x})$ is PDF then we can find an $r > 0$ such that Ω_c is simply connected for all c with $0 \le c < r$. Such an $r > 0$ can in general not be found if the function $V(\boldsymbol{x})$ is only LPDF. The condition that $V(\boldsymbol{x})$ is PDF is therefore essential for measure differential inclusions, when no further assumptions on the system or the form of $V(\boldsymbol{x})$ are made. The importance of this condition has been stated in [180] for hybrid systems, in [34] for measure differential inclusions and in [27,34,162] for mechanical systems with frictionless unilateral constraints.

For special classes of systems and Lyapunov functions, it is possible to relax the condition of positive definiteness to local positive definiteness. Consider the class of systems for which the state vector $\boldsymbol{x}(t) \in \mathbb{R}^n$ consists of time-continuous states $\boldsymbol{x}_1(t) \in \mathbb{R}^m$ and states $\boldsymbol{x}_2(t) \in \mathbb{R}^p$ which are of locally bounded variation:

$$\boldsymbol{x}(t) = \begin{bmatrix} \boldsymbol{x}_1(t) \\ \boldsymbol{x}_2(t) \end{bmatrix}, \quad \boldsymbol{x}_1 \in \mathrm{C}^0(I, \mathbb{R}^m), \quad \boldsymbol{x}_2 \in \mathrm{lbv}(I, \mathbb{R}^p), \quad (6.33)$$

with $n = m + p$ and $I = [t_0, \infty)$. Moreover, we assume that the admissible set $\mathcal{A} = \mathcal{A}_1 \times \mathcal{A}_2$ is such that $\mathcal{A}_2 = \mathbb{R}^p$, i.e. only the states $\boldsymbol{x}_1$ are restricted to an admissible set $\mathcal{A}_1$. Let $V(\boldsymbol{x})$ be of the form

$$V(\boldsymbol{x}) = V_1(\boldsymbol{x}_1) + V_2(\boldsymbol{x}_1, \boldsymbol{x}_2), \quad V_2(\boldsymbol{x}_1, \boldsymbol{x}_2) = \boldsymbol{x}_2^{\mathrm{T}} \boldsymbol{P}(\boldsymbol{x}_1) \boldsymbol{x}_2, \quad (6.34)$$

in which $V(\boldsymbol{x}_1)$ is LPDF and $\boldsymbol{P}(\boldsymbol{x}_1) \in \mathbb{R}^{p \times p}$ is a symmetric positive definite matrix. The differential measure $\mathrm{d}V_1$ can be expressed in a subderivative, because $\boldsymbol{x}_1(t)$ is absolutely continuous and V_1 is an extended lower semi-continuous function. Using (6.28)-(6.30) we obtain the differential measure of $V_1(\boldsymbol{x}_1) = v_1(\boldsymbol{x}_1) + \Psi_{\mathcal{A}_1}(\boldsymbol{x}_1)$

$$
\begin{aligned}
\mathrm{d}V_1 &= \mathrm{d}V_1(\boldsymbol{x}_1)(\mathrm{d}\boldsymbol{x}_1) \\
&= (\nabla v_1(\boldsymbol{x}_1))^{\mathrm{T}} \, \mathrm{d}\boldsymbol{x}_1 + \mathrm{d}\Psi_{\mathcal{A}_1}(\boldsymbol{x}_1)(\mathrm{d}\boldsymbol{x}_1) \\
&= (\nabla v_1(\boldsymbol{x}_1))^{\mathrm{T}} \, \mathrm{d}\boldsymbol{x}_1 + \Psi_{K_{\mathcal{A}_1}(\boldsymbol{x}_1)}(\mathrm{d}\boldsymbol{x}_1) \\
&= (\nabla v_1(\boldsymbol{x}_1))^{\mathrm{T}} \, \dot{\boldsymbol{x}}_1 \mathrm{d}t.
\end{aligned}
\quad (6.35)
$$

Moreover, V_2 is a quadratic form and $\mathrm{d}V_2$ can be expressed using (6.19) and a partial derivative with respect to $\boldsymbol{x}_1$

$$dV_2 = (\boldsymbol{x}_2^+ + \boldsymbol{x}_2^-)^\mathrm{T} \boldsymbol{P}(\boldsymbol{x}_1)d\boldsymbol{x}_2 + \frac{\partial V_2}{\partial \boldsymbol{x}_1}\dot{\boldsymbol{x}}_1 dt. \tag{6.36}$$

In Chapter 7 we will see that mechanical systems with impact belong to this class of systems for which the generalised coordinates $\boldsymbol{q}(t)$ are absolutely continuous and the generalised velocities $\boldsymbol{u}(t)$ are functions of locally bounded variation. Before stating a Lyapunov theorem for this class of systems, we need a proposition on the invariance of level sets of $V(\boldsymbol{x})$ and their connected components.

Proposition 6.24. *Consider the function $V(\boldsymbol{x})$ of the form (6.34) and the measure differential inclusion (6.1) with the property (6.33). Let Ω_{ci} be a bounded connected component of the level set $\Omega_c = \{\boldsymbol{x} \in \mathbb{R}^n \mid V(\boldsymbol{x}) \le c\}$. If $dV \le 0$ for all $\boldsymbol{x} \in \Omega_{ci}$, then the set Ω_{ci} is positively invariant with respect to (6.1).*

Proof: The condition $dV \le 0$ gives $\dot{V} \le 0$ and $V^+ \le V^-$, see (6.16). Using (6.33) and the fact that the system is consistent, the latter can be written as

$$V_1(\boldsymbol{x}_1) + V_2(\boldsymbol{x}_1, \boldsymbol{x}_2^+) \le V_1(\boldsymbol{x}_1) + V_2(\boldsymbol{x}_1, \boldsymbol{x}_2^-), \tag{6.37}$$

with $V_1(\boldsymbol{x}_1) < \infty$ for $\boldsymbol{x}_1 \in \mathcal{A}_1$ and hence

$$V_2(\boldsymbol{x}_1, \boldsymbol{x}_2^+) \le V_2(\boldsymbol{x}_1, \boldsymbol{x}_2^-). \tag{6.38}$$

Let $\Omega_{ci} = \Omega_{ci1} \times \Omega_{ci2}$ with $\Omega_{ci1} \subset \mathbb{R}^m$ and $\Omega_{ci2} \subset \mathbb{R}^p$. Consider the function $V_1^*(\boldsymbol{x}_1)$ to be an extension of $V_1(\boldsymbol{x}_1)$:

$$V_1^*(\boldsymbol{x}_1) = \begin{cases} V_1(\boldsymbol{x}_1), & \boldsymbol{x}_1 \in \Omega_{ci1}, \\ +\infty, & \boldsymbol{x}_1 \notin \Omega_{ci1}. \end{cases} \tag{6.39}$$

It therefore holds that Ω_{ci1} is a level set of $V_1^*(\boldsymbol{x}_1)$. Moreover, the function $V_2(\boldsymbol{x}_1, \boldsymbol{x}_2)$ is a quadratic positive definite form in $\boldsymbol{x}_2$ for fixed $\boldsymbol{x}_1$. The set Ω_{ci} is therefore a level set of $V^*(\boldsymbol{x}) = V_1^*(\boldsymbol{x}_1) + V_2(\boldsymbol{x}_1, \boldsymbol{x}_2)$. It holds that $\dot{V}^* = \dot{V}$ and

$$\begin{aligned} V^{*+} &= V^*(\boldsymbol{x}^+) \\ &= V_1^*(\boldsymbol{x}_1) + V_2(\boldsymbol{x}_1, \boldsymbol{x}_2^+) \\ &\le V_1^*(\boldsymbol{x}_1) + V_2(\boldsymbol{x}_1, \boldsymbol{x}_2^-) \\ &\le V^{*-}, \end{aligned} \tag{6.40}$$

which gives $dV^* \le 0$. Using Proposition 6.5, we conclude that each level set of $V^*(\boldsymbol{x})$ is positively invariant. Consequently, Ω_{ci} is positively invariant. $\square$

The idea behind Proposition 6.24 is illustrated in Figure 6.8. The function $V(\boldsymbol{x})$ is a potential with two wells, such that the level set Ω_c consists of two connected components Ω_{c1} and Ω_{c2}. Consider a solution curve with $\boldsymbol{x}_0 \in \Omega_{c1}$ and $dV \le 0$. Discontinuities in $\boldsymbol{x}(t)$ can only occur in the x_2-direction. The solution curve can therefore not leave the connected component, which makes Ω_{c1} positively invariant.

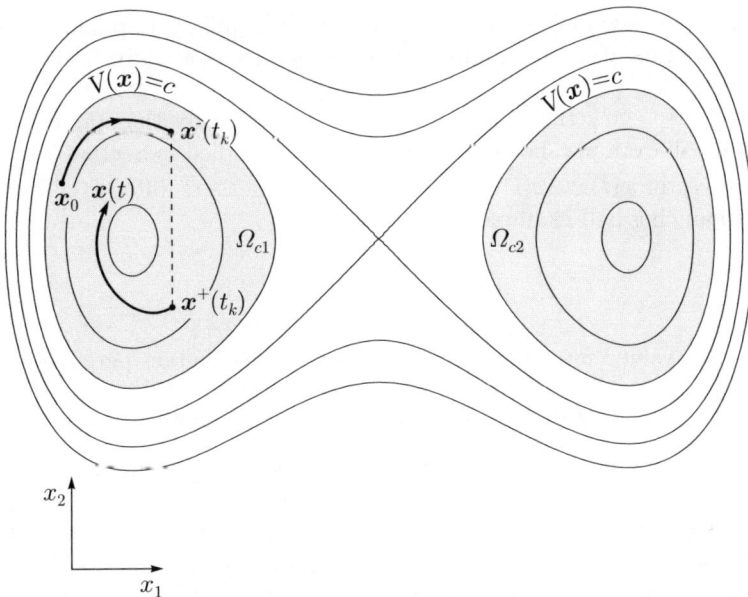

Fig. 6.8. Positive invariance of the connected component Ω_{c1}.

Theorem 6.25 (Lyapunov Stability of an Equilibrium with (6.33)).
Let $\boldsymbol{x}^* = \boldsymbol{0} \in \mathcal{A}$ be an equilibrium point of $\mathrm{d}\boldsymbol{x} \in \mathrm{d}\boldsymbol{\Gamma}(\boldsymbol{x}(t))$, which has the property (6.33) and admits solutions $\boldsymbol{\varphi}(\cdot, t_0, \boldsymbol{x}_0) \in \mathcal{S}(\mathrm{d}\boldsymbol{\Gamma}, t_0, \boldsymbol{x}_0)$ for all $\boldsymbol{x}_0 \in \mathcal{A}$. Let $V(\boldsymbol{x}) \in \mathbb{R} \cup \{\infty\}$ be LPDF and of the form (6.34). The following statements can be made:

a. *If $\mathrm{d}V(\boldsymbol{x}(t)) \leq 0 \; \forall \boldsymbol{x}(t) \in B_h \cap \mathcal{A}$ for some $h > 0$, then the equilibrium point is stable.*

b. *If there exists a function β of class K and a constant $h > 0$ such that $\mathrm{d}V(\boldsymbol{x}(t)) \leq -\beta(\|\boldsymbol{x}\|)\mathrm{d}t \; \forall \boldsymbol{x}(t) \in B_h \cap \mathcal{A}$, then the equilibrium point is locally attractively stable.*

Proof: If $V(\boldsymbol{x})$ is LPDF, then there must exist a bounded ball B_r such that the function $V(\boldsymbol{x})$ has no minima, maxima or saddle-points for $\boldsymbol{x} \in B_r$ other than the origin. The function $V(\boldsymbol{x})$ therefore has the form of a cup within B_r and $\boldsymbol{x} = \boldsymbol{0}$ is the unique minimum in B_r. Let ε^* be the minimum of r and h. Using Proposition 6.24, together with (6.33) and the special form of V (6.34), it follows that each connected component $\Omega_{ci} \subset B_{\varepsilon^*}$ is a positively invariant set. Moreover, the sets $\Omega_{ci} \subset B_{\varepsilon^*}$ are concentric in the sense that $\Omega_{di} \subset \Omega_{ci}$ if $d \leq c \leq c^*$, where Ω_{c^*i} is the largest connected component in B_{ε^*}. The proofs of Theorem 6.25a and b can be continued in the same way as in the proof of Theorem 6.23a and b. $\qquad\square$

In order to prove global attractive stability of an equilibrium point, a positive definite function $V(\boldsymbol{x})$ is required. This theorem has already been given in Theorem 6.23c.

In order to show the use of Theorem 6.23, we study the stability of the bouncing ball example shown in Figure 5.11 and studied in Section 5.4. Using the state-vector $\boldsymbol{x}(t) = \begin{bmatrix} q(t) & u(t) \end{bmatrix}^{\mathrm{T}}$ we write the measure differential inclusion for the bouncing ball example as

$$\mathrm{d}\boldsymbol{x} \in \begin{bmatrix} u\mathrm{d}t \\ -g\mathrm{d}t + \frac{1}{m}\mathrm{d}P_N \end{bmatrix} \tag{6.41}$$

with the set-valued force law for the measure of the contact percussion $\mathrm{d}P_N$

$$-\mathrm{d}P_N \in N_{K_{\mathcal{K}}(q)}(\xi_N), \quad \xi_N = u^+ + e_N u^-, \tag{6.42}$$

and the set $\mathcal{K} = \{q \mid q \geq 0\}$ of admissible positions q. The admissible set of $\boldsymbol{x}(t)$ is the half-space $\mathcal{A} = \{\boldsymbol{x} \in \mathbb{R}^2 \mid q \in \mathcal{K}\}$. We are now interested in the stability of the equilibrium point $\boldsymbol{x}^* = \boldsymbol{0}$. This is the unique equilibrium point of the system and is therefore isolated. As Lyapunov candidate function, we choose the sum of kinetic and potential energy together with the energy potential $\pi_N(q) = \Psi_{C_N}^*(q)$ of the contact force $-\lambda_N \in C_N = \{-\lambda_N \mid \lambda_N \geq 0\}$:

$$V(\boldsymbol{x}) = \frac{1}{2}mu^2 + mgq + \pi_N(q). \tag{6.43}$$

The energy potential $\pi_N(q)$ equals the indicator function on the admissible set $\mathcal{A}$

$$\pi_N(q) = \Psi_{C_N}^*(q) = \Psi_{\mathcal{A}}(\boldsymbol{x}) = \Psi_{\mathcal{K}}(q). \tag{6.44}$$

The function $V(\boldsymbol{x})$ is PDF, because $f(\boldsymbol{x}) = \frac{1}{2}mu^2 + mg|q|$ is PDF and $V(\boldsymbol{x}) \geq f(\boldsymbol{x})$. We see that the term $\pi_N(q) = \Psi_{\mathcal{K}}(q)$ is essential in (6.43) in order to make V a positive definite function. Each level set of V is bordered by a parabola $q = \frac{c}{mg} - \frac{1}{2g}u^2$, $c > 0$, and the line $q = 0$ (see Figure 6.9). Using (3.64), the subderivative of the indicator function (2.54) and the chain rule, we obtain the differential measure

$$\begin{aligned} \mathrm{d}V &= \frac{1}{2}m(u^+ + u^-)\mathrm{d}u + mg\mathrm{d}q + \mathrm{d}\Psi_{\mathcal{K}}(q)(\mathrm{d}q) \\ &= \frac{1}{2}m(u^+ + u^-)\mathrm{d}u + mg\mathrm{d}q + \Psi_{K_{\mathcal{K}}(q)}(\mathrm{d}q), \end{aligned} \tag{6.45}$$

which we evaluate along solution curves $\boldsymbol{x}(t) = \begin{bmatrix} q(t) & u(t) \end{bmatrix}^{\mathrm{T}}$ of the system giving

$$\mathrm{d}V = \frac{1}{2}(u^+ + u^-)(\mathrm{d}P_N - mg\mathrm{d}t) + mgu\mathrm{d}t + \Psi_{K_{\mathcal{K}}(q)}(u\mathrm{d}t). \tag{6.46}$$

The measure of the contact percussion $\mathrm{d}P_N$ obeys the set-valued force law (6.42) with

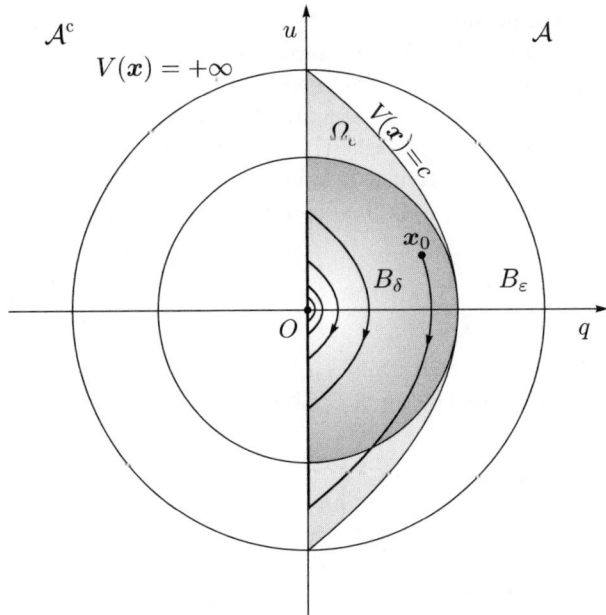

Fig. 6.9. Stability of the bouncing ball example.

$$dP_N = \lambda_N dt + \Lambda_N d\eta, \tag{6.47}$$

in which λ_N is the contact force and Λ_N the contact impulse. Substitution of the decomposition (6.47) in (6.46) and using the fact that $\frac{1}{2}(u^+ + u^-)dt = udt$ holds in front of all Lebesgue measurable terms yields

$$\begin{aligned}
dV &= \frac{1}{2}(u^+ + u^-)(\lambda_N dt + \Lambda_N d\eta - mgdt) + mgudt + \Psi_{K_K(q)}(udt) \\
&= \frac{1}{2}(u^+ + u^-)\Lambda_N d\eta + u\lambda_N dt - mgudt + mgudt + \Psi_{K_K(q)}(udt), \quad (6.48) \\
&= \frac{1}{2}(u^+ + u^-)\Lambda_N d\eta + u\lambda_N dt + \Psi_{K_K(q)}(udt),
\end{aligned}$$

with

$$\begin{aligned}
-\Lambda_N &\in N_{K_K(q)}(u^+ + e_N u^-), \\
-\lambda_N &\in N_{K_K(q)}(u).
\end{aligned} \tag{6.49}$$

If $q(t) > 0$, then it holds that $u(t)$ is locally continuous and $K_K(q) = \mathbb{R}$. As a consequence, the contact force and impulse vanish for an open contact, i.e. $N_{\mathbb{R}}(\cdot) = 0 \Rightarrow \lambda_N = \Lambda_N = 0$. Moreover, the indicator term in the Lyapunov function gives $\Psi_{\mathbb{R}}(\cdot) = 0$. Hence, the differential measure of the Lyapunov function vanishes for $q > 0$

$$q > 0 \Rightarrow dV = 0, \tag{6.50}$$

because the system is conservative in this part of the state-space.

If $q(t) = 0$, then also the contact force λ_N and the impulse Λ_N come into play. It holds that $K_{\mathcal{K}}(0) = \mathbb{R}^+$. The set-valued force laws for the contact force and impulse are such that no penetration can occur. The unilaterality of the contact is preserved. In other words, the system is consistent and it holds that $dq = u dt = u^+ dt \in K_{\mathcal{K}}(q)$. Hence, it holds that $\Psi_{K_{\mathcal{K}}(q)}(dq) = \Psi_{K_{\mathcal{K}}(q)}(u dt) = 0$ and necessarily $u \geq 0$ for $q = 0$. The contact force therefore obeys

$$-\lambda_N \in N_{\mathbb{R}^+}(u) \Rightarrow \begin{cases} u > 0 \Rightarrow -\lambda_N = 0 \\ u = 0 \Rightarrow -\lambda_N \leq 0 \end{cases} \tag{6.51}$$

We conclude that the contact force λ_N has no power, i.e. $u\lambda_N = 0$. Similarly, from the set-valued force law for Λ_N follows

$$-\Lambda_N \in N_{\mathbb{R}^+}(u^+ + e_N u^-) \Rightarrow \begin{cases} u^+ > -e_N u^- \Rightarrow -\Lambda_N = 0 \\ u^+ = -e_N u^- \Rightarrow -\Lambda_N \leq 0 \end{cases} \tag{6.52}$$

and it therefore holds that $dV = 0$ if $u^+ > -e_N u^-$. On the other hand, if $u^+ = -e_N u^-$, then an impulse $\Lambda_N = m(u^+(t) - u^-(t)) \geq 0$ causes the velocity to jump to $u^+ = -e_N u^- \geq 0$ and we obtain the differential measure dV:

$$\begin{aligned} q = 0 \Rightarrow dV &= \frac{1}{2}(u^+ + u^-)\Lambda_N d\eta \\ &= \frac{1}{2}m(u^+ + u^-)(u^+ - u^-)d\eta \\ &= -\frac{1}{2}m(1 - e_N)(1 + e_N)\left(u^-\right)^2 d\eta \\ &= -\frac{1}{2}m(1 - e_N^2)\left(u^-\right)^2 d\eta. \end{aligned} \tag{6.53}$$

Hence, it holds that $dV \leq 0$ if $|e_N| \leq 1$, and because of the consistency requirement $e_N \geq 0$ this condition reduces to $0 \leq e_N \leq 1$. Comparison of the decomposition (6.16)

$$dV = \dot{V} dt + (V^+ - V^-)d\eta$$

of the differential measure dV with (6.53) gives

$$\dot{V} = 0, \qquad V^+ - V^- = -\frac{1}{2}m(1 - e_N^2)\left(u^-\right)^2. \tag{6.54}$$

From the above, it becomes apparent that energy is only dissipated when a not completely elastic impact (i.e. $e_N < 1$) occurs, which agrees with our physical interpretation of the system.

Consequently, $V(\boldsymbol{x})$ is PDF and $dV(\boldsymbol{x}(t)) \leq 0 \ \forall \boldsymbol{x}(t) \in \mathcal{A}$. These conditions are stronger than the conditions required by Theorem 6.23a, but weaker than those in Theorem 6.23b or c. Using Theorem 6.23, we can therefore only prove stability. For each $\varepsilon > 0$ we find the largest level set Ω_c of $V(\boldsymbol{x})$ that fits in the ball B_ε (see Figure 6.9). It holds that $\Omega_c \subset \mathcal{A}$, because $V(\boldsymbol{x}) = +\infty$ for $\boldsymbol{x} \notin \mathcal{A}$. Each level set Ω_c is a positively invariant set, because $\mathrm{d}V \leq 0$ for all $\boldsymbol{x} \subset \mathcal{A}$ (Proposition 6.5). We now choose δ such, that B_δ is the largest ball in Ω_c. Consequently, for each $\varepsilon > 0$ we can find a $\delta > 0$ such that for each $\boldsymbol{x}_0 \in \mathcal{A}$ it holds that

$$\|\boldsymbol{x}_0\| < \delta(\varepsilon), \quad \Rightarrow \|\boldsymbol{\varphi}(t, t_0, \boldsymbol{x}_0)\| < \varepsilon \quad \forall t > t_0$$

with $\boldsymbol{\varphi}(\cdot, t_0, \boldsymbol{x}_0) \in \mathcal{S}(\mathrm{d}\boldsymbol{\Gamma}, t_0, \boldsymbol{x}_0)$, which proves stability of the equilibrium point $\boldsymbol{x}^* = \boldsymbol{0}$ in the sense of Definition 6.13. For $e_N < 1$, all solutions converge to the origin in a finite time and the equilibrium point is even globally symptotically attractively stable, but this can not be proven using Theorem 6.23. To prove attractivity, we need the invariance principle of LaSalle, which will be discussed in Section 6.6. The solution curve drawn in Figure 6.9 corresponds to the time-histories $q(t)$ and $u(t)$ of Figure 5.11. We see that the equilibrium point is symptotically attractive, as it is reached in a finite time. Moreover, the equilibrium point is an accumulation point, because it is reached in a finite time after an infinite number of impacts.

In order to gain more insight in the problem, we replace the rigid unilateral contact with a unilateral spring with spring stiffness k. The total mechanical energy is then composed of the kinetic energy, the potential energy due to gravitation and the potential energy of the unilateral spring: $E(\boldsymbol{x}(t)) = \frac{1}{2}mu(t)^2 + mgq(t) + \frac{1}{2}kd(t)^2$ with the penetration depth $d(t) = \max(-q(t), 0)$. Clearly, when k tends to ∞, then $E(\boldsymbol{x}) \to V(\boldsymbol{x})$. The dissipation in a compliant contact model can be modelled by a viscous damper. For the one-dimensional case, the existence of a sequence of compliant dissipative models which converge to the rigid limiting model with restitution is shown in [23]. This example illustrates that the decomposition of the Lyapunov function in (6.27) (or (6.43) in the example) is a natural choice for mechanical systems with unilateral constraints and will be used in Chapter 7.

6.5.2 Lyapunov Stability of Equilibrium Sets

Theorem 6.23 provides a means to prove the (attractive) stability of equilibrium points. However, (measure) differential inclusions generally have equilibrium sets. Without much effort, we can extend Theorems 6.23 and 6.25 to prove the stability of an equilibrium set. Instead of using a Lyapunov function which is PDF, as has been done in Theorem 6.23 to prove the stability of an equilibrium *point*, we have to use a Lyapunov function which is zero within the equilibrium *set* and positive outside.

Theorem 6.26 (Stability of an Equilibrium Set). *Let $\mathcal{E} \subset \mathcal{A}$ be an equilibrium set of* $\mathrm{d}\boldsymbol{x} \in \mathrm{d}\boldsymbol{\Gamma}(\boldsymbol{x}(t))$, *which admits solutions* $\boldsymbol{\varphi}(\cdot, t_0, \boldsymbol{x}_0) \in \mathcal{S}(\mathrm{d}\boldsymbol{\Gamma}, t_0, \boldsymbol{x}_0)$ *for all* $\boldsymbol{x}_0 \in \mathcal{A}$. *Let $V(\boldsymbol{x}) \in \mathbb{R} \cup \{\infty\}$ be a lower semi-continuous function, such that*

i. $V(\boldsymbol{x}) = 0$ for all $\boldsymbol{x} \in \mathcal{E}$,
ii. $V(\boldsymbol{x}) > 0$ for all $\boldsymbol{x} \in \mathcal{A}\backslash\mathcal{E}$,
iii. $V(\boldsymbol{x}) = +\infty$ for $\boldsymbol{x} \notin \mathcal{A}$ and $V(\boldsymbol{x})$ is radially unbounded.

Then the following statements hold:

a. If $dV(\boldsymbol{x}(t)) \leq 0\ \forall \boldsymbol{x}(t) \in U_h(\mathcal{E}) \cap \mathcal{A}$ for some $h > 0$, then the equilibrium set is stable.
b. If there exists a function β of class K and a constant $h > 0$ such that

$$dV(\boldsymbol{x}(t)) \leq -\beta(\mathrm{dist}_\mathcal{E}(\boldsymbol{x}))dt \quad \forall \boldsymbol{x}(t) \in U_h(\mathcal{E}) \cap \mathcal{A},$$

then the equilibrium set is locally attractively stable.
c. If there exists a function β of class K such that $dV(\boldsymbol{x}(t)) \leq -\beta(\mathrm{dist}_\mathcal{E}(\boldsymbol{x}))dt$ $\forall \boldsymbol{x}(t) \in \mathcal{A}$, then the equilibrium set is globally attractively stable.

Proof:
The proof of Theorem 6.26 is analogous to the proof of Theorem 6.23. Conditions $i - iii$ replace the positive definiteness requirement whereas the neighbourhood U_h replaces the ball B_h of Theorem 6.23. In this way, stability of the equilibrium set can be proven in the sense of Definition 6.18.   □

Clearly, if we let the equilibrium set shrink to an equilibrium point, then the conditions on the Lyapunov function reduce to positive definiteness and that it is of the form (6.27). Theorem 6.26 is therefore a direct generalisation of Theorem 6.23. In the same way, we generalise Theorem 6.25.

Theorem 6.27 (Stability of an Equilibrium Set with (6.33)). *Let $\mathcal{E} \subset \mathcal{A}$ be an equilibrium set of $d\boldsymbol{x} \in d\boldsymbol{\Gamma}(\boldsymbol{x}(t))$, which has the property (6.33) and admits solutions $\boldsymbol{\varphi}(\cdot, t_0, \boldsymbol{x}_0) \in \mathcal{S}(d\boldsymbol{\Gamma}, t_0, \boldsymbol{x}_0)$ for all $\boldsymbol{x}_0 \in \mathcal{A}$. Let $V(\boldsymbol{x}) \in \mathbb{R} \cup \{\infty\}$ be a lower semi-continuous function of the form (6.34), such that*

i. $V(\boldsymbol{x}) = 0$ for all $\boldsymbol{x} \in \mathcal{E}$,
ii. $V(\boldsymbol{x}) > 0$ for all $\boldsymbol{x} \in (U_r(\mathcal{E}) \cap \mathcal{A})\backslash\mathcal{E}$ for some $r > 0$,
iii. $V(\boldsymbol{x}) = +\infty$ for $\boldsymbol{x} \notin \mathcal{A}$.

Then the following statements hold:

a. If $dV(\boldsymbol{x}(t)) \leq 0\ \forall \boldsymbol{x}(t) \in U_h(\mathcal{E}) \cap \mathcal{A}$ for some $h > 0$, then the equilibrium set is stable.
b. If there exists a function β of class K and a constant $h > 0$ such that

$$dV(\boldsymbol{x}(t)) \leq -\beta(\mathrm{dist}_\mathcal{E}(\boldsymbol{x}))dt \quad \forall \boldsymbol{x}(t) \in U_h(\mathcal{E}) \cap \mathcal{A},$$

then the equilibrium set is locally attractively stable.

Proof:
The proof of Theorem 6.27 is analogous to the proof of Theorem 6.25. Conditions i and ii replace the local positive definiteness requirement whereas the

neighbourhood U_h replaces the ball B_h of Theorem 6.25. In this way, stability of the equilibrium set can be proven in the sense of Definition 6.18. $\square$

The main difficulty with Theorem 6.26 and 6.27 is the fact that a suitable Lyapunov function is hard to find. The Lyapunov function V has to be zero within the equilibrium set and be positive outside and still fulfill the conditions i to iii. It might prove to be easier to prove the stability of each equilibrium point $\boldsymbol{x}^* \in \mathcal{E}$ separately using Theorem 6.23a using a different Lyapunov function $V_{\boldsymbol{x}^*}(\boldsymbol{x})$ for each equilibrium point. Subsequently, we can prove the stability of the equilibrium set as a whole [114, 115]:

Theorem 6.28 (Stability of an Equilibrium Set as Union of Equilibrium Points). *Let $\mathcal{E} \subset \mathcal{A}$ be an equilibrium set of $\mathrm{d}\boldsymbol{x} \in \mathrm{d}\boldsymbol{\Gamma}(\boldsymbol{x}(t))$, which admits solutions $\boldsymbol{\varphi}(\cdot, t_0, \boldsymbol{x}_0) \in \mathcal{S}(\mathrm{d}\boldsymbol{\Gamma}, t_0, \boldsymbol{x}_0)$ for all $\boldsymbol{x}_0 \in \mathcal{A}$. If it can be proven that each equilibrium point $\boldsymbol{x}^* \in \mathcal{E}$ is Lyapunov stable using a PDF function $V_{\boldsymbol{x}^*}(\boldsymbol{x})$, then the equilibrium set is stable in the sense of Definition 6.18.*

Proof: Let $\Omega_{\boldsymbol{x}^*, r}$ be the r-level set of the Lyapunov function $V_{\boldsymbol{x}^*}(\boldsymbol{x})$

$$\Omega_{\boldsymbol{x}^*, r} = \{\boldsymbol{x} \in \mathbb{R}^n \mid V_{\boldsymbol{x}^*}(\boldsymbol{x}) \le r\}.$$

Define $r^* > 0$ as the largest value of r such that $\Omega_{\boldsymbol{x}^*, r^*}$ is a positively invariant set for all $\boldsymbol{x}^* \in \mathcal{E}$. Note that such an r^* exists by virtue of the fact that each equilibrium point $\boldsymbol{x}^* \in \mathcal{E}$ is Lyapunov stable. For each $r \in (0, r^*]$, we introduce the set Ω_r as the union of all r-level sets of $V_{\boldsymbol{x}^*}(\boldsymbol{x})$:

$$\Omega_r = \bigcup_{\forall \boldsymbol{x}^* \in \mathcal{E}} \Omega_{\boldsymbol{x}^*, r}, \quad r \in (0, r^*].$$

Note that each set Ω_r is the union of positively invariant sets, which makes it a positively invariant set itself and $\mathcal{E} \subset \text{int } \Omega_r$. For each $r \in (0, r^*]$, we now choose $\varepsilon(r) > 0$ as the smallest value such that $\Omega_r \subset U_\varepsilon(\mathcal{E})$. We can find for each $\varepsilon \in (0, \varepsilon^*]$, with $\varepsilon^* = \varepsilon(r^*)$, a $\delta(\varepsilon) > 0$ such that for any $\boldsymbol{x}_0 \in \mathcal{A}$ with

$$\boldsymbol{x}_0 \in U_\delta(\mathcal{E}) \subset \Omega_r,$$

each solution curve $\boldsymbol{\varphi}(\cdot, t_0, \boldsymbol{x}_0) \in \mathcal{S}(\mathrm{d}\boldsymbol{\Gamma}, t_0, \boldsymbol{x}_0)$ satisfies

$$\boldsymbol{\varphi}(t, t_0, \boldsymbol{x}_0) \in U_\varepsilon(\mathcal{E}).$$

For each $\varepsilon > \varepsilon^*$ we take $\delta(\varepsilon) = \delta(\varepsilon^*)$. Consequently, we can find for each $\varepsilon > 0$ a $\delta(\varepsilon) > 0$ such that for any $\boldsymbol{x}_0 \in \mathcal{A}$ with $\boldsymbol{x}_0 \in U_\delta(\mathcal{E}) \subset \Omega_r$ each solution curve $\boldsymbol{\varphi}(\cdot, t_0, \boldsymbol{x}_0) \in \mathcal{S}(\mathrm{d}\boldsymbol{\Gamma}, t_0, \boldsymbol{x}_0)$ satisfies $\boldsymbol{\varphi}(t, t_0, \boldsymbol{x}_0) \in U_\varepsilon(\mathcal{E})$, which proves stability of the equilibrium set $\mathcal{E}$. $\square$

The different sets used in the proof of Theorem 6.28 are depicted in Figure 6.10. Theorem 6.28 gives an alternative way to prove the stability of an equilibrium set which does not require to find a Lyapunov function for the whole set. However, the proof of Theorem 6.28 introduces a set Ω_r, being

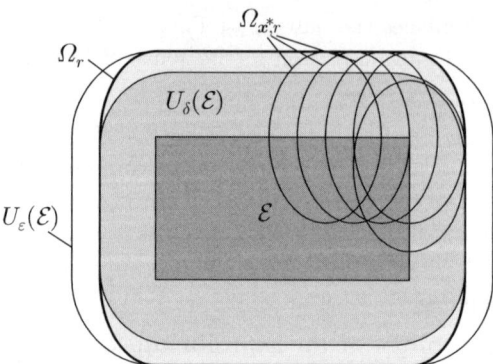

Fig. 6.10. Construction of Ω_r in the proof of Theorem 6.26 as the union of level sets $\Omega_{\boldsymbol{x}^*,r}$.

positively invariant and enclosing the equilibrium set. The question now rises which function V is associated with Ω_r, if we interpret Ω_r as a level set:

$$\Omega_r = \{\boldsymbol{x} \in \mathbb{R}^n \mid V(\boldsymbol{x}) \leq r\}. \tag{6.55}$$

The set Ω_r is the union of all r-level sets of each individual Lyapunov function $V_{\boldsymbol{x}^*}(\boldsymbol{x})$ and it therefore must hold that

$$\operatorname{epi} V = \bigcup_{\forall \boldsymbol{x}^* \in \mathcal{E}} \operatorname{epi} V_{\boldsymbol{x}^*}. \tag{6.56}$$

The function $V(\boldsymbol{x})$ is therefore the pointwise minimum of all functions $V_{\boldsymbol{x}^*}(\boldsymbol{x})$

$$V(\boldsymbol{x}) = \min_{\boldsymbol{x}^* \in \mathcal{E}} V_{\boldsymbol{x}^*}(\boldsymbol{x}). \tag{6.57}$$

The function V defined by (6.56) (or equivalently (6.57)) has the properties

$$V(\boldsymbol{x}) = 0 \ \forall \boldsymbol{x} \in \mathcal{E}, \quad V(\boldsymbol{x}) > 0 \ \forall \boldsymbol{x} \notin \mathcal{E}, \quad V(\boldsymbol{x}) = +\infty \ \forall \boldsymbol{x} \notin \mathcal{A} \tag{6.58}$$

and $V(\boldsymbol{x})$ is radially unbounded since each $V_{\boldsymbol{x}^*}(\boldsymbol{x})$ is a PDF. Consequently, the function V as in (6.57) is a suitable Lyapunov function candidate to prove stability of an equilibrium set using Theorem 6.26. The search for a Lyapunov function V in Theorem 6.26 has therefore been reduced to the search of Lyapunov functions $V_{\boldsymbol{x}^*}(\boldsymbol{x})$ for all equilibrium points $\boldsymbol{x}^*$ in the set $\mathcal{E}$.

The Lyapunov function V in (6.57) can be simplified if the Lyapunov functions $V_{\boldsymbol{x}^*}(\boldsymbol{x})$ have the generic quadratic form

$$V_{\boldsymbol{x}^*}(\boldsymbol{x}) = \frac{1}{2}(\boldsymbol{x} - \boldsymbol{x}^*)^{\mathrm{T}} \boldsymbol{P}(\boldsymbol{x} - \boldsymbol{x}^*) + \Psi_\mathcal{A}(\boldsymbol{x}), \qquad \boldsymbol{P} = \boldsymbol{G}\boldsymbol{G}^{\mathrm{T}} > 0. \tag{6.59}$$

with $\boldsymbol{G}$ being the square root matrix of $\boldsymbol{P} = \boldsymbol{P}^{\mathrm{T}}$. Substitution of this special form of $V_{\boldsymbol{x}^*}(\boldsymbol{x})$ in (6.57) gives

$$V(x) = \min_{x^* \in \mathcal{E}} \frac{1}{2}(x - x^*)^{\mathrm{T}} GG^{\mathrm{T}}(x - x^*) + \Psi_A(x)$$

$$= \min_{x^* \in \mathcal{E}} \frac{1}{2}\|G^{\mathrm{T}}(x - x^*)\|^2 + \Psi_A(x)$$

$$= \min_{z^* \in \mathcal{E}_z} \frac{1}{2}\|z - z^*\|^2 + \Psi_A(x), \tag{6.60}$$

$$= \frac{1}{2}\|z - \mathrm{prox}_{\mathcal{E}_z}(z)\|^2 + \Psi_A(x)$$

$$= \frac{1}{2}\mathrm{dist}^2_{\mathcal{E}_z}(z) + \Psi_A(x),$$

where $z = G^{\mathrm{T}}x$ and $\mathcal{E}_z = \{z \mid z = G^{\mathrm{T}}x, x \in \mathcal{E}\}$. Hence, a generic quadratic form of $V_{x^*}(x)$ allows to express $V(x)$ as the distance to the equilibrium set (see Definition 2.32), but with a metric determined by P. If the state $x(t)$ is absolutely continuous, $dx = \dot{x}dt$, then the differential measure dV can be obtained from (2.44)

$$dV = dV(x)(dx)$$
$$= (z - \mathrm{prox}_{\mathcal{E}_z}(z))^{\mathrm{T}}dz + d\Psi_A(x)(dx)$$
$$= (z - \mathrm{prox}_{\mathcal{E}_z}(z))^{\mathrm{T}}dz + \Psi_{K_A(x)}(dx), \quad dx = \dot{x}dt \in K_A(x) \tag{6.61}$$
$$= x^{\mathrm{T}}Pdx - \mathrm{prox}^{\mathrm{T}}_{\mathcal{E}_z}(G^{\mathrm{T}}x)G^{\mathrm{T}}dx.$$

6.5.3 Stability through Maximal Monotonicity

A special class of systems is formed by measure differential inclusions of the form

$$dx \in -d\mathcal{A}(x^+), \tag{6.62}$$

where $d\mathcal{A}$ is a maximal monotone (set-valued) measure function. This class of systems has already been encountered in Section 4.3. The maximal monotonicity of $d\mathcal{A}$ is a sufficient condition for the stability of an equilibrium set of (6.62).

Theorem 6.29 (Stability of the Equilibrium Set through Monotonicity). *Consider a measure differential inclusion of the form*

$$dx \in -d\mathcal{A}(x^+) = -\mathcal{A}_t(x)dt - \mathcal{A}_\eta(x^+)d\eta,$$

with a non-empty equilibrium set $\mathcal{E} = \{x \in \mathbb{R}^n \mid 0 \in d\mathcal{A}(x)\}$. Let the system be consistent and have existence of solutions for all initial conditions $x_0 \in A$, where the admissible domain A is a closed set. If $d\mathcal{A}(x^+)$ is a maximal monotone set-valued measure function, then the equilibrium set $\mathcal{E}$ is stable. Moreover, if $\mathcal{A}_t(x)$ is a strictly maximal monotone set-valued function, then the equilibrium set[3] is globally attractively stable.

---

[3] If $\mathcal{A}_t(x)$ is strictly maximal monotone, then the equilibrium 'set' is an equilibrium point.

Proof: The set $\mathcal{E} = \mathrm{d}\mathbf{A}^{-1}(\mathbf{0})$ is a closed convex set, because $\mathrm{d}\mathbf{A}^{-1}$ is maximal monotone if $\mathrm{d}\mathbf{A}$ is maximal monotone and the images of a maximal monotone operator are closed convex sets [12]. For the differential measure of $\mathrm{d}\boldsymbol{x}$ we write

$$\mathrm{d}\boldsymbol{x} = -\boldsymbol{a}_t\mathrm{d}t - \boldsymbol{a}_\eta\mathrm{d}\eta, \tag{6.63}$$

where the single-valued densities obey the set-valued force laws

$$-\dot{\boldsymbol{x}} = \boldsymbol{a}_t \in \mathcal{A}_t(\boldsymbol{x}), \qquad -(\boldsymbol{x}^+ - \boldsymbol{x}-) = \boldsymbol{a}_\eta \in \mathcal{A}_\eta(\boldsymbol{x}^+). \tag{6.64}$$

The set-valued operators $\mathcal{A}_t$ and $\mathcal{A}_\eta$ are maximal monotone operators and are related to the set-valued measure function by

$$\mathrm{d}\mathcal{A}(\boldsymbol{x}^+) = \mathcal{A}_t(\boldsymbol{x})\mathrm{d}t + \mathcal{A}_\eta(\boldsymbol{x}^+)\mathrm{d}\eta. \tag{6.65}$$

Consider the Lyapunov function

$$V(\boldsymbol{x}) = \frac{1}{2}\,\mathrm{dist}_\mathcal{E}^2(\boldsymbol{x}) + \Psi_A(\boldsymbol{x}), \tag{6.66}$$

which is a positive definite function with respect to $\mathcal{E}$. Evaluation of the differential measure $\dot{V}$ gives

$$\begin{aligned}
\dot{V} &= (\boldsymbol{x} - \mathrm{prox}_\mathcal{E}(\boldsymbol{x}))^\mathrm{T}\dot{\boldsymbol{x}} \\
&= -(\boldsymbol{x} - \mathrm{prox}_\mathcal{E}(\boldsymbol{x}))^\mathrm{T}\boldsymbol{a}_t.
\end{aligned} \tag{6.67}$$

Using $\mathcal{A}_t(\mathrm{prox}_\mathcal{E}(\boldsymbol{x})) \ni \boldsymbol{a}_{t0} = \boldsymbol{0}$ together with the monotonicity of $\mathcal{A}_t(\boldsymbol{x})$ yields

$$\dot{V} = -(\boldsymbol{x} - \mathrm{prox}_\mathcal{E}(\boldsymbol{x}))^\mathrm{T}(\boldsymbol{a}_t - \boldsymbol{a}_{t0}) \leq 0. \tag{6.68}$$

We now consider the atomic part

$$\begin{aligned}
V^+ - V^- &= \frac{1}{2}\,\mathrm{dist}_\mathcal{E}^2(\boldsymbol{x}^+) - \frac{1}{2}\,\mathrm{dist}_\mathcal{E}^2(\boldsymbol{x}^-) \\
&= \frac{1}{2}\left\|\boldsymbol{x}^+ - \mathrm{prox}_\mathcal{E}(\boldsymbol{x}^+)\right\|^2 - \frac{1}{2}\left\|\boldsymbol{x}^- - \mathrm{prox}_\mathcal{E}(\boldsymbol{x}^-)\right\|^2.
\end{aligned} \tag{6.69}$$

Due to the convexity of $\mathcal{E}$ it holds that

$$\left\|\boldsymbol{x}^+ - \mathrm{prox}_\mathcal{E}(\boldsymbol{x}^+)\right\| \leq \left\|\boldsymbol{x}^+ - \mathrm{prox}_\mathcal{E}(\boldsymbol{x}^-)\right\|, \tag{6.70}$$

which gives

$$V^+ - V^- \leq \frac{1}{2}\left\|\boldsymbol{x}^+ - \mathrm{prox}_\mathcal{E}(\boldsymbol{x}^-)\right\|^2 - \frac{1}{2}\left\|\boldsymbol{x}^- - \mathrm{prox}_\mathcal{E}(\boldsymbol{x}^-)\right\|^2. \tag{6.71}$$

Using $\|\boldsymbol{x}\| = \boldsymbol{x}^\mathrm{T}\boldsymbol{x}$ and rearranging terms gives

$$V^+ - V^- \leq \frac{1}{2} \left(\boldsymbol{x}^+ - \operatorname{prox}_{\mathcal{E}}(\boldsymbol{x}^-) \right)^{\mathrm{T}} \left(\boldsymbol{x}^+ - \operatorname{prox}_{\mathcal{E}}(\boldsymbol{x}^-) \right)$$

$$- \frac{1}{2} \left(\boldsymbol{x}^- - \operatorname{prox}_{\mathcal{E}}(\boldsymbol{x}^-) \right)^{\mathrm{T}} \left(\boldsymbol{x}^- - \operatorname{prox}_{\mathcal{E}}(\boldsymbol{x}^-) \right)$$

$$\leq \frac{1}{2} \left(\|\boldsymbol{x}^+\|^2 - 2 \operatorname{prox}_{\mathcal{E}}(\boldsymbol{x}^-)^{\mathrm{T}} \boldsymbol{x}^+ - \|\boldsymbol{x}^-\|^2 + 2 \operatorname{prox}_{\mathcal{E}}(\boldsymbol{x}^-)^{\mathrm{T}} \boldsymbol{x}^- \right)$$

$$\leq \frac{1}{2} \left(\boldsymbol{x}^+ + \boldsymbol{x}^- - 2 \operatorname{prox}_{\mathcal{E}}(\boldsymbol{x}^-) \right)^{\mathrm{T}} \left(\boldsymbol{x}^+ - \boldsymbol{x}^- \right).$$

$$(6.72)$$

Furthermore, because of the invariance of the equilibrium set $\mathcal{E}$ it holds that $\boldsymbol{0} \in \boldsymbol{A}_\eta(\operatorname{prox}_{\mathcal{E}}(\boldsymbol{x}^-))$. Substitution of $\boldsymbol{a}_\eta = -(\boldsymbol{x}^+ - \boldsymbol{x}^-)$ and $\boldsymbol{a}_{\eta 0} = \boldsymbol{0}$ yields

$$V^+ - V^- \leq -\frac{1}{2} \left(2\boldsymbol{x}^+ + \boldsymbol{a}_\eta - 2 \operatorname{prox}_{\mathcal{E}}(\boldsymbol{x}^-) \right)^{\mathrm{T}} \boldsymbol{a}_\eta$$

$$\leq -\frac{1}{2} \|\boldsymbol{a}_\eta\|^2 - \left(\boldsymbol{x}^+ - \operatorname{prox}_{\mathcal{E}}(\boldsymbol{x}^-) \right)^{\mathrm{T}} \left(\boldsymbol{a}_\eta - \boldsymbol{a}_{\eta 0} \right),$$

$$(6.73)$$

Clearly, using the monotonicity of $\boldsymbol{A}_\eta$, it follows that $V^+ - V^- \leq 0$. Stability of the equilibrium set follows from Theorem 6.26. Moreover, if $\boldsymbol{A}_t(\boldsymbol{x})$ is strictly maximal monotone, then it holds that there exists a KR-function β such that $\dot{V} \leq -\beta(\operatorname{dist}_{\mathcal{E}}(\boldsymbol{x}))$. Furthermore, if $\boldsymbol{A}_t(\boldsymbol{x})$ is strictly maximal monotone, then the closed equilibrium set $\mathcal{E}$ is an equilibrium point and therefore compact. The Lyapunov function (6.66) is therefore a radially unbounded positive definite function with respect to $\mathcal{E}$ and has to decrease as long as the solution $\boldsymbol{x}(t) \notin \mathcal{E}$, which proves global attractivity. $\qquad\square$

Consider for instance the measure differential inclusion

$$\mathrm{d}x \in -(\operatorname{Sign}(x) + cx)\mathrm{d}t, \qquad (6.74)$$

which has the unique equilibrium point $x^* = 0$. The operator $\mathrm{d}A(x) = (\operatorname{Sign}(x) + cx)\mathrm{d}t$ is maximal monotone if $c \geq 0$ and strictly maximal monotone if $c > 0$. Hence, Theorem 6.29 immediately proves stability of $x^* = 0$ for $c \geq 0$ and attractive stability of $x^* = 0$ for $c > 0$. If $c < 0$, then the equilibrium point is still locally attractively stable but this is not proven by Theorem 6.29, because the measure function $\mathrm{d}A(x)$ is no longer maximal monotone for $c < 0$.

We conclude this section with a sufficient condition for incremental stability of a system (see Definition 6.22), which follows directly from the monotonicity property of (6.62).

Theorem 6.30 (Incremental Stability through Monotonicity). *Consider a measure differential inclusion of the form*

$$\mathrm{d}\boldsymbol{x} \in -\mathrm{d}\boldsymbol{A}(\boldsymbol{x}^+) = -\boldsymbol{A}_t(\boldsymbol{x})\mathrm{d}t - \boldsymbol{A}_\eta(\boldsymbol{x}^+)\mathrm{d}\eta.$$

Let the system be consistent and have existence of solutions for all initial conditions $\boldsymbol{x}_0 \in \mathcal{A}$, where the admissible domain $\mathcal{A}$ is a closed set. If $\mathrm{d}\boldsymbol{A}(\boldsymbol{x}^+)$ is *a maximal monotone set-valued measure function, then the system is incrementally stable. Moreover, if $\mathcal{A}_t(\boldsymbol{x})$ is a strictly maximal monotone set-valued function, then the system is attractively incrementally stable.*

Proof: Consider two arbitrary solutions $\boldsymbol{x}_1(t)$ and $\boldsymbol{x}_2(t)$ of the measure differential inclusion with $\boldsymbol{x}_i(t_0) \in \mathcal{A}$, $i = 1, 2$. In order to prove incremental stability, we measure the distance between these two arbitrary solutions and, therefore, consider the Lyapunov function candidate

$$V = \frac{1}{2}\|\boldsymbol{x}_1 - \boldsymbol{x}_2\|^2,$$

which has the differential measure

$$\mathrm{d}V = \dot{V}\mathrm{d}t + (V^+ - V^-)\mathrm{d}\eta, \tag{6.75}$$

with

$$\begin{aligned}
\dot{V} &= (\boldsymbol{x}_2 - \boldsymbol{x}_1)^\mathrm{T}(\dot{\boldsymbol{x}}_2 - \dot{\boldsymbol{x}}_1) \\
&= -(\boldsymbol{x}_2 - \boldsymbol{x}_1)^\mathrm{T}(\boldsymbol{a}_{t2} - \boldsymbol{a}_{t1}) \qquad \boldsymbol{a}_{t1} \in \mathcal{A}_t(\boldsymbol{x}_1), \boldsymbol{a}_{t2} \in \mathcal{A}_t(\boldsymbol{x}_2) \tag{6.76} \\
&\leq 0
\end{aligned}$$

and

$$\begin{aligned}
V^+ - V^- &= \frac{1}{2}(\boldsymbol{x}_2^+ + \boldsymbol{x}_2^- - \boldsymbol{x}_1^+ - \boldsymbol{x}_1^-)^\mathrm{T}(\boldsymbol{x}_2^+ - \boldsymbol{x}_2^- - \boldsymbol{x}_1^+ + \boldsymbol{x}_1^-) \\
&= -\frac{1}{2}(2\boldsymbol{x}_2^+ - 2\boldsymbol{x}_1^+ + \boldsymbol{a}_{\eta 2} - \boldsymbol{a}_{\eta 1})^\mathrm{T}(\boldsymbol{a}_{\eta 2} - \boldsymbol{a}_{\eta 1}) \\
&= -(\boldsymbol{x}_2^+ - \boldsymbol{x}_1^+)^\mathrm{T}(\boldsymbol{a}_{\eta 2} - \boldsymbol{a}_{\eta 1}) - \frac{1}{2}\|\boldsymbol{a}_{\eta 2} - \boldsymbol{a}_{\eta 1}\|^2 \\
&\leq 0,
\end{aligned} \tag{6.77}$$

where $\boldsymbol{a}_{\eta 1} \in \mathcal{A}_\eta(\boldsymbol{x}_1^+)$ and $\boldsymbol{a}_{\eta 2} \in \mathcal{A}_\eta(\boldsymbol{x}_2^+)$, see the proof of Theorem 4.9. Clearly, if $\mathcal{A}_t(\boldsymbol{x})$ and $\mathcal{A}_\eta(\boldsymbol{x}^+)$ are monotone, then it holds that $\mathrm{d}V \leq 0$, from which we deduce that $V^+(t) \leq V(t_0)$ or

$$\|\boldsymbol{x}_1^+(t) - \boldsymbol{x}_2^+(t)\| \leq \|\boldsymbol{x}_1(t_0) - \boldsymbol{x}_2(t_0)\| \quad \forall t \geq t_0.$$

Consequently, the distance between any two solutions can not increase, which proves incremental stability of the system. Moreover, if $\mathcal{A}_t(\boldsymbol{x})$ is strictly monotone then it holds that

$$\begin{aligned}
\mathrm{d}V &\leq \alpha\|\boldsymbol{x}_1 - \boldsymbol{x}_2\|^2\mathrm{d}t \\
&< 0, \quad \text{for } \boldsymbol{x}_1 \neq \boldsymbol{x}_2,
\end{aligned}$$

for some $\alpha > 0$. The solutions therefore have to approach each other as long as they are different from each other, i.e.

$$\lim_{t \to \infty} \|\boldsymbol{x}_1(t) - \boldsymbol{x}_2(t)\| = 0,$$

which proves attractive incremental stability.   □

Incremental stability of the system implies that solutions can not diverge from each other and stability of all its equilibrium points is therefore assured. Consider an equilibrium $x^* \in \mathcal{E}$ and a ball $x^* + B_\varepsilon$ around that equilibrium point. Solutions starting in the ε-ball can not leave this ball, because their distance to x^* can not increase due to incremental stability. For each $\varepsilon > 0$ we can take $\delta = \varepsilon$ and prove Lyapunov stability in the usual way. Theorem 6.29 therefore follows from Theorem 6.30.

Attractive stability excludes the possibility of more than one equilibrium because all solutions have to converge to each other. Hence, if $d\mathcal{A}(x^+)$ is a strictly maximal monotone operator, then the system $dx \in -d\mathcal{A}(x^+)$ does not posses an equilibrium set and at most a unique equilibrium point. Theorems 6.29 and 6.30 will be used in Section 7.2 to prove the stability of the equilibrium set of a mechanical system with friction. In Chapter 8, we will further study the convergence property of maximal monotone measure differential inclusions.

6.6 LaSalle's Invariance Principle

Using the basic Lyapunov theorems, proposed in the previous section, we are able to prove attractive stability if the Lyapunov function is strictly decreasing outside the equilibrium point/set, i.e. $dV < 0$. In many situations, however, we are not able to find such a Lyapunov function (e.g. the bouncing ball example). If the Lyapunov function is not strictly decreasing, then the solution can remain on a level surface of the Lyapunov function and not be attracted towards the equilibrium set. LaSalle introduced a theorem with which we are able to prove the conditional attractivity even when $dV \le 0$. The main idea of LaSalle is, that when the level surfaces of the Lyapunov function outside the equilibrium point/set are not positively invariant, then the equilibrium point/set must be attractive. In this section, we give a generalisation of the invariance principle of LaSalle [89], which has originally been stated for smooth dynamical systems, and formulate it for measure differential inclusions. In the following, we consider the extended lower semi-continuous function $V : \mathbb{R}^n \to \mathbb{R} \cup \{ | \infty \}$ to be of the form

$$V(x) = v(x) + \Psi_A(x), \tag{6.78}$$

in which $v(x)$ is a smooth, continuous, and locally bounded function. The following theorem and proof are adaptations of LaSalle's theorem as given in [85]:

Theorem 6.31 (LaSalle's Invariance Principle). *Let $\Omega \subset A$ be a compact set that is positively invariant with respect to the measure differential inclusion (6.1). Let every limit set $\mathcal{L}^+$ in Ω be positively invariant. Let $V(x)$*

be a function of the form (6.78), with $v(\boldsymbol{x})$ being continuous and bounded from below, such that $\mathrm{d}V \leq 0$ along solution curves in Ω. Let $\mathcal{Z}$ be the set of all points in Ω where $\mathrm{d}V = 0$ can hold, i.e.

$$\mathcal{Z} = \{\boldsymbol{x} \in \Omega \mid \boldsymbol{0} \in \mathrm{d}V(\boldsymbol{x}), \, \mathrm{d}\boldsymbol{x} \in \mathrm{d}\boldsymbol{\Gamma}(\boldsymbol{x})\}.$$

Let $\mathcal{M}$ be the largest positively invariant set in $\mathcal{Z}$. Then every solution curve $\boldsymbol{\varphi}(\cdot, t_0, \boldsymbol{x}_0) \in \mathcal{S}(\mathrm{d}\boldsymbol{\Gamma}, t_0, \boldsymbol{x}_0)$ with $\boldsymbol{x}_0 \in \Omega$ approaches $\mathcal{M}$ as $t \to \infty$.

Proof: Let $\boldsymbol{x}(t) = \boldsymbol{\varphi}(t, t_0, \boldsymbol{x}_0)$ be a solution of (6.1) starting in Ω, i.e. $\boldsymbol{x}_0 \in \Omega$. The solution $\boldsymbol{x}(t)$ remains in Ω for all $t \geq t_0$, because Ω is positively invariant. The function $V(\boldsymbol{x}(t))$ is non-increasing because $\mathrm{d}V(\boldsymbol{x}) \leq 0$ in Ω. Since $V(\boldsymbol{x})$ is bounded from below, $V(\boldsymbol{x}(t))$ has a limit a as $t \to \infty$. The positive limit set $\mathcal{L}^+$ of $\boldsymbol{\varphi}(\cdot, t_0, \boldsymbol{x}_0)$ is in Ω because Ω is a closed set. According to Definition 6.9, for any $\boldsymbol{p} \in \mathcal{L}^+$ there exists a sequence $\{t_j\}$ with $t_j \to \infty$ as $j \to \infty$, such that $\boldsymbol{x}(t_j) \to \boldsymbol{p}$ for $j \to \infty$. It holds that $\boldsymbol{x}(t_j) \in \mathcal{A}$ since $\Omega \subset \mathcal{A}$ is positively invariant. By continuity of $v(\boldsymbol{x})$ it holds that $v(\boldsymbol{p}) = \lim_{j \to \infty} v(\boldsymbol{x}(t_j)) = a$. Moreover, using that $\boldsymbol{x}(t_j) \in \mathcal{A}$ and $\boldsymbol{p} \in \mathcal{A}$ it follows with $\Psi_{\mathcal{A}}(\boldsymbol{p}) = 0$ that $V(\boldsymbol{x}) = a$ for all $\boldsymbol{x} \in \mathcal{L}^+$. The positive limit set $\mathcal{L}^+$ is assumed to be positively invariant. Consequently, $\mathrm{d}V = 0$ on $\mathcal{L}^+$ and

$$\mathcal{L}^+ \subset \mathcal{M} \subset \mathcal{Z} \subset \Omega$$

The compactness of Ω implies boundedness of $\boldsymbol{x}(t)$ (from which we can deduce that $\mathcal{L}^+$ is non-empty [85]) and $\boldsymbol{x}(t)$ approaches $\mathcal{L}^+$ as $t \to \infty$. Consequently, $\boldsymbol{x}(t)$ approaches $\mathcal{M}$ as $t \to \infty$. $\qquad \square$

Note that a sufficient condition for the positive invariance of limit sets is a continuous dependence of solutions $\boldsymbol{x}(t) \in \Omega$ on initial data (Proposition 6.12). The set Ω in Theorem 6.31 has to be positively invariant which can, for instance, be proven with Proposition 6.5. LaSalle's invariance principle does not require the function V to be positive definite, but the function V is usually chosen to be positive definite in order to prove the boundedness (and positive invariance) of the set Ω, which is used in Proposition 6.5.

We briefly discuss an example which shows why the condition that every limit set is positively invariant is essential for LaSalle's theorem to hold. Consider the discontinuous differential equation

$$\dot{x} = \begin{cases} -x + 1 & x > 1, \\ -x & x \leq 1, \end{cases} \tag{6.79}$$

which has a unique equilibrium point at the origin $x = 0$. Note that this system is not a Filippov system. Consider the positive definite function $V(y) = \frac{1}{2}x^2$. The function V does not increase along solution curves, which can be seen from

$$\dot{V} = x\dot{x} = \begin{cases} -x^2 + x < 0 & x > 1, \\ -x^2 < 0 & x \leq 1, x \neq 0, \\ 0 & x = 0. \end{cases} \tag{6.80}$$

Each level set of V is positively invariant and the origin is the largest invariant set within the set for which $\dot{V} = 0$. Still, the origin is not attractive within level sets $V > \frac{1}{2}$. A solution $x(t)$ with $x(0) > 1$ converges asymptotically to the limit point $x = 1$, i.e. $\lim_{t\to\infty} x(t) = 1$, but it never reaches the limit point. The limit point itself is not invariant because $\dot{x} < 0$ for $x - 1$. A solution with $x(0) > 1$ will therefore not be attracted to the origin. Note also that the conditions for global attractivity in Theorem 6.23 are not fulfilled because there does exist a function β of class K such that $\dot{V} \leq -\beta(|x|)$ for all $x \in \mathbb{R}$. In other words, the function $-\dot{V}$ is not positive definite.

We now return to the bouncing ball example which was studied in the previous section. Each level set Ω_c of V, given by (6.43), is bounded because V is PDF. Proposition 6.5 therefore proves that each level set Ω_c is positively invariant. Following Theorem 6.31, let $\mathcal{Z}$ be the set of all points in Ω_c where $dV = 0$ can hold, i.e. $\mathcal{Z}$ is the union of the set $\{x \in \Omega \mid q = 0, u \geq 0\}$ and the set $\{x \in \Omega \mid q > 0\}$:

$$\mathcal{Z} = \{x \in \Omega \mid q = 0, u \geq 0\} \bigcup \{x \in \Omega \mid q > 0\}.$$

The equilibrium $x^* = 0$ is an element of $\mathcal{Z}$. We now seek the largest positively invariant set in $\mathcal{Z}$, which we call $\mathcal{M}$. Because of the gravitation, if the ball is in the air, then it will hit the floor in a finite time. In other words: the solution curve will leave $\operatorname{int} \mathcal{A}$ in a finite time and the set $\operatorname{int} \mathcal{A}$ is therefore not positively invariant. The largest positively invariant set in $\mathcal{Z}$ is therefore $\mathcal{M} = x^* = 0$. This implies that the equilibrium point x^* is globally attractive. Moreover, stability has been proven in the previous section. Consequently, the equilibrium point x^* is globally attractively stable. To prove symptotic attractivity is much more difficult. Symptotic stability of the bouncing ball system for $e < 1$ has been proven in [27,175] using an event-type Poincaré map method. Such an analysis can only be carried out explicitly in very special cases.

6.7 Instability

Lyapunov-type theorems for the stability and attractivity of equilibrium points and sets have been presented in the previous sections. It is also possible to prove instability and/or non-attractivity using a Lyapunov-type argument. We give a generalisation of Chetaev's instability theorem [35,85] for an equilibrium point or set of a measure differential inclusion using an extended lower semi-continuous function $V : \mathbb{R}^n \to \mathbb{R} \cup \{\infty\}$ of the form (6.27).

Theorem 6.32 (Instability and Non-Attractivity of an Equilibrium Point). *Let $x^* = 0 \in \mathcal{A}$ be an equilibrium point of the consistent measure differential inclusions (6.1), which admits solutions $\varphi(\cdot, t_0, x_0) \in \mathcal{S}(d\Gamma, t_0, x_0)$ for all $x_0 \in \mathcal{A}$. Let $V(x)$ be an extended lower semi-continuous function of* *the form (6.27), such that $V(\mathbf{0}) = 0$ and $V(\mathbf{x}_a) = a < 0$ for some $\mathbf{x}_a$ with arbitrarily small $\|\mathbf{x}_a\|$. Moreover, V is such that the set $\mathcal{U}$ defined by*

$$\mathcal{U} = \{\mathbf{x} \in \overline{B}_r \cap \mathcal{A} \mid V(\mathbf{x}) \leq 0\}$$

for some $r > 0$ is a nonempty set. Let every limit set $\mathcal{L}^+$ in $\mathcal{U}$ be positively invariant. Three statements can be made

a. *If $\mathrm{d}V < 0$ for all $\mathbf{x} \in \mathcal{U}\backslash\{\mathbf{0}\}$, then it holds that $\mathbf{x}^* = \mathbf{0}$ is unstable.*
b. *If $\mathrm{d}V \leq 0$ for all $\mathbf{x} \in \mathcal{U}$ and the solution can not stay in $\mathcal{Z} = \{\mathbf{x} \in \mathcal{U}\backslash\{\mathbf{0}\} \mid \mathrm{d}V = 0\}$, then $\mathbf{x}^* = \mathbf{0}$ is unstable.*
c. *If $\mathrm{d}V \leq 0$ for all $\mathbf{x} \in \mathcal{U}$ and $\mathbf{0} \in \overline{\mathcal{U}\backslash(\mathrm{bdry}\,\mathcal{U}\backslash\mathrm{bdry}\,\mathcal{A})}$, then it holds that $\mathbf{x}^* = \mathbf{0}$ is not attractive.*

Proof: Theorem 6.32a: First of all, remark that $\mathcal{U} \subset \mathcal{A}$ is compact and all solution curves $\varphi(\cdot, t_0, \mathbf{x}_0) \in \mathcal{S}(\mathrm{d}\mathbf{\Gamma}, t_0, \mathbf{x}_0)$ remain in $\mathcal{A}$ for $\mathbf{x}_0 \in \mathcal{A}$ due to the consistency assumption (Definition 4.8). From (6.27) we see that $V(\mathbf{x}) = v(\mathbf{x})$ for all $\mathbf{x} \in \mathcal{U}$ and $V(\mathbf{x})$ is therefore bounded on $\mathcal{U}$. It holds that $\mathbf{x}_a \in \mathrm{int}\,\mathcal{U}$ because $V(\mathbf{x}_a) = a < 0$ and $\|\mathbf{x}_a\| < r$ is arbitrarily small. Moreover, for every solution curve $\varphi(\cdot, t_0, \mathbf{x}_a) \in \mathcal{S}(\mathrm{d}\mathbf{\Gamma}, t_0, \mathbf{x}_a)$ it holds that the function $V(\varphi(t, t_0, \mathbf{x}_a))$ with

$$V(\varphi(t, t_0, \mathbf{x}_a)) = V(\mathbf{x}_a) + \int_{[t_0, t]} \mathrm{d}V \tag{6.81}$$

is strictly decreasing as long as $\varphi(t, t_0, \mathbf{x}_a) \in \mathcal{U}$ since $\mathrm{d}V < 0$ in $\mathcal{U}\backslash\{\mathbf{0}\}$. This means that if a solution curve $\varphi(t, t_0, \mathbf{x}_a)$ would stay forever in $\mathcal{U}$ then $V(\varphi(t, t_0, \mathbf{x}_a))$ would tend to $-\infty$. This is not possible because $V(\mathbf{x})$ is bounded from below on $\mathcal{U}$ and all solution curves $\varphi(t, t_0, \mathbf{x}_a)$ must therefore leave the set $\mathcal{U}$. The solution curves $\varphi(\cdot, t_0, \mathbf{x}_a) \in \mathcal{S}(\mathrm{d}\mathbf{\Gamma}, t_0, \mathbf{x}_a)$ can not leave through $\mathrm{bdry}\,\mathcal{A}$ because of the viability of solution curves. The solution curves can also not leave through the surface $V(\mathbf{x}) = 0$ because $V(\varphi(t, t_0, \mathbf{x}_a)) \leq a < 0$ for almost all $t \geq t_0$. Consequently, all solution curves $\varphi(\cdot, t_0, \mathbf{x}_a) \in \mathcal{S}(\mathrm{d}\mathbf{\Gamma}, t_0, \mathbf{x}_a)$ leave through the sphere $\|\mathbf{x}\| = r$. Notice that $\mathrm{bdry}\,B_r \cap \mathrm{bdry}\,\mathcal{U} \neq \emptyset$ because the solution curves have to escape from $\mathcal{U}$. We can take $\mathbf{x}_a$ arbitrarily close to the origin and the origin is therefore unstable.

Theorem 6.32b: We extend Theorem 6.32a with a LaSalle-type of argumentation. The function V is non-increasing along solution curves of the system within $\mathcal{U}$ because $\mathrm{d}V \leq 0$. No solution curve $\varphi(t, t_0, \mathbf{x}_a) \in \mathcal{U}$ therefore exists which is attracted to the origin. Solution curves might remain temporarily on a level surface of V if $\mathrm{d}V = 0$, i.e. if $\mathbf{x} \in \mathcal{Z}$. However, the solution can not stay in $\mathcal{Z}$, i.e. there exists no positively invariant subset of $\mathcal{Z}$. This means that solution curves ultimately will leave this set and enter $\mathcal{U}\backslash\mathcal{Z}$ for which $\mathrm{d}V < 0$. So, the function V will decrease again. The rest of the reasoning is similar to the proof of Theorem 6.32a. All solution curves may enter and leave $\mathcal{Z}$ a couple of times, but finally leave $\mathcal{U}$ through the sphere $\|\mathbf{x}\| = r$ which proves instability.

Theorem 6.32c: As in b it holds that the function V is non-increasing along solution curves of the system within $\mathcal{U}$ because $dV \le 0$. No solution curve $\varphi(t, t_0, \boldsymbol{x}_a) \in \mathcal{U}$ therefore exists which is attracted to the origin. Moreover, since $\boldsymbol{0} \in \mathcal{U} \backslash \overline{(\mathrm{bdry}\,\mathcal{U} \backslash \mathrm{bdry}\,\mathcal{A})}$ it holds that either $\boldsymbol{0} \in \mathrm{int}\,\mathcal{U}$ or the origin is on the boundary between $\mathcal{U}$ and $\mathcal{A}^c$, i.e. it is surrounded by $\mathcal{U} \sqcup \mathcal{A}^c$ (see Figure 6.12). If $\boldsymbol{0} \in \mathrm{int}\,\mathcal{U}$ then the set $\mathcal{U}$ is a neighbourhood of the origin. Consequently, there exists no solution which is attracted to the origin. Similarly, if $\boldsymbol{0}$ is on the boundary between $\mathcal{U}$ and $\mathcal{A}^c$, then it can not be attracted through $\mathcal{U}$ nor through $\mathcal{A}^c$. □

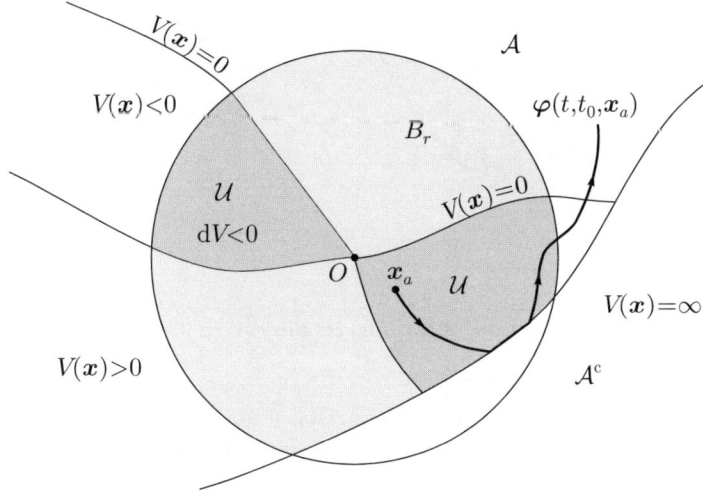

Fig. 6.11. Sets used in the proof of Theorem 6.32a.

The various sets used in the proof of Theorem 6.32a have been elucidated in Figure 6.11. A solution curve is drawn which starts in $\mathcal{U}$ at point $\boldsymbol{x}_a$, slides along the boundary of $\mathcal{A}$ and finally leaves $\mathcal{U}$ through the sphere $\|\boldsymbol{x}\| = r$. Notice that $\mathcal{U}$ can consist of more than one part (e.g. two, as drawn in the Figure 6.11), but has to contain the origin in its interior or boundary.

Theorem 6.32a and b prove the conditional instability of an equilibrium point. All solution curves $\varphi(t, t_0, \boldsymbol{x}_a)$ with $\boldsymbol{x}_a \in \mathcal{U}$ have to leave $\mathcal{U}$ through the sphere $\|\boldsymbol{x}\| = r > 0$. So there exists at least one solution curve which starts arbitrarily close to the origin and which arrives at a distance $r > 0$ from the origin. Non-attractivity is however not proven in a and b because solution curves which have left $\mathcal{U}$ might still be attracted to the origin in $\mathcal{A} \backslash \mathcal{U}$. Such a 'return-loop' of solution curves can of course not exist when $\boldsymbol{x}^* \in \mathrm{int}\,\mathcal{U}$ and this is the main idea behind Theorem 6.32c. Also, when $\boldsymbol{x}^*$ is on the boundary of $\mathcal{A}$ but within $\mathcal{A}$ surrounded by $\mathcal{U}$, then such a return-loop can not exist, see Figure 6.12. Therefore, a return-loop can not exist if $\boldsymbol{x}^*$ is in the relative interior of $\mathcal{U}$ with respect to $\mathcal{A}$:

$$\exists r > 0 \ B_r(\boldsymbol{x}^*) \subset (\mathcal{U} \cup \mathcal{A}^{\mathrm{c}}) \Longleftrightarrow \boldsymbol{x}^* \in \mathcal{U} \backslash \overline{(\mathrm{bdry}\,\mathcal{U} \backslash \mathrm{bdry}\,\mathcal{A})}. \qquad (6.82)$$

An equilibrium set $\mathcal{E}$ is called stable when all equilibrium points $\boldsymbol{x}^* \in \mathcal{E}$ are stable. To prove instability of an equilibrium set, it therefore suffices to prove the instability of *one* equilibrium point $\boldsymbol{x}^* \in \mathcal{E}$, for instance by using Theorem 6.32.

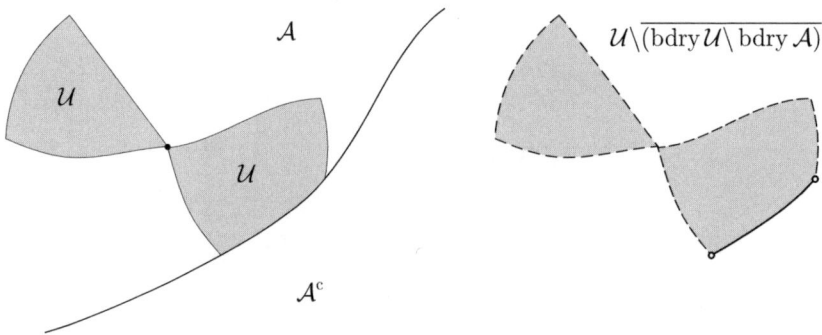

Fig. 6.12. Meaning of $\mathcal{U} \backslash \overline{(\mathrm{bdry}\,\mathcal{U} \backslash \mathrm{bdry}\,\mathcal{A})}$ in Theorem 6.32c.

An equilibrium set $\mathcal{E}$ is called attractive if all solution curves in a neighbourhood of $\mathcal{E}$ are attracted to the equilibrium set. To prove non-attractivity of an equilibrium set it does therefore *not* suffice to prove the non-attractivity of one equilibrium point $\boldsymbol{x}_1^* \in \mathcal{E}$, because repelling solution curves of $\boldsymbol{x}_1^*$ might still be attracted to $\boldsymbol{x}_2^* \in \mathcal{E}$. If we want to prove non-attractivity of $\mathcal{E}$, then we have to prove the non-attractivity of the set $\mathcal{E}$ as a whole. We can do this for instance with a function $V(\boldsymbol{x})$ with the following properties

$$V(\boldsymbol{x}) = 0 \ \forall \boldsymbol{x} \in \mathcal{E}, \quad V(\boldsymbol{x}) < 0 \, \forall \boldsymbol{x} \in U_\delta(\mathcal{E}) \backslash \mathcal{E} \qquad (6.83)$$

for some $\delta > 0$. If $\mathrm{d}V \leq 0$ for all $\boldsymbol{x} \in U_\delta(\mathcal{E})$ then the equilibrium set is non-attractive and even unstable if $\mathrm{d}V < 0$ for all $\boldsymbol{x} \in U_\delta(\mathcal{E}) \backslash \mathcal{E}$. However, in the following theorem we will propose less stringent conditions for instability and non-attractivity of an equilibrium set.

Theorem 6.33 (Instability and Non-Attractivity of an Equilibrium Set). *Let $\mathcal{E} \in \mathcal{A}$ be a compact equilibrium set of the consistent measure differential inclusion (6.1), which admits solutions $\boldsymbol{\varphi}(\cdot, t_0, \boldsymbol{x}_0) \in \mathcal{S}(\mathrm{d}\boldsymbol{\Gamma}, t_0, \boldsymbol{x}_0)$ for all $\boldsymbol{x}_0 \in \mathcal{A}$. Let $V(\boldsymbol{x})$ be an extended lower semi-continuous function of the form (6.27), such that $V(\boldsymbol{x}_a) < V_{\mathcal{E}} \leq 0$ for some $\boldsymbol{x}_a$, for which $\mathrm{dist}_{\mathcal{E}}(\boldsymbol{x}_a) > 0$ is arbitrarily small, with $V_{\mathcal{E}} = \min_{\boldsymbol{x} \in \mathcal{E}} V(\boldsymbol{x})$. Moreover, let V be such that the set $\mathcal{U}$ defined by*

$$\mathcal{U} = \{\boldsymbol{x} \in \overline{U}_r(\mathcal{E}) \cap \mathcal{A} \mid V(\boldsymbol{x}) \leq 0\}$$

for some $r > 0$ is a nonempty set. Let every limit set $\mathcal{L}^+$ in $\mathcal{U}$ be positively invariant. Three statements can be made:

a. *If $dV < 0$ for all $\boldsymbol{x} \in \mathcal{U}\backslash\mathcal{E}$, then $\mathcal{E}$ is unstable.*
b. *If $dV \leq 0$ for all $\boldsymbol{x} \in \mathcal{U}$ and the solution can not stay in $\mathcal{Z} = \{\boldsymbol{x} \in \mathcal{U}\backslash\mathcal{E} \mid dV = 0\}$, then it holds that $\mathcal{E}$ is unstable.*
c. *If $dV \leq 0$ for all $\boldsymbol{x} \in \mathcal{U}$ and $\mathcal{E} \subset \mathcal{U}\backslash\overline{(\text{bdry}\,\mathcal{U}\backslash\text{bdry}\,\mathcal{A})}$, then $\mathcal{E}$ is not attractive.*

Proof: Theorem 6.33a: The set $\mathcal{U}$ is compact as it is the intersection of compact sets. The point $\boldsymbol{x}_a \notin \mathcal{E}$ is in the interior of $\mathcal{U}$ and can be taken arbitrarily close to an equilibrium point $\boldsymbol{x}^* \in \text{bdry}\,\mathcal{E}$. The local continuity of $V(\boldsymbol{x})$ for $\boldsymbol{x} \in \text{int}\,\mathcal{A}$ implies that the function $V(\boldsymbol{x})$ attains its minimum $V_\mathcal{E}$ in $\mathcal{E}$ for a point $\boldsymbol{x}^* \in \text{bdry}\,\mathcal{E}$, because $V(\boldsymbol{x}_a) < V_\mathcal{E}$ with $\boldsymbol{x}_a$ taken arbitrarily close to $\mathcal{E}$. Figure 6.13 shows, similarly to Figure 6.11, the various sets used in the proof of Theorem 6.33a. Using a very similar reasoning as in the proof of Theorem 6.32a, we can prove that $\boldsymbol{x}^*$ is an unstable equilibrium point from which follows the instability of $\mathcal{E}$. Each solution curve $\boldsymbol{\varphi}(\cdot, t_0, \boldsymbol{x}_a) \in \mathcal{S}(d\boldsymbol{\Gamma}, t_0, \boldsymbol{x}_a)$ can not be attracted to $\mathcal{E} \cap \mathcal{U}$ because $V(\boldsymbol{\varphi}(t, t_0, \boldsymbol{x}_a)) < V_\mathcal{E}$ will decrease as long as $\boldsymbol{\varphi}(t, t_0, \boldsymbol{x}_a) \in \mathcal{U}$. Moreover, each solution curve can not remain in $\mathcal{U}$ because V would then tend to $-\infty$, which is not possible due to the boundedness of $\mathcal{U}$. Hence, $\boldsymbol{x}^*$ is unstable, and so is $\mathcal{E}$.

Theorem 6.33b: We use again the reasoning of LaSalle. A solution curve $\boldsymbol{\varphi}(t, t_0, \boldsymbol{x}_a)$ might temporarily remain on a level surface $V(\boldsymbol{\varphi}(t, t_0, \boldsymbol{x}_a)) = c$ when $\boldsymbol{\varphi}(t, t_0, \boldsymbol{x}_a) \in \mathcal{Z}$, but finally will have to leave $\mathcal{Z}$ because there exists no positively invariant subset of $\mathcal{Z}$.

Theorem 6.33c: Each solution curve $\boldsymbol{\varphi}(t, t_0, \boldsymbol{x}_a) \in \mathcal{U}$ can not be attracted to $\mathcal{E}$ because $V(\boldsymbol{\varphi}(t, t_0, \boldsymbol{x}_a)) < V_\mathcal{E}$ will not increase as long as $\boldsymbol{\varphi}(t, t_0, \boldsymbol{x}_a) \in \mathcal{U}$. If a solution curve $\boldsymbol{\varphi}(t, t_0, \boldsymbol{x}_a)$ has left $\mathcal{U}$ it can not be attracted to $\mathcal{E}$ because $\mathcal{E} \subset \text{int}\,\mathcal{U}$ or $\mathcal{E}$ is surrounded by $\mathcal{U}$ and $\mathcal{A}^c$. Consequently, the equilibrium point is not only not attractive, but its repelling neighbouring solution curves can not return to $\mathcal{E}$. The equilibrium set $\mathcal{E}$ is therefore not attractive in the sense of Definition 6.18. $\square$

6.8 Summary

Lyapunov's direct method, LaSalle's invariance principle and Chetaev's instability theorem have been generalised to prove stability properties of autonomous measure differential inclusions of the form (6.1). All theorems rely on a suitable choice of a Lyapunov-type function $V(\boldsymbol{x})$. This is exactly the weak spot of theorems of this type. How do we find a suitable Lyapunov-type function $V(\boldsymbol{x})$? For mechanical systems with unilateral constraints, however,

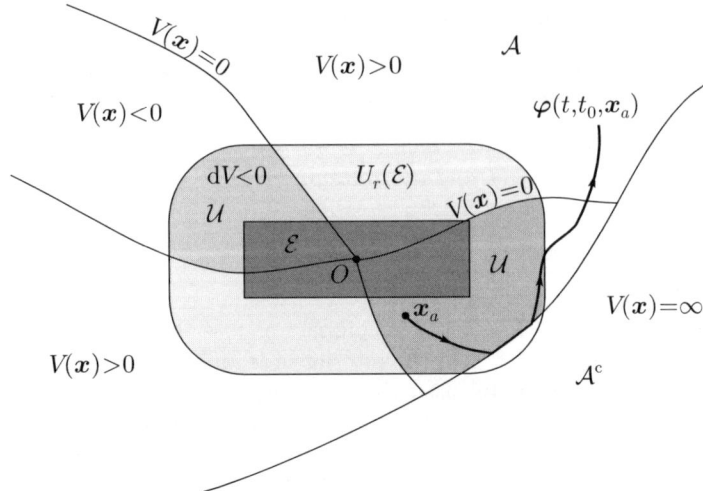

Fig. 6.13. Sets used in the proof of Theorem 6.33a.

there is a natural choice for the Lyapunov function: the total mechanical energy which includes the support function of the unilateral constraints.

The Lyapunov-type theorems, which have been presented in this chapter, will be used in Chapter 7 to derive theorems for the stability and attractivity of equilibrium sets in mechanical systems with frictional unilateral contact.

7

Stability Properties in Mechanical Systems

In this chapter, we apply the stability results of Chapter 6 for measure differential inclusions to Lagrangian mechanical systems with set-valued force laws, which have been formulated in Chapter 5. The special structure of Lagrangian mechanical systems allows for a natural choice of the Lyapunov function, a systematic derivation of the proof for this large class of systems as well as a physical interpretation of the results.

In the following, we study stability properties of equilibrium sets of the measure differential inclusion (5.96)

$$M(q)\mathrm{d}u - h(q, u)\mathrm{d}t = W_N(q)\mathrm{d}P_N + W_D(q)\mathrm{d}P_D \quad \forall t,$$

where $\mathrm{d}P_N$ and $\mathrm{d}P_D$ obey the set-valued constitutive laws (5.98). We assume existence of solutions of (5.96) for all admissible initial conditions. We denote an equilibrium position of (5.96) by q^*. The generalised velocities u vanish at an equilibrium point as we assume all contacts to be scleronomic (i.e. the normal contact distances $g_N(q)$ as well as all relative velocities $\gamma_D(q, u)$ do not explicitly depend on time). With $\mathcal{E}$ we denote an equilibrium set of the measure differential inclusion in first-order form, while $\mathcal{E}_q$ is reserved for the union of equilibrium positions q^*, i.e. $\mathcal{E} = \{(q, u) \in \mathbb{R}^n \times \mathbb{R}^n \mid q \in \mathcal{E}_q, u = 0\}$.

7.1 Total Mechanical Energy

The basic Lyapunov theorems stated in Theorem 6.23 for the (attractive) stability of an equilibrium point demand the choice of a suitable Lyapunov function V. The function V should be of the form (6.27)

$$V(x) = v(x) + \Psi_A(x), \tag{7.1}$$

which contains an indicator function on the set $\mathcal{A}$, which is the set of admissible states. When studying mechanical systems it is natural to choose the total mechanical energy of the system as a Lyapunov candidate function.

Motivated by the bouncing ball example of Section 6.5, it will sometimes be useful to choose the sum of the kinetic energy $T(\boldsymbol{q}, \boldsymbol{u})$, the potential energy of the smooth conservative forces $U(\boldsymbol{q})$ and the energy potential of the normal contact forces as Lyapunov candidate function.

Let the potential $Q(\boldsymbol{q})$ be the *total* potential energy of the system

$$Q(\boldsymbol{q}) = U(\boldsymbol{q}) - U(\boldsymbol{0}) + \sum_{i \in I_G} \pi_{Ni}(g_{Ni}(\boldsymbol{q}))$$
$$= U(\boldsymbol{q}) - U(\boldsymbol{0}) + \sum_{i \in I_G} \Psi_{\mathbb{R}+}(g_{Ni}(\boldsymbol{q})). \tag{7.2}$$

which is the sum of the potential energy $U(\boldsymbol{q})$ of all smooth potential forces and the support functions $\pi_{Ni}(g_{Ni}(\boldsymbol{q}))$ of the normal contact forces. The support function $\pi_{Ni}(g_{Ni}(\boldsymbol{q}))$ equal indicator functions $\Psi_{\mathbb{R}+}(g_{Ni}(\boldsymbol{q}))$ which express the unilaterality of the constraints $g_{Ni}(\boldsymbol{q}) \geq 0$, which restrict the generalised coordinates $\boldsymbol{q}$ to the set $\mathcal{K}$

$$\mathcal{K} = \{\boldsymbol{q} \in \mathbb{R}^n \mid g_{Ni}(\boldsymbol{q}) \geq 0 \,\forall i\}, \tag{7.3}$$

which we call the set of admissible generalised coordinates. We can therefore write the total potential energy function as

$$Q(\boldsymbol{q}) = U(\boldsymbol{q}) - U(\boldsymbol{0}) + \Psi_{\mathcal{K}}(\boldsymbol{q}). \tag{7.4}$$

Moreover, in the stability proofs of Section 7.2 we will assume the total potential energy $Q(\boldsymbol{q})$ to be a locally positive definite function such that

$$Q(\boldsymbol{0}) = 0, \qquad Q(\boldsymbol{q}) > 0 \,\forall \boldsymbol{q} \in \mathcal{U} \backslash \{\boldsymbol{0}\}, \tag{7.5}$$

where $\mathcal{U}$ is some neighbourhood of the origin.

The kinetic energy $T(\boldsymbol{q}, \boldsymbol{u})$ is assumed to be a symmetric positive definite quadratic form in $\boldsymbol{u}$

$$T(\boldsymbol{q}, \boldsymbol{u}) = \frac{1}{2} \boldsymbol{u}^{\mathrm{T}} \boldsymbol{M}(\boldsymbol{q}) \boldsymbol{u} \tag{7.6}$$

with $\boldsymbol{M}(\boldsymbol{q}) = \boldsymbol{M}(\boldsymbol{q})^{\mathrm{T}} > 0$.

In the following, we will use the total mechanical energy function

$$E_{\mathrm{tot}} = T(\boldsymbol{q}, \boldsymbol{u}) + Q(\boldsymbol{q}) = T(\boldsymbol{q}, \boldsymbol{u}) + U(\boldsymbol{q}) - U(\boldsymbol{0}) + \Psi_{\mathcal{K}}(\boldsymbol{q}), \tag{7.7}$$

being the sum of kinetic and total potential energy. The total mechanical energy function E_{tot} (7.7) is locally positive definite if the total potential energy is locally positive definite and E_{tot} has the property

$$E_{\mathrm{tot}}(\boldsymbol{q}, \boldsymbol{u}) = \infty \quad \forall \boldsymbol{q} \notin \mathcal{K}, \tag{7.8}$$

where $\mathcal{K}$ is defined by (7.3). If we gather the generalised positions $\boldsymbol{q}$ and velocities $\boldsymbol{u}$ in the state-vector $\boldsymbol{x} = \begin{bmatrix} \boldsymbol{q}^{\mathrm{T}} & \boldsymbol{u}^{\mathrm{T}} \end{bmatrix}^{\mathrm{T}}$, then the set $\mathcal{K}$ induces a set

$\mathcal{A} = \{x \mid q \in \mathcal{K}\}$ of admissible states. Clearly, the total mechanical energy function $E_{\text{tot}}(q, u)$ is an extended lower semi-continuous function of the form (6.27). Moreover, the states (q, u) of a mechanical system consist of generalised positions $q(t)$, which are assumed to be absolutely continuous, and generalised velocities $u(t)$, which are of locally bounded variation. Only the generalised positions are restricted to an admissible set $\mathcal{K}$, whereas no restriction exists on the velocities u. The total mechanical energy function $E_{\text{tot}}(q, u)$ consists of a (possibly) locally positive definite function $Q(q)$ which is solely dependent on q (being absolutely continuous) and a function $T(q, u)$ which is a PDF in u for fixed values of q. Consequently, a mechanical system (5.96) is a measure differential inclusion with the property (6.33) and the total mechanical energy $E_{\text{tot}}(q, u)$ is a function of the form (6.34) if $Q(q)$ is a LPDF.

We compute the differential measure of E_{tot} along solution curves of (5.96):

$$\mathrm{d}F_{\text{tot}} = \mathrm{d}T + \mathrm{d}Q. \tag{7.9}$$

The total potential energy is only a function of the generalised displacements and it holds that

$$\begin{aligned} \mathrm{d}Q &= dQ(q)(\mathrm{d}q) \\ &= U_{,q}\mathrm{d}q + d\Psi_{\mathcal{K}}(q)(\mathrm{d}q) \end{aligned} \tag{7.10}$$

where $dQ(q)(\mathrm{d}q)$ is the subderivative (see Section 2.6) of Q at q in the direction $\mathrm{d}q$. The subderivative $d\Psi_{\mathcal{K}}(q)(\mathrm{d}q)$ of the indicator function $\Psi_{\mathcal{K}}(q)$ equals the indicator function on the associated contingent cone $K_{\mathcal{K}}(q)$ (see (2.54))

$$d\Psi_{\mathcal{K}}(q)(\mathrm{d}q) = \Psi_{K_{\mathcal{K}}(q)}(\mathrm{d}q). \tag{7.11}$$

It holds that $\mathrm{d}q = u\mathrm{d}t$ with $u = u^+ \in K_{\mathcal{K}}(q)$ due to the consistency of the system and the indicator function on the contingent cone therefore vanishes, i.e. $\Psi_{K_{\mathcal{K}}(q)}(u\mathrm{d}t) = 0$. The differential measure of Q simplifies to

$$\begin{aligned} \mathrm{d}Q &= U_{,q}\mathrm{d}q + \Psi_{K_{\mathcal{K}}(q)}(\mathrm{d}q) \\ &= U_{,q}u\mathrm{d}t + \Psi_{K_{\mathcal{K}}(q)}(u\mathrm{d}t), \quad u \in K_{\mathcal{K}}(q) \\ &= U_{,q}u\mathrm{d}t. \end{aligned} \tag{7.12}$$

Recalling that the kinetic energy $T(q, u)$ (7.6) is a quadratic form and using the results of Section 3.6, we deduce that the differential measure of T is

$$\mathrm{d}T = \frac{1}{2}(u^+ + u^-)^{\mathrm{T}}M(q)\mathrm{d}u + T_{,q}\mathrm{d}q. \tag{7.13}$$

The differential measure of E_{tot} becomes by (7.9), (7.12) and (7.13)

$$\begin{aligned} \mathrm{d}E_{\text{tot}} &= \frac{1}{2}(u^+ + u^-)^{\mathrm{T}}M(q)\mathrm{d}u + (T_{,q} + U_{,q})\,\mathrm{d}q \\ &\stackrel{(5.101)}{=} \frac{1}{2}(u^+ + u^-)^{\mathrm{T}}(h(q, u)\mathrm{d}t + W\mathrm{d}P) + (T_{,q} + U_{,q})\,u\mathrm{d}t \end{aligned} \tag{7.14}$$

A term $\frac{1}{2}(u^+ + u^-)^{\mathrm{T}}\mathrm{d}t$ in front of a Lebesgue measurable term equals $u^{\mathrm{T}}\mathrm{d}t$. Together with (5.86), yielding $h = f^{\mathrm{nc}} + f^{\mathrm{gyr}} - (T_{,q} + U_{,q})^{\mathrm{T}}$ and (5.97) with (5.102) we obtain

$$\mathrm{d}E_{\mathrm{tot}} = u^{\mathrm{T}}f^{\mathrm{nc}}\mathrm{d}t + u^{\mathrm{T}}f^{\mathrm{gyr}}\mathrm{d}t + u^{\mathrm{T}}W\lambda\mathrm{d}t + \frac{1}{2}(u^+ + u^-)^{\mathrm{T}}W\Lambda\mathrm{d}\eta \quad (7.15)$$

The gyroscopic forces have zero power, i.e. $u^{\mathrm{T}}f^{\mathrm{gyr}} = 0$ as motivated by (5.84). Moreover, the constraints are assumed to be scleronomic and it therefore holds according to (5.91) that $\gamma = W^{\mathrm{T}}u$, which gives

$$\mathrm{d}E_{\mathrm{tot}} = u^{\mathrm{T}}f^{\mathrm{nc}}\mathrm{d}t + \gamma^{\mathrm{T}}\lambda\mathrm{d}t + \frac{1}{2}(\gamma^+ + \gamma^-)^{\mathrm{T}}\Lambda\mathrm{d}\eta$$

$$\overset{(5.104)}{=} u^{\mathrm{T}}f^{\mathrm{nc}}\mathrm{d}t + \gamma^{\mathrm{T}}\lambda\mathrm{d}t + \frac{1}{2}\left((I + E)^{-1}(2\xi - (I - E)\delta)\right)^{\mathrm{T}}\Lambda\mathrm{d}\eta$$

$$= u^{\mathrm{T}}f^{\mathrm{nc}}\mathrm{d}t + \gamma^{\mathrm{T}}\lambda\mathrm{d}t + \xi^{\mathrm{T}}(I + E)^{-1}\Lambda\mathrm{d}\eta - \frac{1}{2}\delta^{\mathrm{T}}(I - E)(I + E)^{-1}\Lambda\mathrm{d}\eta$$

$$\overset{(5.97)+(5.103)}{=} u^{\mathrm{T}}f^{\mathrm{nc}}\mathrm{d}t + \xi^{\mathrm{T}}(I + E)^{-1}\mathrm{d}P - \frac{1}{2}\delta^{\mathrm{T}}(I - E)(I + E)^{-1}\Lambda\mathrm{d}\eta$$

$$= u^{\mathrm{T}}f^{\mathrm{nc}}\mathrm{d}t + \sum_{i \in I_N}\left(\frac{\xi_{Ni}\mathrm{d}P_{Ni}}{1 + e_{Ni}} + \frac{\xi_{Di}^{\mathrm{T}}\mathrm{d}P_{Di}}{1 + e_{Di}}\right) - \frac{1}{2}\delta^{\mathrm{T}}\Delta\Lambda\mathrm{d}\eta,$$
$$(7.16)$$

where $\Delta := (I - E)(I + E)^{-1}$ is the dissipation index matrix (5.105). Using (5.98) and (2.33), we obtain

$$\begin{aligned} \xi_{Ni}\mathrm{d}P_{Ni} &= -\Psi_{C_N}^*(\xi_{Ni})(\mathrm{d}t + \mathrm{d}\eta) = 0 \\ \xi_{Di}^{\mathrm{T}}\mathrm{d}P_{Di} &= -\Psi_{C_{Di}(\lambda_{Ni})}^*(\xi_{Di})\mathrm{d}t - \Psi_{C_{Di}(\Lambda_{Ni})}^*(\xi_{Di})\mathrm{d}\eta \leq 0, \end{aligned} \quad (7.17)$$

because of (2.37) and $\Psi_{C_N}^*(\xi_{Ni}) = \Psi_{\mathbb{R}^+}(\xi_{Ni}) = 0$ for admissible $\xi_{Ni} \geq 0$. Moreover, applying (5.91) to (5.103) and using the impact equation (5.100) gives

$$\delta := \gamma^+ - \gamma^- = W^{\mathrm{T}}(u^+ - u^-) = W^{\mathrm{T}}M^{-1}W\Lambda = G\Lambda, \quad (7.18)$$

in which we used the abbreviation

$$G := W^{\mathrm{T}}M^{-1}W, \quad (7.19)$$

which is known as the Delassus matrix. The matrix G is positive definite when W has full rank, because $M > 0$. The matrix G is only positive semi-definite if the matrix W does not have full rank, meaning that the generalised force directions of the contact forces are linearly dependent. However, we assume that the matrix W only contains the generalised force directions of unilateral constraints, and that these unilateral constraints do not constitute a bilateral constraint. It therefore holds that there exists no $\Lambda_N \neq 0$ such that $\boldsymbol{W}_N \boldsymbol{\Lambda}_N = \boldsymbol{0}$. The impact law requires that $\boldsymbol{\Lambda}_N \geq \boldsymbol{0}$. Hence, it holds that $\boldsymbol{\Lambda}_N^{\mathrm{T}} \boldsymbol{W}_N^{\mathrm{T}} \boldsymbol{M}^{-1} \boldsymbol{W}_N \boldsymbol{\Lambda}_N > 0$ for all $\boldsymbol{\Lambda}_N$ such that $\boldsymbol{\Lambda}_N \neq \boldsymbol{0}$ and $\boldsymbol{\Lambda}_N \geq \boldsymbol{0}$, even if the unilateral constraints are linearly dependent. Moreover, $\boldsymbol{\Lambda}_T \neq \boldsymbol{0}$ implies $\boldsymbol{\Lambda}_N \neq \boldsymbol{0}$. The inequality $\boldsymbol{\Lambda}^{\mathrm{T}} \boldsymbol{G} \boldsymbol{\Lambda} > 0$ therefore holds for all $\boldsymbol{\Lambda} \neq \boldsymbol{0}$ which obey the impact law (5.74), even if dependent unilateral constraints are considered

Using (7.18), we can put the last term in (7.16) in the quadratic form

$$\frac{1}{2} \boldsymbol{\delta}^{\mathrm{T}} \boldsymbol{\Delta} \boldsymbol{\Lambda} \mathrm{d}\eta = \frac{1}{2} \boldsymbol{\Lambda}^{\mathrm{T}} \boldsymbol{G} \boldsymbol{\Delta} \boldsymbol{\Lambda} \mathrm{d}\eta. \tag{7.20}$$

in which $\boldsymbol{G} \boldsymbol{\Delta}$ is a square matrix. The dissipation index matrix $\boldsymbol{\Delta}$ (5.105) is a diagonal matrix which is positive definite if the contacts are not purely elastic, i.e. $0 \leq e_{Ni} < 1$ and $|e_{Di}| < 1$ for all i. However, the matrix $\boldsymbol{G} \boldsymbol{\Delta}$ is in general not a positive definite matrix. We first present two propositions which guarantee the positive definiteness of the product of two square positive definite matrices.

Proposition 7.1. *Let $\boldsymbol{A} = \boldsymbol{A}^{\mathrm{T}} \in \mathbb{R}^{n \times n}$ and $\boldsymbol{B} = \boldsymbol{B}^{\mathrm{T}} \in \mathbb{R}^{n \times n}$ be two square symmetric matrices which commute, i.e. $\boldsymbol{A} \boldsymbol{B} = \boldsymbol{B} \boldsymbol{A}$. If $\boldsymbol{A} > 0$ and $\boldsymbol{B} > 0$, then it holds that $\boldsymbol{A} \boldsymbol{B} > 0$. If $\boldsymbol{A} > 0$ and $\boldsymbol{B} \geq 0$, then it holds that $\boldsymbol{A} \boldsymbol{B} \geq 0$.*

Proof: The matrix $\boldsymbol{A} \boldsymbol{B}$ is symmetric, because $\boldsymbol{A} \boldsymbol{B} = (\boldsymbol{A} \boldsymbol{B})^{\mathrm{T}}$. The matrix $\boldsymbol{A}$ is non-singular because it is positive definite and symmetric. Let λ be an eigenvalue of $\boldsymbol{A} \boldsymbol{B}$ with eigenvector $\boldsymbol{v}$,

$$\boldsymbol{A} \boldsymbol{B} \boldsymbol{v} = \lambda \boldsymbol{v},$$

which we can write as $\boldsymbol{B} \boldsymbol{v} = \lambda \boldsymbol{A}^{-1} \boldsymbol{v}$. The Rayleigh coefficient shows that the matrix $\boldsymbol{A} \boldsymbol{B}$ has real eigenvalues,

$$\lambda = \frac{\boldsymbol{v}^{\mathrm{T}} \boldsymbol{B} \boldsymbol{v}}{\boldsymbol{v}^{\mathrm{T}} \boldsymbol{A}^{-1} \boldsymbol{v}},$$

which are positive if $\boldsymbol{B} > 0$ and non-negative if $\boldsymbol{B} \geq 0$. Moreover, because $\boldsymbol{A} \boldsymbol{B}$ is symmetric and has positive or non-negative real eigenvalues it holds that

$$\boldsymbol{x}^{\mathrm{T}} \boldsymbol{A} \boldsymbol{B} \boldsymbol{x} \geq \lambda_{\min}(\boldsymbol{A} \boldsymbol{B}) \|\boldsymbol{x}\|^2,$$

where $\lambda_{\min}(\boldsymbol{A} \boldsymbol{B})$ is the smallest eigenvalue of $\boldsymbol{A} \boldsymbol{B}$. Hence, if $\boldsymbol{B} > 0$, then $\boldsymbol{x}^{\mathrm{T}} \boldsymbol{A} \boldsymbol{B} \boldsymbol{x} > 0$ for all $\boldsymbol{x} \neq \boldsymbol{0}$ and if $\boldsymbol{B} \geq 0$, then $\boldsymbol{x}^{\mathrm{T}} \boldsymbol{A} \boldsymbol{B} \boldsymbol{x} \geq 0$ for all $\boldsymbol{x}$.   □

Proposition 7.2. *Let $\boldsymbol{A} \in \mathbb{R}^{n \times n}$ be a symmetric positive definite matrix and $\boldsymbol{B} \in \mathbb{R}^{n \times n}$ be a diagonal positive definite matrix with the diagonal elements b_{ii} which fulfil $1 \geq b_{ii} \geq b_{\min} > 0$, $i = 1, \ldots, n$. If $1 - b_{\min} < \frac{1}{\mathrm{cond}(A)}$ then it holds that the matrix $\boldsymbol{A} \boldsymbol{B}$ is positive definite.*

Proof: The matrix $\boldsymbol{A} = \boldsymbol{A}^{\mathrm{T}} > 0$ has real positive eigenvalues and it therefore holds that

$$\boldsymbol{x}^{\mathrm{T}} \boldsymbol{A} \boldsymbol{x} \geq \lambda_{\min}(\boldsymbol{A}) \, \|\boldsymbol{x}\|^2 \,, \tag{7.21}$$

where $\lambda_{\min}(\boldsymbol{A})$ is the smallest eigenvalue of $\boldsymbol{A}$. Moreover, it holds that

$$
\begin{aligned}
\boldsymbol{x}^{\mathrm{T}} \boldsymbol{A}(\boldsymbol{I} - \boldsymbol{B})\boldsymbol{x} &\leq |\boldsymbol{x}^{\mathrm{T}} \boldsymbol{A}(\boldsymbol{I} - \boldsymbol{B})\boldsymbol{x}| \\
&\leq |\boldsymbol{A}| \, |\boldsymbol{I} - \boldsymbol{B}| \, \|\boldsymbol{x}\|^2 \\
&\leq \lambda_{\max}(\boldsymbol{A}) \, (1 - b_{\min}) \|\boldsymbol{x}\|^2 \,,
\end{aligned}
\tag{7.22}
$$

where $\lambda_{\max}(\boldsymbol{A})$ is the largest eigenvalue of $\boldsymbol{A}$ and $b_{\min} \leq 1$ is the smallest diagonal element of $\boldsymbol{B}$. Using the above inequalities, we deduce that

$$
\begin{aligned}
\boldsymbol{x}^{\mathrm{T}} \boldsymbol{A} \boldsymbol{B} \boldsymbol{x} &= \boldsymbol{x}^{\mathrm{T}}(\boldsymbol{A} - \boldsymbol{A}(\boldsymbol{I} - \boldsymbol{B}))\boldsymbol{x} \\
&\geq (\lambda_{\min}(\boldsymbol{A}) - \lambda_{\max}(\boldsymbol{A}) \, (1 - b_{\min})) \, \|\boldsymbol{x}\|^2 \,.
\end{aligned}
$$

Hence, if it holds that

$$1 - b_{\min} < \frac{\lambda_{\min}(\boldsymbol{A})}{\lambda_{\max}(\boldsymbol{A})} =: \frac{1}{\operatorname{cond}(\boldsymbol{A})} \,,$$

then it follows that $\boldsymbol{x}^{\mathrm{T}} \boldsymbol{A} \boldsymbol{B} \boldsymbol{x} > 0$ holds for all $\boldsymbol{x} \neq \boldsymbol{0}$. $\qquad \square$

Proposition 7.3. *Let $\boldsymbol{A} \in \mathbb{R}^{n \times n}$ be a symmetric positive definite matrix and $\boldsymbol{B} \in \mathbb{R}^{n \times n}$ be a diagonal positive definite matrix with the diagonal elements b_{ii} which fulfil $1 \geq b_{\max} \geq b_{ii} \geq b_{\min} > 0$, $i = 1, \dots, n$. If it holds that*

$$\frac{\operatorname{cond}(\boldsymbol{B}) - 1}{\operatorname{cond}(\boldsymbol{B}) + 1} < \frac{1}{\operatorname{cond}(\boldsymbol{A})} \,,$$

then the matrix $\boldsymbol{A} \boldsymbol{B}$ is positive definite.

Proof: The matrix $\boldsymbol{A} = \boldsymbol{A}^{\mathrm{T}} > 0$ has real positive eigenvalues and it therefore holds that

$$\boldsymbol{x}^{\mathrm{T}} \boldsymbol{A} \bar{b} \, \boldsymbol{x} \geq \bar{b} \lambda_{\min}(\boldsymbol{A}) \, \|\boldsymbol{x}\|^2 \,,$$

where $\lambda_{\min}(\boldsymbol{A})$ is the smallest eigenvalue of $\boldsymbol{A}$ and $\bar{b} = \frac{1}{2}(b_{\max} + b_{\min})$ is the mean of the largest and smallest diagonal element of $\boldsymbol{B}$. Moreover, it holds that

$$
\begin{aligned}
\boldsymbol{x}^{\mathrm{T}} \boldsymbol{A}(\bar{b}\boldsymbol{I} - \boldsymbol{B})\boldsymbol{x} &\leq |\boldsymbol{x}^{\mathrm{T}} \boldsymbol{A}(\bar{b}\boldsymbol{I} - \boldsymbol{B})\boldsymbol{x}| \\
&\leq |\boldsymbol{A}| \, |\bar{b}\boldsymbol{I} - \boldsymbol{B}| \, \|\boldsymbol{x}\|^2 \\
&\leq \lambda_{\max}(\boldsymbol{A}) \, (b_{\max} - \bar{b}) \|\boldsymbol{x}\|^2 \,,
\end{aligned}
$$

where $\lambda_{\max}(\boldsymbol{A})$ is the largest eigenvalue of $\boldsymbol{A}$. Using the above inequalities, we deduce that

$$
\begin{aligned}
\boldsymbol{x}^{\mathrm{T}} \boldsymbol{A} \boldsymbol{B} \boldsymbol{x} &= \boldsymbol{x}^{\mathrm{T}}(\boldsymbol{A}\bar{b} - \boldsymbol{A}(\bar{b}\boldsymbol{I} - \boldsymbol{B}))\boldsymbol{x} \\
&\geq \left(\bar{b}\lambda_{\min}(\boldsymbol{A}) - \lambda_{\max}(\boldsymbol{A}) \, (b_{\max} - \bar{b})\right) \|\boldsymbol{x}\|^2 \,. \\
&= \frac{1}{2} \left((b_{\max} + b_{\min})\lambda_{\min}(\boldsymbol{A}) - (b_{\max} - b_{\min})\lambda_{\max}(\boldsymbol{A})\right) \|\boldsymbol{x}\|^2 \,.
\end{aligned}
$$

Hence, if it holds that

$$\frac{b_{\max} - b_{\min}}{b_{\max} + b_{\min}} < \frac{\lambda_{\min}(\boldsymbol{A})}{\lambda_{\max}(\boldsymbol{A})} ,$$

which is equivalent to

$$\frac{\operatorname{cond}(\boldsymbol{B}) - 1}{\operatorname{cond}(\boldsymbol{B}) + 1} < \frac{1}{\operatorname{cond}(\boldsymbol{A})} ,$$

then it follows that $\boldsymbol{x}^{\mathrm{T}} \boldsymbol{A} \boldsymbol{B} \boldsymbol{x} > 0$ holds for all $\boldsymbol{x} \neq \boldsymbol{0}$. $\square$

In the theorems which are to be presented in the remainder of this work, the assumption will be made that the system is energetically consistent in the sense that the contact impulses $\boldsymbol{\Lambda}$ are dissipative, i.e.

$$\frac{1}{2}(\boldsymbol{\gamma}^+ + \boldsymbol{\gamma}^-)^{\mathrm{T}} \boldsymbol{\Lambda} \leq 0. \tag{7.23}$$

If no conditions on the restitution coefficients exist (other than $0 \leq e_{Ni} < 1$ and $|e_{Di}| < 1 \forall i$) and if friction is present, then the impact laws (5.74) and (5.75) can, under circumstances, lead to an energy increase. Such an energetic inconsistency has been reported by Kane and Levinson [84] (see also [23]). We will give sufficient conditions for the system parameters such that energetic consistency (7.23) can be guaranteed:

1. *Frictionless contacts*. If friction is absent then there exist only normal contact impulses. Let $I_A = \{i \in I_N \mid \Lambda_{Ni} > 0\}$ be the set of contact points which participate in the contact. The dissipation due to the contact impulses becomes

$$\begin{aligned}
\frac{1}{2}(\boldsymbol{\gamma}^+ + \boldsymbol{\gamma}^-)^{\mathrm{T}} \boldsymbol{\Lambda} &= \frac{1}{2} \sum_{i \in I_A} (\gamma_{Ni}^+ + \gamma_{Ni}^-) \Lambda_{Ni} \\
&= \frac{1}{2} \sum_{i \in I_A} (1 - e_{Ni}) \gamma_{Ni}^- \Lambda_{Ni}
\end{aligned} \tag{7.24}$$

Unilaterality of the contact implies $\Lambda_{Ni} \geq 0$, whereas kinematic compatibility requires $\gamma_{Ni} \leq 0$. Hence, if $e_{Ni} < 1$ for all i, then each term in the above sum is negative, implying that each normal contact dissipates energy.

2. *Commuting matrices* $\boldsymbol{G}$ *and* $\boldsymbol{\Delta}$. If the Delassus matrix $\boldsymbol{G} > 0$ is positive definite (no dependent constraints) and if $\boldsymbol{\Delta} > 0$ (it holds that $0 < e_{Ni} < 1$ and $|e_{Di}| < 1$), then the matrix $\boldsymbol{G}\boldsymbol{\Delta}$ is positive definite (see Proposition 7.1) and energetic consistency is guaranteed. Moreover, if one of these matrices is positive definite and the other is positive semi-definite while they still commute, then energetic consistency is still guaranteed. There are some special cases for which these matrices commute:

a) *Equal restitution coefficients.* If all restitution coefficients are equal, i.e. $e_{Ni} = e_{Di} = e$ for all i, then the matrix $\boldsymbol{\Delta}$ simplifies to $\frac{1-e}{1+e}\boldsymbol{I}$, which commutes with $\boldsymbol{G}$. This case is equivalent to Moreau's global dissipation index and includes the special case of completely inelastic contacts $e = 0$.

b) *Decoupling.* If $\boldsymbol{G}$ is diagonal, then the contacts are decoupled and $\boldsymbol{G}$ and $\boldsymbol{\Delta}$ commute as they are both diagonal.

3. *Small restitution coefficients.* The smallest diagonal element of $\boldsymbol{\Delta}$ is $(1 - e_{\max})/(1 + e_{\max})$ where $e_{\max}$ is the largest diagonal element of $\boldsymbol{E}$. Using Proposition 7.1 we derive that if the restitution coefficients are small in the sense that

$$1 - \Delta_{\min} = \frac{2e_{\max}}{1 + e_{\max}} < \frac{1}{\mathrm{cond}(\boldsymbol{G}(\boldsymbol{q}))} \quad \forall \boldsymbol{q} \in \mathcal{K}, \qquad (7.25)$$

then the matrix $\boldsymbol{G}\boldsymbol{\Delta}$ is positive definite and energetic consistency is therefore guaranteed.

4. *Almost equal restitution coefficients.* Using Proposition 7.3 we derive that if the restitution coefficients are almost equal in the sense that

$$\frac{\mathrm{cond}(\boldsymbol{\Delta}) - 1}{\mathrm{cond}(\boldsymbol{\Delta}) + 1} < \frac{1}{\mathrm{cond}(\boldsymbol{G}(\boldsymbol{q}))} \quad \forall \boldsymbol{q} \in \mathcal{K}, \qquad (7.26)$$

then the matrix $\boldsymbol{G}\boldsymbol{\Delta}$ is positive definite and energetic consistency is therefore guaranteed. This case includes the special case 2(a) of equal restitution coefficients.

Looking again at the differential measure of the total energy (7.16), we realise that if the above conditions are met then all terms related to the contact forces and impulses are dissipative or passive. Moreover, if we consider not purely elastic contacts, then nonzero contact impulses $\boldsymbol{\Lambda}$ strictly dissipate energy.

We define the following nonlinear functionals $\mathbb{R}^n \to \mathbb{R}$ on $\boldsymbol{u} \in \mathbb{R}^n$:

- $D_{\boldsymbol{q}}^{\mathrm{nc}}(\boldsymbol{u}) := -\boldsymbol{u}^{\mathrm{T}}\boldsymbol{f}^{\mathrm{nc}}(\boldsymbol{q}, \boldsymbol{u})$ is the (rate of) dissipation function[1] of the smooth non-conservative forces,
- $D_{\boldsymbol{q}}^{\lambda_D}(\boldsymbol{u}) := \sum_{i \in I_N} \frac{1}{1+e_{Di}} \Psi^*_{C_{Di}(\lambda_{Ni})}(\boldsymbol{\xi}_{Di}(\boldsymbol{q}, \boldsymbol{u})) \geq 0$ is the dissipation function of the frictional contact forces,
- $D_{\boldsymbol{q}}^{\Lambda_D}(\boldsymbol{u}) := \sum_{i \in I_N} \frac{1}{1+e_{Di}} \Psi^*_{C_{Di}(\Lambda_{Ni})}(\boldsymbol{\xi}_{Di}(\boldsymbol{q}, \boldsymbol{u})) \geq 0$ is the dissipation function of the frictional contact impulses.

The dissipation functions $D_{\boldsymbol{q}}^{\lambda_D}(\boldsymbol{u})$ and $D_{\boldsymbol{q}}^{\Lambda_D}(\boldsymbol{u})$ are non-negative because they consist of convex support functions whereas *a priori* nothing can be said of the dissipation function $D_{\boldsymbol{q}}^{\mathrm{nc}}(\boldsymbol{u})$ of the smooth non-conservative forces. For

---

[1] Usually, the term 'dissipation function' is used to denote the pseudo-potential of the non-conservative forces. Here, we use the term dissipation function for the function $D_{\boldsymbol{q}}^{\mathrm{nc}}(\boldsymbol{u})$, which gives the rate of dissipation.

non-impulsive motion, it holds that $\gamma_D = \gamma_D^+ = \gamma_D^-$ and $\boldsymbol{\xi}_D = (1 + e_D)\gamma_D$. Due to the fact that the support function is positively homogeneous it follows that

$$D_q^{\lambda_D}(\boldsymbol{u}) = \sum_{i \in I_N} \varPsi^*_{C_{Di}(\lambda_{Ni})}(\gamma_{Di}(\boldsymbol{q}, \boldsymbol{u})) - \sum_{i \in I_N} -\lambda_{Di}\gamma_{Di}(\boldsymbol{q}, \boldsymbol{u}), \qquad (7.27)$$

from which we see that the dissipation function of the frictional contact forces does not depend on the restitution coefficient e_D. The above dissipation functions are of course functions of $(\boldsymbol{q}, \boldsymbol{u})$, but we write them as nonlinear functionals on $\boldsymbol{u}$ so that we can speak of the zero set of the functional $D_q(\boldsymbol{u})$:

$$D_q^{-1}(0) = \{\boldsymbol{u} \in \mathbb{R}^n \mid D_q(\boldsymbol{u}) = 0\}. \qquad (7.28)$$

We can now decompose the differential measure dE_{tot} in a Lebesgue part and an atomic part

$$dE_{\text{tot}} = \dot{E}_{\text{tot}}dt + (E_{\text{tot}}^+ - E_{\text{tot}}^-)d\eta, \qquad (7.29)$$

with

$$\begin{aligned} \dot{E}_{\text{tot}} &= \boldsymbol{u}^{\text{T}}\boldsymbol{f}^{\text{nc}} - \sum_{i \in I_N} \frac{1}{1 + e_{Di}}\varPsi^*_{C_{Di}(\lambda_{Ni})}(\boldsymbol{\xi}_{Di}) \\ &= -D_q^{\text{nc}}(\boldsymbol{u}) - D_q^{\lambda_D}(\boldsymbol{u}) \end{aligned} \qquad (7.30)$$

and

$$\begin{aligned} E_{\text{tot}}^+ - E_{\text{tot}}^- &= -\sum_{i \in I_N} \left(\frac{1}{1 + e_{Di}}\varPsi^*_{C_{Di}(\Lambda_{Ni})}(\boldsymbol{\xi}_{Di})\right) - \frac{1}{2}\boldsymbol{\Lambda}^{\text{T}}\boldsymbol{G}\boldsymbol{\Delta}\boldsymbol{\Lambda} \\ &= -D_q^{\Lambda_D}(\boldsymbol{u}) - \frac{1}{2}\boldsymbol{\Lambda}^{\text{T}}\boldsymbol{G}\boldsymbol{\Delta}\boldsymbol{\Lambda}. \end{aligned} \qquad (7.31)$$

We see that the dissipation rate during non-impulsive motion is governed by the dissipation of the non-conservative smooth forces and the dissipation of the (Lebesgue measurable) friction forces, whereas the jumps in the total energy are due to the dissipation of the frictional impulses and the dissipation caused by impact. Energetic consistency of the system requires that $E_{\text{tot}}^+ - E_{\text{tot}}^- \leq 0$. A sufficient condition for energetic consistency is the positiveness of the term $\frac{1}{2}\boldsymbol{\Lambda}^{\text{T}}\boldsymbol{G}\boldsymbol{\Delta}\boldsymbol{\Lambda}$. Sufficient conditions for $\frac{1}{2}\boldsymbol{\Lambda}^{\text{T}}\boldsymbol{G}\boldsymbol{\Delta}\boldsymbol{\Lambda} \geq 0$ have been derived in the conditions 1-4 on page 159.

Remark: In Section 5.4 (Equation (5.81) or (7.6)), we assumed that the the kinetic energy is a purely quadratic form in the generalised velocities: $T = \frac{1}{2}\boldsymbol{u}^{\text{T}}\boldsymbol{M}(\boldsymbol{q})\boldsymbol{u}$. Subsequently, the analysis in Section 5.4 has been based on this assumption, as are the expressions for the differential measure of the total mechanical energy function in (7.29)-(7.31). The latter expression will prove to be of central importance in exploiting E_{tot} as a Lyapunov function candidate in the next section. One could consider the more general case of a kinetic energy of the form

$$T(\boldsymbol{q}, \boldsymbol{u}) = T_2(\boldsymbol{q}, \boldsymbol{u}) + T_1(\boldsymbol{q}, \boldsymbol{u}) + T_0(\boldsymbol{q}), \qquad (7.32)$$

with

$$T_2(\boldsymbol{q}, \boldsymbol{u}) := \frac{1}{2}\boldsymbol{u}^{\mathrm{T}} M(\boldsymbol{q})\boldsymbol{u},$$
$$T_1(\boldsymbol{q}, \boldsymbol{u}) := \boldsymbol{m}^T(\boldsymbol{q})\boldsymbol{u}. \qquad (7.33)$$

In this more general case, one should use an adapted form of the Lyapunov function candidate, which involves the function $\bar{E}_{\mathrm{tot}} = T_2(\boldsymbol{q}, \boldsymbol{u}) - T_0(\boldsymbol{q}) + Q(\boldsymbol{q})$ (instead of $E_{\mathrm{tot}} = T_2(\boldsymbol{q}, \boldsymbol{u}) + Q(\boldsymbol{q})$). The function $\bar{E}_{\mathrm{tot}}$ is called the Jacobi-Painlevé generalised energy [129]. Under the assumption that T_2, T_1 and T_0 do not depend explicitly on time, similar expressions for $\mathrm{d}\bar{E}_{\mathrm{tot}}$ can be derived (as in (7.29)-(7.31) for $\mathrm{d}E_{\mathrm{tot}}$). For the sake of simplicity, we restrict ourselves to the case of a purely quadratic kinetic energy in the next section. However, we mention that, using the Jacobi-Painlevé generalised energy function, the more general case (i.e. for systems characterised by a kinetic energy as in (7.32)) can be treated in an analogous fashion.

7.2 Stability Results for Mechanical Systems

The stability of the equilibrium position of the bouncing ball example has been proven in Section 6.5 by using Theorem 6.23a. Mechanical systems, for which the generalised position $\boldsymbol{q}(t)$ are absolutely continuous and the generalised velocities $\boldsymbol{u}(t)$ are of locally bounded variation, are of the form (6.33). We can therefore use Theorem 6.25a to prove the stability of an equilibrium point of a general mechanical system (5.96), taking the total mechanical energy function (7.7) as a Lyapunov candidate function.

Theorem 7.4 (Lyapunov Stability of an Equilibrium Position at the Origin). *Let $\boldsymbol{q}^* = \boldsymbol{0} \in \mathcal{K}$ be an equilibrium position of (5.96) and let the total potential energy Q as in (7.4) be a locally positive definite function. If $D_{\boldsymbol{q}}^{\mathrm{nc}}(\boldsymbol{u}) + D_{\boldsymbol{q}}^{\lambda_D}(\boldsymbol{u}) \geq 0 \; \forall (\boldsymbol{q}, \boldsymbol{u}) \in B_h \cap (\mathcal{K} \times \mathbb{R}^n)$ for some $h > 0$, and if the system is energetically consistent, then the equilibrium position is stable.*

Proof: Consider the Lyapunov function candidate $V = E_{\mathrm{tot}}$ given by (7.7) which is a locally positive definite function with $V(\boldsymbol{0}, \boldsymbol{0}) = 0$ and $V(\boldsymbol{q}, \boldsymbol{u}) > 0 \; \forall (\boldsymbol{q}, \boldsymbol{u}) \in (\mathcal{U} \times \mathbb{R}^n) \setminus (\boldsymbol{0}, \boldsymbol{0})$. The differential measure $\mathrm{d}V$ consists of a Lebesgue part and an atomic part

$$\mathrm{d}V = \dot{V}\mathrm{d}t + (V^+ - V^-)\mathrm{d}\eta. \qquad (7.34)$$

Using $D_q^{nc}(u) + D_q^{\lambda_D}(u) \geq 0$ and (7.29) it follows that $\dot{V} \leq 0 \ \forall(q, u) \in B_h \cap (\mathcal{K} \times \mathbb{R}^n)$ for some $h > 0$. Moreover, it holds that $V^+ - V^- \leq 0$ because of the system is assumed to be energetically consistent. The total mechanical energy V does therefore not increase along solution curves of the system in the neighbourhood of the origin. Lyapunov stability in the sense of Definition 6.13 follows from Theorem 6.25a. $\qquad\qquad\square$

Notice that it is not useful to state Theorem 6.25b or Theorem 6.23c for mechanical systems using the total energy function as Lyapunov function V, because the dissipation is absent for $u = 0$ and $dV(q, 0)$ therefore vanishes for all $q \in \mathcal{K}$. Hence, attractivity has to be proven separately using LaSalle-type of arguments.

Up to now, we only considered the stability of an equilibrium position $q^* = 0$. If we are interested in the stability of an equilibrium position q^* which is not located at the origin, then we we have to consider an altered potential energy function $U_a(q)$ such that the altered total potential energy function

$$Q_a(q) = U_a(q) + \sum_{i \in I_G} \pi_{Ni}(g_{Ni}(q)), \qquad (7.35)$$

fulfills the conditions

$$Q_a(q^*) = 0, \quad Q_a(q) > 0 \ \forall q \in \mathcal{U}\backslash\{q^*\} \qquad (7.36)$$

where $\mathcal{U}$ is some neighbourhood of q^*. The latter condition is a local positive definiteness requirement for $Q_a(q)$ with respect to the point q^*. Using $Q_a(q)$, we can construct the Lyapunov function $V_{q^*}(q, u) = T(q, u) + Q_a(q)$ which can be used to prove the stability of q^*. The big question is: how do we find an altered total potential energy function $U_a(q)$ which fulfills (7.36)? We are not able to give one function $U_a(q)$ that can be used to prove the stability of equilibrium positions of a general mechanical system (5.96), but we will study some special cases.

If the total potential energy function $Q(q)$ (7.4) has a local minimum at q^* and is locally strictly convex, then the altered total potential energy function $Q_a(q)$ based on the shifted potential energy function $U_a(q)$, with

$$U_a(q) = U(q) - U(q^*), \qquad (7.37)$$

fulfills the conditions (7.36). Using $dU_a = dU$ it follows that the differential measure of $V_{q^*}(q, u) = T(q, u) + Q_a(q)$ agrees with the differential measure of the total mechanical energy function V, i.e. $dV_{q^*} = dV = dE_{tot}$, and the stability therefore depends on the sign of the term $D_q^{nc}(u) + D_q^{\lambda_D}(u)$. This special case may occur for instance when only frictionless contacts are considered [33]. An example is the bouncing ball example, but with a floor on a non-zero height.

Another special case occurs when the system only contains frictional bilateral contacts with an associated Coulomb friction law (5.50) with a constant normal contact force F_N, i.e. the system is described by the differential inclusion

$$M(q)\dot{u} - h(q, u) = W_D(q)\lambda_D, \qquad -\lambda_D \in \Psi^*_{C_D}(\gamma_D), \qquad (7.38)$$

where C_D is a constant set. Generally, an equilibrium q^* is accompanied by a nonzero static friction force λ^*_D which makes equilibrium with the vector h of smooth forces:

$$-h(q^*, 0) = W_D(q^*)\lambda^*_D. \qquad (7.39)$$

The vector h (5.86) consists of non-conservative forces $f^{nc}(q, u)$, which are assumed to vanish for $u = 0$, forces $f^{gyr}(q, u) + T_{,q}(q, u)$ derived from the kinetic energy which also vanish at an equilibrium point and forces $U_{,q}(q)$ derived from the potential energy. The force equilibrium (7.39) therefore simplifies to

$$U_{,q}(q^*)^T = W_D(q^*)\lambda^*_D. \qquad (7.40)$$

In order to prove the (conditional) stability of an equilibrium position $q^* \neq 0$ of (7.38), we consider the altered potential energy function

$$U_a(q) = U(q) - (q - q^*)^T W_D(q^*)\lambda^*_D - U(q^*). \qquad (7.41)$$

If $U(q)$ consists of a linear and quadratic form (e.g. the system is linear)

$$U(q) = \frac{1}{2}q^T K q - q^T f_0 \qquad (7.42)$$

then it holds that $U_{,q}(q^*)^T = Kq^* - f_0$ which gives

$$\begin{aligned}
U_a(q) &= \frac{1}{2}q^T K q - q^T f_0 - (q - q^*)^T W_D(q^*)\lambda^*_D - \frac{1}{2}q^{*T} K q^* + q^{*T} f_0 \\
&= \frac{1}{2}q^T K q - q^T f_0 - (q - q^*)^T (K q^* - f_0) - \frac{1}{2}q^{*T} K q^* + q^{*T} f_0 \\
&= \frac{1}{2}q^T K q - q^T K q^* + \frac{1}{2}q^{*T} K q^* \\
&= \frac{1}{2}(q - q^*)^T K (q - q^*).
\end{aligned} \qquad (7.43)$$

This shows that the altered potential energy function of a linear system with $K > 0$ is positive definite with respect to q^*, i.e. (7.36) is satisfied. Moreover, it holds that $U_a(q) = U(q - q^*)$ if $f_0 = 0$. If we now consider the differential measure of the Lyapunov function candidate

$$\begin{aligned}
V_{q^*}(q, u) &= T(q, u) + U_a(q) \\
&= E_{tot}(q, u) - (q - q^*)^T W_D(q^*)\lambda^*_D - U(q^*),
\end{aligned} \qquad (7.44)$$

then the additional terms only affect the Lebesgue part $\dot{V}_{q^*}$ of the differential measure dV_{q^*}:

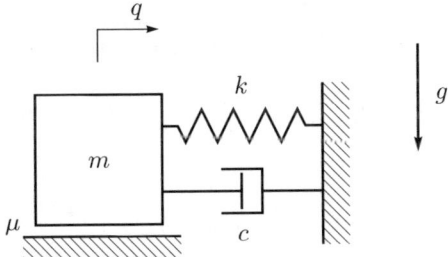

Fig. 7.1. One-DOF system with friction.

$$\dot{V}_{q^*} = \dot{E}_{\text{tot}} - u^{\mathrm{T}} W_D(q^*) \lambda_D^*$$
$$= -D_q^{\text{nc}}(u) - D_q^{\lambda_D}(u) - u^{\mathrm{T}} W_D(q^*) \lambda_D^* \tag{7.45}$$

If W_D is constant, then $\dot{V}_{q^*}$ simplifies to

$$\dot{V}_{q^*} = -D_q^{\text{nc}}(u) - D_q^{\lambda_D}(u) - \gamma_D^{\mathrm{T}} \lambda_D^*. \tag{7.46}$$

It holds that $\gamma_D \in \partial \Psi_{C_D}(-\lambda_D)$. Using (2.26) it immediately follows that $0 \geq -\gamma_D^{\mathrm{T}}(\lambda_D^* - \lambda_D)$ for all $-\lambda_D^* \in C_D$. The dissipation of the friction forces

$$D_q^{\lambda_D}(u) + \gamma_D^{\mathrm{T}} \lambda_D^* = \gamma_D^{\mathrm{T}}(\lambda_D^* - \lambda_D) \geq 0 \tag{7.47}$$

is therefore always non-negative.

Theorem 7.5 (Lyapunov Stability of an Equilibrium Position). *Let $q^* \in \mathcal{K}$ be an equilibrium position of the differential inclusion (7.38) and let the altered potential energy*

$$U_a(q) = U(q) - (q - q^*)^{\mathrm{T}} W_D(q^*) \lambda_D^* - U(q^*)$$

be a locally positive definite function with respect to q^*, i.e. (7.36) is satisfied. If $D_q^{\text{nc}}(u) + D_q^{\lambda_D}(u) + u^{\mathrm{T}} W_D(q^*) \lambda_D^* \geq 0 \,\forall (q, u) \in B_h \cap (\mathcal{K} \times \mathbb{R}^n)$ for some $h > 0$, and if the system is energetically consistent, then the equilibrium position q^* is stable.

Proof: The proof is identical to the proof of Theorem 7.4, but we now use the Lyapunov function $V_{q^*}(q, u)$ given by (7.44). □

If we are able to prove the stability of each equilibrium position in an equilibrium set $\mathcal{E}_q$, then stability of the equilibrium set as a whole follows from Theorem 6.28, or alternatively from Theorem 6.27 using a global Lyapunov function (6.57).

We now discuss a one-degree-of-freedom example which illustrates the use of Theorem 7.5. Consider a block with mass m on a floor which is attached to a wall by a spring k and a dashpot c (see Figure 7.1). The spring is unstressed for $q = 0$. Associated Coulomb friction with friction coefficient μ is present between the block and the floor. The contact force F_N has the constant value mg. The dynamics of the system is described by the differential inclusion

$$\dot{q} = u, \quad m\dot{u} + cu + kq = \lambda_T, \qquad -\lambda_T \in \mu F_N \,\text{Sign}(u) \quad \text{a.e.} \qquad (7.48)$$

and has the equilibrium set

$$\mathcal{E}_q = \{q \in \mathbb{R} \mid kq \in -\mu F_N \,\text{Sign}(0)\} = \left[-\frac{\mu mg}{k}, \frac{\mu mg}{k}\right]. \qquad (7.49)$$

An equilibrium position q^* is accompanied by a friction force $\lambda_T^* = kq^* \in -C_T$ with $C_T = [-\mu mg, \mu mg]$. We consider the stability of an equilibrium position $q^* \in \mathcal{E}_q$ using the Lyapunov function

$$\begin{aligned} V_{q^*}(q, u) &= T(u) + U_a(q) \\ &= \frac{1}{2}m\dot{u}^2 + \frac{1}{2}k(q - q^*)^2 \end{aligned} \qquad (7.50)$$

where $U_a(q)$ is the altered potential energy function (7.41) which can be simplified as in (7.43). It holds that $V_{q^*}(q^*, 0) = 0$ and $V_{q^*}(q, u) > 0$ for all $(q, u) \neq (q^*, 0)$ and the Lyapunov function is therefore positive definite with respect to the equilibrium point $(q^*, 0)$. The time derivative of V_{q^*} along solution curves of the system is given by

$$\begin{aligned} \dot{V}_{q^*} &= m u \dot{u} + k(q - q^*)u \\ &= -cu^2 + \lambda_T u - kq^* u \\ &= -cu^2 + (\lambda_T - \lambda_T^*)u \\ &= -cu^2 - \mu mg|u| - \lambda_T^* u, \quad |\lambda_T^*| \leq \mu mg \\ &\leq 0 \end{aligned} \qquad (7.51)$$

and is non-positive for $c \geq 0$ for all equilibrium positions $q^* \in \mathcal{E}_q$. Consequently, each equilibrium position $q^* \in \mathcal{E}_q$ is stable. Stability of $\mathcal{E}_q$ as a whole follows from Theorem 6.28. Alternatively, we can use Theorem 6.26 to prove the stability of the equilibrium set taking a Lyapunov function V which is zero on the whole equilibrium set (see [170]). To this end, we introduce a coordinate transformation $z = \begin{bmatrix} z_1 & z_2 \end{bmatrix}^T = \begin{bmatrix} \sqrt{k}q & \sqrt{m}u \end{bmatrix}^T$. The equilibrium set $\mathcal{E}$ in z-coordinates is given by the convex set

$$\mathcal{E} = \left\{z \in \mathbb{R}^2 \mid |z_1| \leq \frac{\mu F_N}{\sqrt{k}}, \ z_2 = 0\right\}. \qquad (7.52)$$

Consider the Lyapunov candidate function V, as defined by

$$V(z) = \frac{1}{2} \,\text{dist}_{\mathcal{E}}^2(z), \qquad (7.53)$$

where Definition 2.32 has been used. This Lyapunov function V has the desirable property that $V(z) = 0$ for $z \in \mathcal{E}$, which allows for the stability analysis of the equilibrium set as a whole. Using $\dot{V} = (z - \text{prox}_{\mathcal{E}}(z))^{\text{T}} \dot{z}$ it is shown in [170] that $\dot{V} \leq 0$ along solution curves of the system for $c \geq 0$. Consequently, Lyapunov stability of the equilibrium set $\mathcal{E}$ follows from Theorem 6.26 for $c \geq 0$. Moreover, if we evaluate the distance function using $\text{prox}_{\mathcal{E}}(z) = [q_p\ 0]^{\text{T}}$, with $q_p = \text{prox}_{\mathcal{E}_q}(q)$, as

$$\text{dist}_{\mathcal{E}}(z) = \|z - \text{prox}_{\mathcal{E}}(z)\|$$
$$= \sqrt{k(q - q_p)^2 + mu^2}, \tag{7.54}$$

then we see that the Lyapunov function V (7.53) has the form

$$V(z) = \frac{1}{2}\,\text{dist}_{\mathcal{E}}^2(z)$$
$$= \frac{1}{2}mu^2 + \frac{1}{2}k(q - q_p)^2 \tag{7.55}$$
$$= \min_{q_p \in \mathcal{E}_q}\left(\frac{1}{2}mu^2 + \frac{1}{2}k(q - q_p)^2\right),$$

which we recognise to be a Lyapunov function of the form (6.57)

$$V(q, u) = \min_{q^* \in \mathcal{E}_q} V_{q^*}(q, u), \tag{7.56}$$

where $V_{q^*}(q, u)$ is given by (7.50). It is also of interest to consider the system in z-coordinates in the form

$$\dot{z} \in -\mathcal{F}(z), \tag{7.57}$$

with

$$\mathcal{F}(z) = \begin{bmatrix} -\sqrt{k}u \\ \frac{1}{\sqrt{m}}(cu + kq + \mu F_N\,\text{Sign}(u)) \end{bmatrix}. \tag{7.58}$$

It holds that

$$(z - z^*)^{\text{T}}(\mathcal{F}(z) - \mathcal{F}(z^*))$$
$$= -k(q - q^*)(u - u^*) + (u - u^*)(cu + kq + \mu F_N\,\text{Sign}(u)$$
$$\quad - cu^* - kq^* - \mu F_N\,\text{Sign}(u^*)) \tag{7.59}$$
$$= (u - u^*)(cu + \mu F_N\,\text{Sign}(u) - cu^* - \mu F_N\,\text{Sign}(u^*))$$
$$\geq 0$$

because $\mu F_N\,\text{Sign}(u) + cu$ is maximal monotone in u for $c \geq 0$ and $\mu F_N \geq 0$. Consequently, the operator $\mathcal{F}(z)$ is maximal monotone, but not strictly maximal monotone because of its dependence on q. Stability of the equilibrium set $\mathcal{E}$ immediately follows from Theorem 6.29. Moreover, the system is incrementally stable (see Theorem 6.30). Similar stability results for Lagrangian systems described by differential inclusions with maximal monotone operators can be found in [3].

7.3 Attractivity of Equilibrium Sets

In this section, we will investigate the attractivity properties of equilibrium
sets of Lagrangian mechanical systems with impact and/or friction using a
generalisation of LaSalle's invariance principle (Theorem 6.31). We first discuss attractivity of equilibrium sets in systems with frictional unilateral constraints in Subsection 7.3.1 and subsequently discuss systems with frictional
bilateral constraints as a special case in Subsection 7.3.2.

7.3.1 Systems with Frictional Unilateral Constraints

As stated before, the type of systems under investigation described by (5.96),
(5.98) may exhibit multiple equilibrium sets. Here, we will study the attractivity properties of a specific given equilibrium set $\mathcal{E}$ satisfying (5.109). The
set $\mathcal{E}$ may be the total equilibrium set of the system or just only a simply
connected subset of the total equilibrium set. By $\boldsymbol{q}_e$ we denote an equilibrium
point of the system with unilateral *frictionless* contacts

$$\boldsymbol{M}(\boldsymbol{q})\mathrm{d}\boldsymbol{u} - \boldsymbol{h}(\boldsymbol{q}, \boldsymbol{u})\mathrm{d}t - \boldsymbol{W}_N(\boldsymbol{q})\mathrm{d}\boldsymbol{P}_N = \boldsymbol{0}, \tag{7.60}$$

from which follows that the equilibrium is determined by the inclusion

$$\boldsymbol{h}(\boldsymbol{q}_e, \boldsymbol{0}) - \sum_{i \in I_G} \boldsymbol{w}_{Ni}(\boldsymbol{q}_e)\partial\Psi_{C_N}^*(g_{Ni}(\boldsymbol{q}_e)) \ni \boldsymbol{0} \tag{7.61}$$

or

$$\boldsymbol{h}(\boldsymbol{q}_e, \boldsymbol{0}) - \sum_{i \in I_N} \boldsymbol{w}_{Ni}(\boldsymbol{q}_e)\partial\Psi_{C_N}^*(\underbrace{\gamma_{Ni}(\boldsymbol{q}_e, \boldsymbol{0})}_{=0}) \ni \boldsymbol{0}, \tag{7.62}$$

which is equivalent to

$$\boldsymbol{h}(\boldsymbol{q}_e, \boldsymbol{0}) + \boldsymbol{W}_N(\boldsymbol{q}_e)\mathbb{R}^+ \ni \boldsymbol{0}, \qquad \boldsymbol{W}_N = \{\boldsymbol{w}_{Ni}\}, \ i \in I_N. \tag{7.63}$$

We assume that the equilibrium position $\boldsymbol{q}_e$ is a local minimum of the total
potential energy function $Q(\boldsymbol{q})$ (7.4), i.e.

$$Q(\boldsymbol{q}) - Q(\boldsymbol{q}_e) = \begin{cases} 0 & \boldsymbol{q} = \boldsymbol{q}_e \\ >0 & \forall \boldsymbol{q} \in \mathcal{U} \backslash \{\boldsymbol{q}_e\} \end{cases}, \qquad \boldsymbol{0} \notin \partial Q(\boldsymbol{q}), \forall \boldsymbol{q} \in \mathcal{U} \backslash \{\boldsymbol{q}_e\}. \tag{7.64}$$

The subset $\mathcal{U}$ is assumed to enclose the equilibrium set $\mathcal{E}_q$ of the equilibrium
positions under investigation, where $\mathcal{E}_q = \{\boldsymbol{q} \mid (\boldsymbol{q}, \boldsymbol{0}) \in \mathcal{E}\}$. Notice that the
equilibrium point $\boldsymbol{q}_e$ of the system without friction is also an equilibrium point
of the system with friction, $\boldsymbol{q}_e \in \mathcal{E}_q$. In case the system does exhibit multiple
equilibrium sets, the attractivity of $\mathcal{E}$ will be only local for obvious reasons.

In the following, we will use the total mechanical energy (7.7) shifted with
respect to $\boldsymbol{q}_e$ as Lyapunov candidate function $V(\boldsymbol{q}, \boldsymbol{u}) = E_{\text{tot}}(\boldsymbol{q}, \boldsymbol{u}) - Q(\boldsymbol{q}_e)$.
We now formulate a theorem which states sufficient conditions under which
an equilibrium set $\mathcal{E}$ can be proven to be (locally) attractive.

Theorem 7.6 (Attractivity of the equilibrium set).
Consider system (5.96), with constitutive laws (5.98) and with $M(q) = M^{\mathrm{T}}(q) > 0$. If

1. *the equilibrium position q_e is a local minimum of the total potential energy $Q(q)$ and $Q(q)$ has a non-vanishing generalised gradient for all $q \in \mathcal{U} \backslash \{q_e\}$, i.e. $0 \notin \partial Q(q) \forall q \in \mathcal{U} \backslash \{q_e\}$, and the equilibrium set $\mathcal{E}_q$ is contained in $\mathcal{U}$, i.e. $\mathcal{E}_q \subset \mathcal{U}$,*
2. *$D_q^{\mathrm{nc}}(u) = -u^{\mathrm{T}} f^{\mathrm{nc}} \geq 0$, i.e. the smooth non-conservative forces are dissipative, and $f^{\mathrm{nc}} = 0$ for $u = 0$,*
3. *there exists a non-empty set $I_C \subset I_G$ and an open neighbourhood $\mathcal{V} \subset \mathbb{R}^n \times \mathbb{R}^n$ of the equilibrium set, such that $\dot{\gamma}_{Ni}(q, u) < 0$ (a.e.) for $\forall i \in I_C \backslash I_N$ and $(q, u) \in \mathcal{V}$ (see (5.89) for the definition of the index sets),*
4. *$D_q^{\mathrm{nc}^{-1}}(0) \cap D_q^{\lambda_{DC}^{-1}}(0) \cap \ker W_{NC}^{\mathrm{T}}(q) = \{0\} \quad \forall q \in \mathcal{C}$ with*

$$g_{NC} = \{g_{Ni}\}, W_{NC} - \{w_{Ni}\} \text{ for } i \in I_C \text{ as defined in 3.,}$$

$$\mathcal{C} = \{q \mid g_{NC}(q) = 0\}, \qquad D_q^{\lambda_{DC}}(u) = \sum_{i \in I_C \cap I_N} \Psi_{C_{Di}(\lambda_{Ni})}^*(\gamma_{Di}(q, u)),$$

5. *$0 \leq e_{Ni} < 1, |e_{Di}| < 1 \forall i \in I_G$,*
6. *$\mathcal{E} \subset \Omega_{\rho^*}$ in which the set Ω_{ρ^*} is the largest level set of $V(q, u) = E_{\mathrm{tot}}(q, u) - Q(q_e)$, that is contained in $\mathcal{V}$ and $\mathcal{Q} = \{(q, u) \in \mathbb{R}^n \times \mathbb{R}^n \mid q \in \mathcal{U}\}$, i.e.*

$$\rho^* = \max_{\{\rho: \Omega_\rho \subset (\mathcal{V} \cap \mathcal{Q})\}} \rho,$$

7. *the system is energetically consistent,*
8. *$\mathcal{E}$ is the largest set of equilibrium points in Ω_{ρ^*},*
9. *each limit set in Ω_{ρ^*} is positively invariant,*

then the equilibrium set $\mathcal{E} \supset (q_e, 0)$ is locally attractive and Ω_{ρ^*} is a conservative estimate for the region of attraction.

Proof: Note that V is positive definite around the equilibrium point $(q, u) = (q_e, 0)$ due to condition 1 in the theorem. The differential measure $dV = dE_{\mathrm{tot}}$ has been computed in Section 7.1. We can decompose the differential measure dV in a Lebesgue part and an atomic part:

$$dV = \dot{V}dt + (V^+ - V^-)d\eta, \tag{7.65}$$

with (7.30)

$$\dot{V} = -D_q^{\mathrm{nc}}(u) - D_q^{\lambda_D}(u), \tag{7.66}$$

which is non-positive due to condition 2, and (7.31) gives

$$V^+ - V^- = -D_q^{\Lambda_D}(u) - \frac{1}{2}\Lambda^{\mathrm{T}}G\Delta\Lambda, \tag{7.67}$$

which is non-positive due to condition 5 and 7. For positive differential measures dt and $d\eta$ we deduce that the differential measure of V (7.65) is non-positive, $dV \leq 0$. There are a number of cases for dV to distinguish:

- **Case $u = 0$:** It directly follows that $dV = 0$.
- **Case $g_{Ni} = 0$ and $\gamma_{Ni}^- < 0$ for some $i \in I_N$:** One or more contacts are closing, i.e. there are impacts. It follows from (7.67) that $V^+ - V^- < 0$ and $dV < 0$.
- **Case $g_{NC} = 0$, $u \in \ker W_{NC}^T$ and $u = u^- = u^+$ with $g_{NC} = \{g_{Ni}\}$ for** $i \in I_C$: It then holds that all contacts in I_C are closed and remain closed, $I_C \subset I_N$. We now consider $\dot{V}$ as a nonlinear functional on u and write

$$
\begin{aligned}
\dot{V}(q, u) &= 0, \; u \in \dot{V}_q^{-1}(0), \\
\dot{V}(q, u) &< 0, \; u \notin \dot{V}_q^{-1}(0),
\end{aligned}
\tag{7.68}
$$

with

$$
\begin{aligned}
\dot{V}_q^{-1}(0) &= D_q^{\text{nc}\,-1}(0) \cap D_q^{\lambda_D\,-1}(0) \\
&\subset D_q^{\text{nc}\,-1}(0) \cap D_q^{\lambda_{DC}\,-1}(0).
\end{aligned}
\tag{7.69}
$$

Condition 4 states that, if the contacts in I_C are persistent ($W_{NC}^T u = 0$), then dissipation can only vanish if $u = 0$, i.e. $D_q^{\text{nc}\,-1}(0) \cap D_q^{\lambda_{DC}\,-1}(0) = \{0\}$. In other words, if all contacts in I_C are closed and remain closed and $u \neq 0$ then dissipation is present. Using condition 4 and $u \in \ker W_{NC}^T \setminus \{0\}$, it follows that $\dot{V}_q^{-1}(0) = \{0\}$ and hence

$$
\begin{aligned}
\dot{V} &= 0, \; u = 0, \\
\dot{V} &< 0, \; u \neq 0.
\end{aligned}
\tag{7.70}
$$

Impulsive motion for this case is excluded. For a strictly positive differential measure dt we obtain the differential measure of V as given in (7.65)

$$
\begin{aligned}
dV &= 0, \; u = 0, \\
dV &< 0, \; u \neq 0.
\end{aligned}
\tag{7.71}
$$

- **Case $g_{NC} = 0$, $u \notin \ker W_{NC}^T$:** It then holds that $w_{Ni}^T u > 0$ for some $i \in I_C$ and one or more contacts will open. All we can say is that $dV \leq 0$.
- **Case $g_{Ni} > 0$ for some $i \in I_C$:** One or more contacts are open. All we can say is that $dV \leq 0$.

We conclude that

$$
\begin{aligned}
dV &= 0 \text{ for } u = 0, \\
dV &\leq 0 \text{ for } g_{NC} \neq 0, \\
dV &< 0 \text{ for } g_{NC} = 0, u^- \neq 0.
\end{aligned}
\tag{7.72}
$$

We now apply LaSalle's invariance principle (Theorem 6.31), which is valid when every limit set is a positively invariant set as stated in Condition 9 (see the remark below). Let us hereto consider the set Ω_{ρ^*} where ρ^* is chosen such that $\Omega_{\rho^*} \subset (\mathcal{V} \cap \mathcal{Q})$. Note that Ω_{ρ^*} is a positively invariant set due to the choice of V. Moreover, the set $\mathcal{Z}$ is defined as

$$\mathcal{Z} = \{(\boldsymbol{q}, \boldsymbol{u}) \mid \mathrm{d}V = 0\}, \tag{7.73}$$

which generally has a nonzero intersection with $\mathcal{P} = \{(\boldsymbol{q}, \boldsymbol{u}) \mid \boldsymbol{g}_{NC} \neq \boldsymbol{0}, \boldsymbol{g}_{NC} \geq \boldsymbol{0}\}$.

Consider a solution curve with an arbitrary initial condition in $\mathcal{P}$ for $t = t_0$. Due to condition 3, which requires that $\dot{\gamma}_{Ni} < 0$ (a.e.) for $\forall i \in I_C \backslash I_N$, at least one impact will occur for some $t > t_0$. The impact does not necessarily occur at a contact in I_C. In any case, the impact will cause $\mathrm{d}V < 0$ at the impact time. Therefore, there exists no solution curve with initial condition in $\mathcal{P}$ that remains in the intersection $\mathcal{P} \cap \mathcal{Z}$. Hence, it holds that the intersection $\mathcal{P} \cap \mathcal{Z}$ does not contain any invariant subset. Therefore, we seek the largest invariant set in $\mathcal{T} = \{(\boldsymbol{q}, \boldsymbol{u}) \mid \boldsymbol{g}_{NC}(\boldsymbol{q}) = \boldsymbol{0}, \boldsymbol{u} = \boldsymbol{0}\}$. Using the fact that $\boldsymbol{u}$ should be zero, and that this implies that no impulsive forces can occur in the measure differential inclusion describing the dynamics of the system, yields:

$$
\begin{aligned}
&\boldsymbol{M}(\boldsymbol{q})\mathrm{d}\boldsymbol{u} - \boldsymbol{h}(\boldsymbol{q}, \boldsymbol{0})\mathrm{d}t = \boldsymbol{W}_N(\boldsymbol{q})\mathrm{d}\boldsymbol{P}_N + \boldsymbol{W}_D(\boldsymbol{q})\mathrm{d}\boldsymbol{P}_D \\
&\Rightarrow \boldsymbol{h}(\boldsymbol{q}, \boldsymbol{0})\mathrm{d}t + \boldsymbol{W}_N(\boldsymbol{q})\boldsymbol{\lambda}_N\mathrm{d}t + \boldsymbol{W}_D(\boldsymbol{q})\boldsymbol{\lambda}_D\mathrm{d}t = \boldsymbol{0} \\
&\Rightarrow \boldsymbol{h}(\boldsymbol{q}, \boldsymbol{0}) + \boldsymbol{W}_N(\boldsymbol{q})\boldsymbol{\lambda}_N + \boldsymbol{W}_D(\boldsymbol{q})\boldsymbol{\lambda}_D = \boldsymbol{0} \\
&\Rightarrow \boldsymbol{h}(\boldsymbol{q}, \boldsymbol{0}) - \sum_i \boldsymbol{W}_{N_i}(\boldsymbol{q})\partial \Psi^*_{C_{N_i}}(0) - \sum_i \boldsymbol{W}_{D_i}(\boldsymbol{q})\partial \Psi^*_{C_{D_i}}(0) \ni \boldsymbol{0} \\
&\Rightarrow \boldsymbol{h}(\boldsymbol{q}, \boldsymbol{0}) + \sum_i \boldsymbol{W}_{N_i}(\boldsymbol{q})\mathbb{R}^+ - \sum_i \boldsymbol{W}_{D_i}(\boldsymbol{q})C_{D_i} \ni \boldsymbol{0}.
\end{aligned}
\tag{7.74}
$$

The equilibrium set $\mathcal{E}$ is the largest set of equilibria in Ω_{ρ^*} (Condition 8). Consequently, we can conclude that the largest invariant set in $\mathcal{Z}$ is the equilibrium set $\mathcal{E}$. Therefore, it can be concluded from Theorem 6.31 that $\mathcal{E}$ is an attractive set.   □

Remark: Theorem 6.31, being a generalisation of LaSalle's invariance principle, is valid when every limit set is a positively invariant set [33]. A sufficient condition for the latter is continuity of the solution with respect to the initial condition (Proposition 6.12). Non-smooth mechanical systems with multiple simultaneous impacts do generally not possess continuity with respect to the initial condition. Hence, the positive invariance of limit sets has been explicitly assumed in Condition 9 of Theorem 7.6. However, the assumptions of the above theorem make it *likely* that all contacts close in finite time (just as in the bouncing ball system). When all contacts are closed, then the system behaves after this time-instance like a Filippov system (a differential inclusion of Filippov-type), for which continuity with respect to the initial condition is guaranteed. It then would hold that that every limit set is a positively invariant set and the generalisation of LaSalle's invariance principle can be applied.

In the following propositions, we derive some sufficient conditions for condition 2 and 4 of Theorem 7.6 if the system obeys some additional properties.

Proposition 7.7. *If $\boldsymbol{f}^{\mathrm{nc}} = -\boldsymbol{C}\boldsymbol{u}$, then it holds that $D_{\boldsymbol{q}}^{\mathrm{nc}\,-1}(0) = \ker \boldsymbol{C}$, i.e. the zero set of $D_{\boldsymbol{q}}^{\mathrm{nc}}(\boldsymbol{u})$ is the nullspace of $\boldsymbol{C}$.*

Proof: Substitution gives $D_{\boldsymbol{q}}^{\mathrm{nc}}(\boldsymbol{u}) = \boldsymbol{u}^{\mathrm{T}}\boldsymbol{C}\boldsymbol{u}$. The proof is immediate.   □

The forces $\boldsymbol{\lambda}_{Di}$ (and impulses $\boldsymbol{\Lambda}_{Di}$), which are derived from a support function on the set C_{Di}, have in the above almost always been associated with friction forces, but can also be forces from a one-way clutch. Friction and the one-way clutch are described by the same inclusion on velocity level, but they are different in the sense that $\boldsymbol{0} \in \mathrm{bdry}\, C_{Di}$ holds for the one-way clutch and $\boldsymbol{0} \in \mathrm{int}\, C_{Di}$ holds for friction. The dissipation function of friction is a PDF, meaning that friction is dissipative when a relative sliding velocity is present, whereas no dissipation occurs in the one-way clutch. This insight leads to the following proposition:

Proposition 7.8. *If $\boldsymbol{0} \in \mathrm{int}\, C_{Di}\,\forall i \in I_G$, then it holds that $D_{\boldsymbol{q}}^{\boldsymbol{\lambda}_D\,-1}(0) = \ker \boldsymbol{W}_D^{\mathrm{T}}(\boldsymbol{q})$, i.e. the zero set of $D_{\boldsymbol{q}}^{\boldsymbol{\lambda}_D}(\boldsymbol{u})$ is the nullspace of $\boldsymbol{W}_D^{\mathrm{T}}(\boldsymbol{q})$.*

Proof: Because of $\boldsymbol{0} \in \mathrm{int}\, C_{Di}\,\forall i \in I_G$, it follows from (2.38) that $\Psi_{C_{Di}}^{*}(\boldsymbol{\gamma}_{Di}) > 0$ for $\boldsymbol{\gamma}_{Di} \neq \boldsymbol{0}$, i.e. $\Psi_{C_{Di}}^{*}(\boldsymbol{\gamma}_{Di}(\boldsymbol{q},\boldsymbol{u})) = 0 \Leftrightarrow \boldsymbol{\gamma}_{Di}(\boldsymbol{q},\boldsymbol{u}) = \boldsymbol{0}$. Moreover, it follows from assumption (5.91) that $\boldsymbol{\gamma}_{Di}(\boldsymbol{q},\boldsymbol{u}) = \boldsymbol{0} \Leftrightarrow \boldsymbol{u} \in \ker \boldsymbol{W}_{Di}^{\mathrm{T}}(\boldsymbol{q})$. The proof follows from the definition (7.27) of $D_{\boldsymbol{q}}^{\boldsymbol{\lambda}_D}(\boldsymbol{u})$.   □

If the above propositions are fulfilled then we can simplify condition 2 and 4 of Theorem 7.6.

Corollary 7.9. *If $\boldsymbol{f}^{\mathrm{nc}} = -\boldsymbol{C}\boldsymbol{u}$ and $\boldsymbol{0} \in \mathrm{int}\, C_{Di}\,\forall i \in I_G$, then condition 2 is equivalent to $\boldsymbol{C} > 0$ and condition 4 is equivalent to $\ker \boldsymbol{C} \cap \ker \boldsymbol{W}_D^{\mathrm{T}}(\boldsymbol{q}) \cap \ker \boldsymbol{W}_N^{\mathrm{T}}(\boldsymbol{q}) = \{\boldsymbol{0}\}$.*

7.3.2 Systems with Frictional Bilateral Constraints

In this section, we focus on systems with bilateral constraints with Coulomb friction (frictional sliders). The restriction to bilateral constraints excludes unilateral contact phenomena such as impact and detachment. This kind of systems is very common in engineering practice; think for example of industrial robots with play-free joints. We assume that a set of independent generalised coordinates is known (denoted by $\boldsymbol{q} \in \mathbb{R}^n$ in this section), for which these bilateral constraints are eliminated from the formulation of the dynamics of the system. We formulate the dynamics of the system using a Lagrangian approach, resulting in

$$\left(\frac{\mathrm{d}}{\mathrm{d}t}\,(T_{,\boldsymbol{u}}) - T_{,\boldsymbol{q}} + U_{,\boldsymbol{q}}\right)^{\mathrm{T}} = \boldsymbol{f}^{\mathrm{nc}} + \boldsymbol{W}_D(\boldsymbol{q})\boldsymbol{\lambda}_D, \tag{7.75}$$

or, alternatively,

$$\boldsymbol{M}(\boldsymbol{q})\dot{\boldsymbol{u}} - \boldsymbol{h}(\boldsymbol{q},\boldsymbol{u}) = \boldsymbol{W}_D(\boldsymbol{q})\boldsymbol{\lambda}_D. \tag{7.76}$$

The friction forces are assumed to obey a set-valued force law (5.70). Note that no unilateral contact forces are present in this formulation. Since impacts are excluded, there is no need to formulate the dynamics on momentum level, as no impulsive forces occur. Consequently, the equation (7.75) or (7.76) together with the set-valued force law (5.70) represent a differential inclusion on force level. An equilibrium set of (7.76) obeys

$$\mathcal{E} \subset \left\{ (q, u) \in \mathbb{R}^n \times \mathbb{R}^n \,|\, (u = 0) \wedge h(q, 0) - \sum_{i \in I_G} W_{Di}(q) C_{Di} \ni 0 \right\}, \quad (7.77)$$

where I_G is the set of all frictional bilateral contact points (frictional sliders). An equilibrium set is positively invariant if we assume uniqueness of solutions in forward time. By q_e we denote an equilibrium position of the system without friction, i.e. $h(q_e, 0) = 0$. We consider the attractivity of an equilibrium set $\mathcal{E} \supset (q_e, 0)$ in the following theorem. Theorem 7.10 is almost a corollary of Theorem 7.6, but the smooth non-conservative forces $f^{nc}(q, u)$ are not necessarily dissipative.

Theorem 7.10 (Attractivity of the equilibrium set). *Consider system (7.76) with friction law (5.70). If*

1. *the equilibrium position q_e is a local minimum of the potential energy $U(q)$ and $U(q)$ has a non-vanishing generalised gradient for all $q \in \mathcal{U}\backslash\{q_e\}$, i.e. $0 \notin \partial U(q) \,\forall q \in \mathcal{U}\backslash\{q_e\}$, and the equilibrium set $\mathcal{E}_q$ is contained in $\mathcal{U}$, i.e. $\mathcal{E}_q \subset \mathcal{U}$,*
2. *$D_q^{nc}(u) = -u^T f^{nc}$ and $f^{nc} = 0$ for $u = 0$,*
3. *there exists a set $\mathcal{D} \subset \mathbb{R}^n \times \mathbb{R}^n$, with $\mathcal{E} \subset int\,\mathcal{D}$, such that $D_q^{nc}(u) + D_q^{\lambda_D}(u) > 0 \quad \forall (q, u) \in \mathcal{D}\backslash\mathcal{N}$, where $\mathcal{N} = \{(q, u) \in \mathbb{R}^n \times \mathbb{R}^n \,|\, u = 0\}$,*
4. *$\mathcal{E} \subset \Omega_{\rho^*}$ in which the set Ω_{ρ^*} is the largest level set of $V(q, u) = T(q, u) + U(q) - U(q_e)$, that is contained in $\mathcal{D}$ and $\mathcal{Q} = \{(q, u) \in \mathbb{R}^n \times \mathbb{R}^n \,|\, q \in \mathcal{U}\}$, i.e.*

$$\rho^* = \max_{\{\rho : \Omega_\rho \subset (\mathcal{D} \cap \mathcal{Q})\}} \rho, \quad (7.78)$$

5. *$\mathcal{E}$ is the largest set of equilibrium points in Ω_{ρ^*},*

then the equilibrium set $\mathcal{E} \supset (q_e, 0)$ is locally attractive and Ω_{ρ^*} is a conservative estimate for the region of attraction.

Proof: The Lyapunov function $V(q, u) = T(q, u) + U(q) - U(q_e)$ is positive definite around the equilibrium point $(q, u) = (q_e, 0)$ due to condition 1 in the theorem. The function V is absolutely continuous along solutions curves of the system because the system is of Filippov-type. The differential measure dV only consists of a Lebesgue part $\dot{V}dt$ with

$$\dot{V} = -D_q^{nc}(u) - D_q^{\lambda_D}(u). \quad (7.79)$$

Conditions 2 and 3 assure that the Lyapunov function does not increase along solution curves within the set $\mathcal{D}$:

$$\dot{V} = 0 \; \boldsymbol{u} = \boldsymbol{0}$$
$$\dot{V} < 0 \; (\boldsymbol{q}, \boldsymbol{u}) \in \mathcal{D} \backslash \mathcal{N} \tag{7.80}$$

The proof now continues in the same way as the proof of Theorem 7.6.   □

The above theorem shows that the equilibrium set $\mathcal{E}$ of a mechanical system with dry friction can be (locally) attractive even when the equilibrium position $\boldsymbol{q}_e$ of the mechanical system without dry friction is unstable due to negative damping, i.e. the smooth non-conservative forces are non-dissipative in certain generalised force directions. For instance, fluid, aeroelastic, control and gyroscopic forces can lead to a non-positive definite damping matrix of the linearised system and can therefore pump energy in the system and destabilise an equilibrium through a Hopf bifurcation. The fact that the presence of dry friction can have a 'stabilising' effect can be explained by pointing out that the dry friction forces are of zero-th order (in terms of generalised velocities) whereas the 'destabilising' damping forces are of first order or higher. Consequently, the 'stabilising' effect of the dry friction forces can locally dominate the 'destabilising' smooth non-conservative forces leading to the local attractivity of the equilibrium set. In [170], these facts have been proved rigorously for linear mechanical systems with bilateral frictional constraints. The requirement $D_{\boldsymbol{q}}^{\mathrm{nc}}(\boldsymbol{u}) + D_{\boldsymbol{q}}^{\lambda_D}(\boldsymbol{u}) > 0$ of condition 3 simplifies for a linear mechanical system with a symmetric damping matrix $\boldsymbol{C} \not> 0$ and a constant matrix $\boldsymbol{W}_D$ to

$$\boldsymbol{U}_{ci} \in \mathrm{span}\, \boldsymbol{W}_D, \quad i = 1 \ldots n_q \tag{7.81}$$

where $\boldsymbol{U}_c = \{\boldsymbol{U}_{ci}\}$ is a matrix containing the n_q eigencolumns of the eigenvalues of $\boldsymbol{C}$ which lie in the left-half complex plane [170]. This condition can be interpreted as follows: the space spanned by the eigendirections of the damping matrix, related to negative eigenvalues, lies in the space spanned by the generalised force directions of the friction force.

Resuming, we can conclude that, in this subsection, we have formulated sufficient conditions for the (local) attractivity of equilibrium sets of a rather wide class of nonlinear mechanical systems with bilateral frictional sliders. The nonlinearities may involve: nonlinearities in the mass-matrix, both nonlinear conservative forces and non-conservative forces (possibly even non-dissipative). Moreover, the generalised force directions of the dry friction forces may depend on the generalised coordinates and the normal forces in the friction sliders may depend on both the generalised coordinates and the generalised velocities.

7.4 Instability of Equilibrium Positions and Sets

Instability of equilibrium points and sets of measure differential inclusions has been discussed in Section 6.7. When speaking of mechanical systems, at least two kinds of instability of equilibrium positions and sets can be distinguished.

The first kind is due to the fact that the total potential energy does not have a local minimum at the equilibrium point (or at an equilibrium point in an equilibrium set). Such an equilibrium point is a saddle-point in the phase-space when the system is described by a smooth differential equation. For a linear mechanical system this would mean that the stiffness matrix is not positive semi-definite. Examples of such equilibrium points can be found in buckling phenomena caused by static bifurcations (pitchfork, transcritical or saddle-node bifurcations) of an equilibrium point.

The second kind of instability is caused by non-dissipating non-conservative forces. Roughly speaking, we could say that the system has 'negative damping', i.e. some non-conservative forces are pumping energy into the system. A linear mechanical system would in this case have a damping matrix which is not positive semi-definite. This phenomenon occurs in nonlinear smooth dynamical systems when a Hopf bifurcation occurs, i.e. a stable focus transforms into an unstable focus and periodic solutions are created.

If we want to prove the instability of an equilibrium set of a mechanical system with frictional unilateral contacts using Theorem 6.33, then the choice of the V has to be based on the kind of instability in hand. We first present a theorem that can be used to prove instability of the first kind, i.e. when the total potential energy is not a positive definite function.

Theorem 7.11 (Instability Type I of an Isolated Equilibrium at the Origin). *Let $q^* = 0 \in \mathcal{K}$ be an isolated equilibrium position of (5.96), (5.98) and let the total potential energy $Q(q)$ be such that $Q(q_a) < 0$ for some q_a with arbitrary small $\|q_a\|$. We assume the system to be energetically consistent. If it holds that $D_q^{nc}(u) + D_q^{\lambda_D}(u) > 0 \; \forall q \in \mathcal{K}, u \neq 0$, then the equilibrium position is unstable.*

Proof: It holds that $q_a \in \mathcal{K}$ because $Q(q_a) < 0$ and $Q(q) = +\infty$ for all $q \notin \mathcal{K}$. Consider the function V to be the total energy function

$$V(q, u) = E_{tot}(q, u) = T(q, u) + U(q) - U(0) + \sum_{i \in I_G} \pi_{Ni}(g_{Ni}(q)),$$

which has the properties

$$V(0, 0) = 0, \quad V(q_a, 0) < 0$$

for some q_a with arbitrary small $\|q_a\|$. The set

$$\mathcal{U} = \{(q, u) \in \overline{B}_r \mid V(q, u) \leq 0\}$$

is a nonempty set because $(q_a, 0) \in \mathcal{U}$ and it holds that $\mathcal{U} \subset \mathcal{A} = \{(q, u) \in \mathbb{R}^n \times \mathbb{R}^n \mid q \in \mathcal{K}\}$ because $V(q, u) = \infty$ for $q \notin \mathcal{K}$. Because $q^* = 0$ is an isolated equilibrium position, we can take r such that $q^* = 0$ is the unique equilibrium position in $\mathcal{U}$. The differential measure dV consists of a Lebesgue part and an atomic part:

$$dV = \dot{V}dt + (V^+ - V^-)d\eta.$$

Using $D_q^{nc}(u) + D_q^{\lambda_P}(u) \geq 0$, (7.29) and (7.30) it follows that $\dot{V} \leq 0 \ \forall (q, u) \in \mathcal{U} \subset \mathcal{K} \times \mathbb{R}^n$. Moreover, it holds that $V^+ - V^- \leq 0$ because of the energetic consistency of the system. The function V does therefore not increase along solution curves of the system in $\mathcal{U}$. The set

$$\mathcal{Z} = \{(q, u) \in \mathcal{U} \backslash \{0\} \mid dV = 0\}$$

can be written as $\mathcal{Z} = \{(q, u) \in \mathcal{U} \backslash \{0\} \mid u = 0\}$. An invariant set in $\mathcal{Z}$ necessarily consists of equilibrium points. However, the origin is the unique equilibrium point in $\mathcal{U}$ and $\mathcal{Z} \subset \mathcal{U} \backslash \{0\}$ does therefore not contain an equilibrium point. Consequently, $\mathcal{Z}$ is not positively invariant. Instability of $q^* = 0$ follows from Theorem 6.32b. □

Theorem 7.11 can be applied when we can indeed find an arbitrary small q_a for which $Q(q_a) < 0$. This means that the total potential energy does not have a local minimum at the origin and corresponds therefore to an instability of the first kind. We now generalise this theorem to prove instability of an equilibrium set.

Theorem 7.12 (Instability Type I of an Equilibrium Set). *Consider an equilibrium set $\mathcal{E}$ of (5.96), (5.98) which is assumed to be energetically consistent. Let $q_m^* \in$ bdry $\mathcal{E}_q$ be an equilibrium position for which the total potential energy $Q(q)$ is minimal on the boundary, i.e.*

$$Q(q_m^*) \leq Q(q^*) \quad \forall q^* \in \text{bdry } \mathcal{E}_q. \tag{7.82}$$

Assume that $Q(q_a) < Q(q_m^*)$ for some $q_a \notin \mathcal{E}_q$ with arbitrarily small $\|q_a - q_m^*\|$. Moreover, if $D_q^{nc}(u) + D_q^{\lambda_P}(u) > 0 \ \forall q \in \mathcal{K}, \ u \neq 0$, then the equilibrium set $\mathcal{E}$ is unstable.

Proof: It holds that $q_a \in \mathcal{K}$ because $Q(q_a) < Q(q_m^*) < \infty$ and $Q(q) = +\infty$ for all $q \notin \mathcal{K}$. Consider the function V to be the total energy function

$$V(q, u) = E_{\text{tot}}(q, u) = T(q, u) + U(q) - U(0) + \sum_{i \in I_G} \pi_{Ni}(g_{Ni}(q)),$$

which has the properties

$$V(0, 0) = 0, \quad V(q_a, 0) < V(q_m^*, 0) \implies V(q_a, 0) < V(q^*, 0) \ \forall q^* \in \text{bdry } \mathcal{E}$$

for some $q_a \notin \mathcal{E}_q$ with arbitrary small $\|q_a - q_m^*\|$. The set

$$\mathcal{U} = \{(q, u) \in (q_m^*, 0) + \overline{B}_r \mid V(q, u) \leq V(q_m^*, 0)\}$$

is a nonempty set because $(q_a, 0) \in \mathcal{U}$ and it holds that $\mathcal{U} \subset \mathcal{A} = \{(q, u) \in \mathbb{R}^n \times \mathbb{R}^n \mid q \in \mathcal{K}\}$ because $V(q, u) = \infty$ for $q \notin \mathcal{K}$. We can take r such that $\mathcal{U}$ does not contain equilibrium points other than those in $\mathcal{E}$, being the equilibrium set under investigation. The differential measure dV consists of a Lebesgue part and an atomic part

$$dV = \dot{V}dt + (V^+ - V^-)d\eta.$$

Using $D_q^{nc}(u) + D_q^{\lambda_D}(u) \geq 0$, (7.29) and (7.30) it follows that $\dot{V} \leq 0 \; \forall (q, u) \in \mathcal{U} \subset \mathcal{K} \times \mathbb{R}^n$. Moreover, it holds that $V^+ - V^- \leq 0$ because of the energetic consistency requirement. The function V does therefore not increase along solution curves of the system in $\mathcal{U}$. This means that the forward solution $\varphi(t, t_0, q_a)$ can not cross the boundary of $\mathcal{E}$ and therefore not enter $\bar{\mathcal{E}}$. The set

$$\mathcal{Z} = \{(q, u) \in \mathcal{U} \backslash \mathcal{E} \mid dV = 0\}$$

can be written as $\mathcal{Z} = \{(q, u) \in \mathcal{U} \backslash \mathcal{E} \mid u = 0\}$. An invariant set in $\mathcal{Z}$ necessarily consists of equilibrium points, but the value of r is chosen such that the set $\mathcal{U} \backslash \mathcal{E}$ does not contain an equilibrium point. The set $\mathcal{Z} \subset \mathcal{U} \backslash \mathcal{E}$ is therefore not positively invariant. Instability of the equilibrium point $(q_m^*, 0)$ follows from Theorem 6.32b. Consequently, the equilibrium set $\mathcal{E}$ is unstable.
□

We now focus on instability of the second type, i.e. Hopf bifucation instability caused by energy creating non-conservative forces. The dissipation $D_q^{nc}(u)$ due to the non-conservative forces is therefore by definition negative for non-zero generalised velocities. As Lyapunov function we choose again the total mechanical energy. In order to prove instability using Theorem 6.32, we have to pose conditions which force the total dissipation rate to be non-positive. The 'atomic' dissipation change due to impacts can only be non-positive if the contacts are assumed to be completely elastic (all restitution coefficients equal unity) or if impacts are absent in the system. The dissipation rate during impact-free motion is governed by the non-conservative forces as well as the friction forces. This implies that we have to state the condition $D_q^{nc}(u) + D_q^{\lambda_D}(u) \leq 0$. However, the friction forces have a set-valued nature and dominate over the smooth non-conservative forces for small generalised velocities. Hence, we have to exclude friction in order to prove instability of the second type using Theorem 6.32.

Theorem 7.13 (Instability Type II of an Isolated Equilibrium at the Origin). *Let $q^* = 0 \in \mathcal{K}$ be an isolated equilibrium position of (5.96), (5.98) and let the total potential energy $Q(q)$ (7.4) be locally positive definite. Let all contacts be frictionless and be completely elastic or bilateral. If it holds that $D_q^{nc}(u) := -u^T f^{nc}(q, u) < 0 \; \forall q \in \mathcal{K}, \; u \neq 0$, then the equilibrium position is unstable.*

Proof: Consider the function V to be the total mechanical energy function

$$V(q, u) = E_{tot}(q, u) = T(q, u) + U(q) - U(0) + \sum_{i \in I_G} \pi_{Ni}(g_{Ni}(q)),$$

which is a locally positive definite function. Because $\boldsymbol{q}^* = \boldsymbol{0}$ is an isolated equilibrium position, we find an $r > 0$ such that $\boldsymbol{q}^* = \boldsymbol{0}$ is the unique equilibrium position in $\mathcal{U} = \overline{B}_r \cap \mathcal{A}$. The function V is bounded on $\mathcal{U}$ because $\mathcal{U}$ is compact and a subset of $\mathcal{A}$. The differential measure $\mathrm{d}V$ consists of a Lebesgue part and an atomic part: $\mathrm{d}V = \dot{V}\mathrm{d}t + (V^+ - V^-)\mathrm{d}\eta$. Using $D_{\boldsymbol{q}}^{\mathrm{nc}}(\boldsymbol{u}) < 0$ for $\boldsymbol{u} \neq \boldsymbol{0}$, $D_{\boldsymbol{q}}^{\lambda_D}(\boldsymbol{u}) = 0$ and (7.29), (7.30) it follows that $\dot{V} > 0$ when $\boldsymbol{u} \neq \boldsymbol{0}$ and $\dot{V} = 0$ for $\boldsymbol{u} = \boldsymbol{0}$. Moreover, it holds that $V^+ - V^- = 0$ because all contacts are assumed to be completely elastic or bilateral, which implies that no dissipation can occur due to impacts. Hence, it holds that $\mathrm{d}V = 0$ for $(\boldsymbol{q}, \boldsymbol{u}) \in \mathcal{U}_0 = \{(\boldsymbol{q}, \boldsymbol{u}) \in \mathcal{U} \mid \boldsymbol{u} = \boldsymbol{0}\}$ and $\mathrm{d}V = \dot{V}\mathrm{d}t > 0$ for $(\boldsymbol{q}, \boldsymbol{u}) \in \mathcal{U}\backslash\mathcal{U}_0$. The set $\mathcal{U}_0\backslash\{\boldsymbol{0}\}$ does not contain any positively invariant subset because the origin is the unique equilibrium in $\mathcal{U}$. The equilibrium is not attractive because V is locally positive definite and can not decrease in $\mathcal{U}$. We will prove that $\mathcal{U}\backslash\{\boldsymbol{0}\}$ does not contain any positively invariant subset by a reductio ad absurdum. Imagine that $\mathcal{U}\backslash\{\boldsymbol{0}\}$ has the positively invariant subset $\mathcal{M}$. Consider the solution curve $(\boldsymbol{q}(t), \boldsymbol{u}(t))$ starting from $(\boldsymbol{q}_0, \boldsymbol{u}_0) \in \mathcal{M}$. It holds that $(\boldsymbol{q}(t), \boldsymbol{u}(t))$ can only stay for a Lebesgue negligible set of time-instances on $\mathcal{U}_0\backslash\{\boldsymbol{0}\}$. As a consequence, the mean value of $\dot{V}(t)$ will be strictly positive for any non-empty time-interval. Consequently, if the solution remains within $\mathcal{M}$ for all future times, then V will grow unboundedly. This is in contradiction with the boundedness of V on $\mathcal{U}$. Instability of $(\boldsymbol{q}^*, \boldsymbol{0}) = (\boldsymbol{0}, \boldsymbol{0})$ follows from $\mathcal{M} = \emptyset$, i.e. there does not exist a $\delta > 0$ such that all solutions starting in the δ-neighbourhood of the origin remain in $\mathcal{U}$. $\square$

The exclusion of friction and inelastic impact makes the above theorem rather weak. An instability theorem for equilibrium sets for this kind of instability can be pursued in the same way, but the limited use of the theorem remains. More elaborate instability theorems have to be considered when dealing with instability of the second kind.

7.5 Examples

In this section we show how the results of the previous sections can be used to prove the stability, attractivity or instability of an equilibrium set of a number of mechanical systems. Sections 7.5.1 and 7.5.2 involve examples of mechanical systems with unilateral contact, impact and friction and are of increasing complexity. Section 7.5.3 treats an example of a mechanical system with bilateral frictional constraints.

7.5.1 Falling Block

Consider a planar rigid block (see Figure 7.2) with mass m under the action of gravity (gravitational acceleration g), which is attached to a vertical wall with a spring. The block can freely move in the vertical direction but is not able

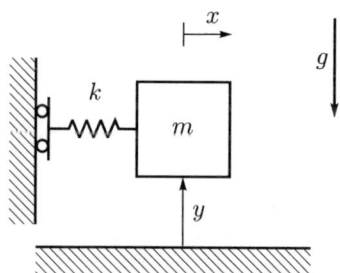

Fig. 7.2. Falling block.

to undergo a rotation. The coordinates x and y describe the position of the block. The spring, with stiffness k, is unstressed for $x = 0$. The block comes into contact with a horizontal floor when the contact distance $g_N = y$ becomes zero. The constitutive properties of the contact are the friction coefficient μ and the restitution coefficients $0 \le e_N < 1$ and $e_T = 0$. The equations of motion for impact free motion read as

$$
\begin{aligned}
m\ddot{x} + kx &= \lambda_T, \\
m\ddot{y} &= -mg + \lambda_N.
\end{aligned}
\tag{7.83}
$$

Using the generalised coordinates $\boldsymbol{q} = \begin{bmatrix} x & y \end{bmatrix}^{\mathrm{T}}$, we can describe the system in the form (5.96) with

$$
\boldsymbol{M} = \begin{bmatrix} m & 0 \\ 0 & m \end{bmatrix}, \quad
\boldsymbol{h} = \begin{bmatrix} -kx \\ -mg \end{bmatrix}, \quad
\boldsymbol{W}_N = \begin{bmatrix} 0 \\ 1 \end{bmatrix}, \quad
\boldsymbol{W}_T = \begin{bmatrix} 1 \\ 0 \end{bmatrix}.
\tag{7.84}
$$

The system for $\mu = 0$ admits a unique equilibrium position $\boldsymbol{q}_e = \boldsymbol{0}$. For $\mu > 0$ there exists an equilibrium set $\mathcal{E} = \{(x, y, \dot{x}, \dot{y}) \mid k|x| \le \mu mg, \, y = 0, \, \dot{x} = \dot{y} = 0\}$ and it holds that $(\boldsymbol{q}_e, \boldsymbol{0}) \in \mathcal{E}$.

The total potential energy function consists of the potential energy of the spring, the gravitation and the support function of the unilateral contact

$$
\begin{aligned}
Q(\boldsymbol{q}) &= U(\boldsymbol{q}) + \Psi^*_{C_N}(g_N(\boldsymbol{q})) \\
&= \frac{1}{2}kx^2 + mgy + \Psi^*_{\mathbb{R}^-}(y) \\
&= \frac{1}{2}kx^2 + mgy + \Psi_{\mathbb{R}^+}(y).
\end{aligned}
\tag{7.85}
$$

Notice that the term $mgy + \Psi_{\mathbb{R}^+}(y)$ is a positive definite term in y. It holds that Q is a positive definite function in $\boldsymbol{q}$, because it is bounded from below by another positive definite function $Q(\boldsymbol{q}) \ge \frac{1}{2}kx^2 + mg|y|$. Moreover, the global minimum of Q is located at the equilibrium point $\boldsymbol{q}_e = \boldsymbol{0}$, because $\partial Q(\boldsymbol{q}_e) \ni \boldsymbol{0}$ and is unique because of the strict convexity of Q. In order to study the altered potential energy of an equilibrium point $(\boldsymbol{q}^*, \boldsymbol{0}) = ((x^*, 0), \boldsymbol{0}) \in \mathcal{E}$, we use the altered potential energy function (7.43)

$$Q_a(q) = U(q - q^*) + \Psi^*_{C_N}(g_N(q))$$
$$= \frac{1}{2}k(x - x^*)^2 + mgy + \Psi_{\mathbb{R}^+}(y), \tag{7.86}$$

which is positive definite with respect to the equilibrium position q^*. The system does not contain smooth non-conservative forces, i.e. $f^{nc} = 0$. Using the fact that the dissipation due to friction (7.47), $D_q^{\lambda_T}(u) + \gamma_T^T \lambda_T^*$, is always non-negative, it follows from Theorem 7.5 that each equilibrium position $q^* \in \mathcal{E}_q$ is stable. Henceforth, we are able to prove that the equilibrium set $\mathcal{E}$ is stable.

Attractivity of the equilibrium set $\mathcal{E}$ can be proven with Theorem 7.6. The total potential energy function $Q(q)$ (7.85) fulfills condition 1 of Theorem 7.6 for all $q \in \mathbb{R}^2$. The system does not contain smooth non-conservative forces, i.e. $f^{nc} = 0$, which fulfills condition 2 of Theorem 7.6. Denote the contact between block and floor as contact 1 and take $I_C = I_G = \{1\}$. It holds that $\dot{\gamma}_N = -g$ for $g_N = y > 0$, which guarantees that if the block is not in contact with the floor, then it will come again in contact with the floor in a finite time. Condition 3 of Theorem 7.6 is therefore satisfied. Furthermore, it holds that $D_q^{nc^{-1}}(0) = \mathbb{R}^n$ and $D_q^{\lambda_{TC}^{-1}}(0) = \ker W_T^T$. Because the vectors W_N and W_T are linearly independent it holds that $\ker W_T^T \cap \ker W_N^T = \{0\}$ and condition 4 of Theorem 7.6 is therefore fulfilled. The restitution coefficients fulfill condition 5. Condition 6 is fulfilled because $\mathcal{V} = \mathcal{U} = \mathcal{A} = \{(q, u) \mid y \geq 0\}$. Condition 7 is fulfilled because the system has only one contact point and $e_N < 1$. Condition 8 is fulfilled by definition. The only limit set is the equilibrium set $\mathcal{E}$, which is positively invariant by definition. Hence, condition 9 is fulfilled. Consequently, Theorem 7.6 proves that the equilibrium set $\mathcal{E}$ is attractive. Moreover, the Lyapunov function $V(q, u) = E_{tot}(q, u) - Q(q_e)$ is radially unbounded and the equilibrium set is therefore globally attractively stable. The equilibrium set is clearly also symptotically attractively stable, but this can not be proven with the presented theorems.

7.5.2 Rocking Bar

Consider a planar rigid bar with mass m and inertia J_S with respect to the centre of mass S, which is attached to a vertical wall with a spring (Figure 7.3). The gravitational acceleration is denoted by g. The position and orientation of the bar are described by the generalised coordinates

$$q = \begin{bmatrix} x & y & \varphi \end{bmatrix}^T, \tag{7.87}$$

where x and y are the displacements of the centre of mass S with respect to the coordinate frame (e_x^I, e_y^I) and φ is the inclination angle. The spring is unstressed for $x = 0$. The bar has length $2a$. The two endpoints 1 and 2 can come into contact with the floor. The contact between bar and floor is described by a friction coefficient $\mu > 0$ and a normal restitution coefficient

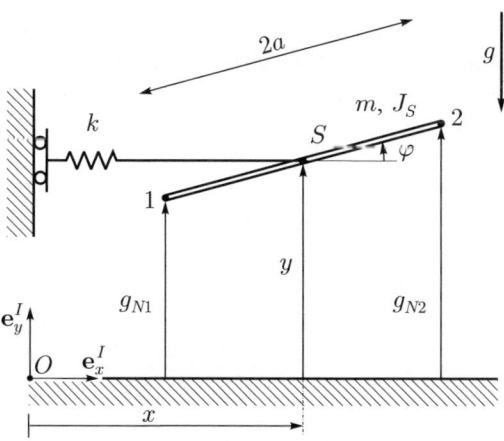

Fig. 7.3. Rocking bar

$0 \leq e_N < 1$ which is equal to the tangential restitution $e_T = e_N$. The contact distances, indicated in Figure 7.3, are

$$\begin{aligned} g_{N1} &= y - a\sin\varphi, \\ g_{N2} &= y + a\sin\varphi. \end{aligned} \qquad (7.88)$$

The relative velocities of contact points 1 and 2 with respect to the floor read as

$$\begin{aligned} \gamma_{T1} &= \dot{x} + a\dot{\varphi}\sin\varphi, \\ \gamma_{T2} &= \dot{x} - a\dot{\varphi}\sin\varphi. \end{aligned} \qquad (7.89)$$

We can describe the system in the form (5.96) with

$$\boldsymbol{M} = \begin{bmatrix} m & 0 & 0 \\ 0 & m & 0 \\ 0 & 0 & J_S \end{bmatrix}, \quad \boldsymbol{h} = \begin{bmatrix} -kx \\ -mg \\ 0 \end{bmatrix}, \qquad (7.90)$$

$$\boldsymbol{W}_N^{\mathrm{T}} = \begin{bmatrix} 0 & 1 & -a\cos\varphi \\ 0 & 1 & a\cos\varphi \end{bmatrix}, \quad \boldsymbol{W}_T^{\mathrm{T}} = \begin{bmatrix} 1 & 0 & a\sin\varphi \\ 1 & 0 & -a\sin\varphi \end{bmatrix}. \qquad (7.91)$$

The system contains a number of equilibrium sets. We will consider the equilibrium set

$$\mathcal{E} = \{(x, y, \varphi, \dot{x}, \dot{y}, \dot{\varphi}) \mid k|x| \leq \mu mg, \, y = 0, \, \varphi = 0, \, \dot{x} = \dot{y} = \dot{\varphi} = 0\}, \qquad (7.92)$$

for which $g_{N1} = g_{N2} = 0$. The total potential energy function

$$\begin{aligned} Q(\boldsymbol{q}) &= U(\boldsymbol{q}) + \Psi_{C_N}^*(g_{N1}(\boldsymbol{q})) + \Psi_{C_N}^*(g_{N2}(\boldsymbol{q})) \\ &= \frac{1}{2}kx^2 + mgy + \Psi_{\mathbb{R}^-}^*(g_{N1}) + \Psi_{\mathbb{R}^-}^*(g_{N2}) \\ &= \frac{1}{2}kx^2 + mgy + \Psi_{\mathbb{R}^+}(g_{N1}) + \Psi_{\mathbb{R}^+}(g_{N2}) \end{aligned} \qquad (7.93)$$

contains a quadratic term in x, a linear term in y and two indicator functions on the contact distances. Notice that $Q(\boldsymbol{q}) = 0$ for $\boldsymbol{q} = \boldsymbol{q}_e = \boldsymbol{0}$, where $(\boldsymbol{q}_e, \boldsymbol{0}) \ni \mathcal{E}$ is the equilibrium point for $\mu = 0$. Moreover, it holds that if $g_{N1} \geq 0$ and $g_{N2} \geq 0$ then $y \geq 0$ and $a|\sin\varphi| \leq y$. We therefore deduce that

$$g_{N1} \geq 0 \wedge g_{N2} \geq 0 \Longrightarrow Q(\boldsymbol{q}) = \frac{1}{2}kx^2 + mgy$$

$$Q(\boldsymbol{q}) = \frac{1}{2}kx^2 + \frac{mg}{2}(|y| + y) \tag{7.94}$$

$$Q(\boldsymbol{q}) \geq \frac{1}{2}kx^2 + \frac{mg}{2}(|y| + a|\sin\varphi|)$$

and

$$g_{N1} < 0 \vee g_{N2} < 0 \Longrightarrow Q(\boldsymbol{q}) = +\infty$$

$$Q(\boldsymbol{q}) > \frac{1}{2}kx^2 + \frac{mg}{2}(|y| + a|\sin\varphi|). \tag{7.95}$$

The function $f(\boldsymbol{q}) = \frac{1}{2}kx^2 + \frac{mg}{2}(|y| + a|\sin\varphi|)$ is locally positive definite in the set $\mathcal{U} = \{\boldsymbol{q} \in \mathbb{R}^n \mid |\varphi| < \frac{\pi}{2}\}$. Consequently, the total potential energy function $Q(\boldsymbol{q}) \geq f(\boldsymbol{q})$ is locally positive definite in the set $\mathcal{U}$ as well. It can be easily checked that the generalised gradient

$$\partial Q(\boldsymbol{q}) = \begin{bmatrix} kx \\ mg + \partial\Psi_{\mathbb{R}^+}(g_{N1}) + \partial\Psi_{\mathbb{R}^+}(g_{N2}) \\ -\partial\Psi_{\mathbb{R}^+}(g_{N1})a\cos\varphi + \partial\Psi_{\mathbb{R}^+}(g_{N2})a\cos\varphi \end{bmatrix} \tag{7.96}$$

can only vanish in the set $\mathcal{U}$ for $\boldsymbol{q} = \boldsymbol{q}_e$, i.e. $\boldsymbol{0} \notin \partial Q(\boldsymbol{q}) \,\forall \boldsymbol{q} \in \mathcal{U} \backslash \{\boldsymbol{q}_e\}$ and $\boldsymbol{0} \in \partial Q(\boldsymbol{q}_e)$. Similarly, we can prove that the altered total potential energy function

$$Q_a(\boldsymbol{q}) = Q(\boldsymbol{q} - \boldsymbol{q}^*) = U(\boldsymbol{q} - \boldsymbol{q}^*) + \Psi^*_{C_N}(-g_{N1}(\boldsymbol{q})) + \Psi^*_{C_N}(-g_{N2}(\boldsymbol{q}))$$

$$= \frac{1}{2}k(x - x^*)^2 + mgy + \Psi_{\mathbb{R}^+}(g_{N1}) + \Psi_{\mathbb{R}^+}(g_{N2}), \tag{7.97}$$

with $\boldsymbol{q}^* \in \mathcal{E}_q$, is locally positive definite with respect to $\boldsymbol{q}^*$.

Smooth non-conservative forces are absent in this system, i.e. $\boldsymbol{f}^{\mathrm{nc}} = \boldsymbol{0}$ and $D_{\boldsymbol{q}}^{\mathrm{nc}}(\boldsymbol{u}) = 0$. Using the fact that the dissipation due to friction (7.47), $D_{\boldsymbol{q}}^{\lambda_T}(\boldsymbol{u}) + \boldsymbol{\gamma}_T^{\mathrm{T}}\boldsymbol{\lambda}_T^*$, is always non-negative, it follows from Theorem 7.5 that each equilibrium position $\boldsymbol{q}^* \in \mathcal{E}_q$ is stable. Henceforth, we are able to prove that the equilibrium set $\mathcal{E}$ is stable.

Proving that the equilibrium set $\mathcal{E}$ is also attractive, is much more complicated. We now have to prove that condition 3 of Theorem 7.6 holds with $I_C = \{1, 2\}$. Consider the open subset $\mathcal{V} = \{(\boldsymbol{q}, \boldsymbol{u}) \in \mathbb{R}^n \times \mathbb{R}^n \mid \mu|\tan\varphi| < 1, a\dot{\varphi}^2 < g\}$ which contains the equilibrium set, i.e. $\mathcal{E} \subset \mathcal{V}$. We consider the following cases with $(\boldsymbol{q}, \boldsymbol{u}) \in \mathcal{V}$:

- $I_N = \emptyset$: both contacts are open, i.e. $g_{N1} > 0$ and $g_{N2} > 0$. It holds for $(\boldsymbol{q}, \boldsymbol{u}) \in \mathcal{V}$ that

$$
\begin{aligned}
\dot{\gamma}_{N1} &= \ddot{y} - a\ddot{\varphi}\cos\varphi + a\dot{\varphi}^2\sin\varphi \\
&= -g + a\dot{\varphi}^2\sin\varphi \\
&< 0 \\
\dot{\gamma}_{N2} &= \ddot{y} + a\ddot{\varphi}\cos\varphi - a\dot{\varphi}^2\sin\varphi \\
&= -g - a\dot{\varphi}^2\sin\varphi \\
&< 0.
\end{aligned} \tag{7.98}
$$

- $I_N = \{1\}$: contact 1 is closed and contact 2 is open, i.e. $g_{N1} = 0$ and $g_{N2} > 0$. We consider contact 1 to be closed for a nonzero time-interval. The normal contact acceleration of the closed contact 1 must vanish:

$$
\dot{\gamma}_{N1} = \ddot{y} - a\ddot{\varphi}\cos\varphi + a\dot{\varphi}^2\sin\varphi
$$

$$
0 = -g + \frac{1}{m}\lambda_{N1} + \frac{a^2}{J_S}\cos^2\varphi\,\lambda_{N1} - \frac{a^2}{J_S}\cos\varphi\sin\varphi\,\lambda_{T1} + a\dot{\varphi}^2\sin\varphi
$$

$$
0 = -g + \left(\frac{1}{m} + \frac{a^2}{J_S}\cos\varphi(\cos\varphi - \bar{\mu}\sin\varphi)\right)\lambda_{N1} + a\dot{\varphi}^2\sin\varphi, \tag{7.99}
$$

with $\lambda_{T1} = \bar{\mu}\lambda_{N1}$, i.e. $\bar{\mu} \in -\mu\operatorname{Sign}(\gamma_{T1})$. It follows from (7.99) that the normal contact force λ_{N1} is a function of φ and $\dot{\varphi}$. The contact acceleration of contact 2 therefore becomes

$$
\begin{aligned}
\dot{\gamma}_{N2} &= \ddot{y} + a\ddot{\varphi}\cos\varphi - a\dot{\varphi}^2\sin\varphi \\
&= -g + \frac{1}{m}\lambda_{N1} - \frac{a^2}{J_S}\cos^2\varphi\,\lambda_{N1} + \frac{a^2}{J_S}\cos\varphi\sin\varphi\,\lambda_{T1} - a\dot{\varphi}^2\sin\varphi \\
&= -g + \left(\frac{1}{m} - \frac{a^2}{J_S}\cos\varphi(\cos\varphi - \bar{\mu}\sin\varphi)\right)\lambda_{N1} - a\dot{\varphi}^2\sin\varphi \\
&= \frac{\frac{1}{m} - \frac{a^2}{J_S}\cos\varphi(\cos\varphi - \bar{\mu}\sin\varphi)}{\frac{1}{m} + \frac{a^2}{J_S}\cos\varphi(\cos\varphi - \bar{\mu}\sin\varphi)}(g - a\dot{\varphi}^2\sin\varphi) - g - a\dot{\varphi}^2\sin\varphi \\
&= \frac{-2ga^2\frac{m}{J_S}\cos\varphi(\cos\varphi - \bar{\mu}\sin\varphi) - 2a\dot{\varphi}^2\sin\varphi}{1 + a^2\frac{m}{J_S}\cos\varphi(\cos\varphi - \bar{\mu}\sin\varphi)}.
\end{aligned} \tag{7.100}
$$

Using $|\bar{\mu}| \le \mu$ and $(\boldsymbol{q}, \boldsymbol{u}) \in \mathcal{V}$ it follows that $\dot{\gamma}_{N2} < 0$.
- $I_N = \{2\}$: contact 1 is open and contact 2 is closed, i.e. $g_{N1} > 0$ and $g_{N2} = 0$. Similar to the previous case we can prove that $\dot{\gamma}_{N1} < 0$.

Hence, there exists a non-empty set $I_C = \{1, 2\}$, such that $\dot{\gamma}_{Ni}(\boldsymbol{q}, \boldsymbol{u}) < 0$ (a.e.) for $\forall i \in I_C \backslash I_N$ and $\forall (\boldsymbol{q}, \boldsymbol{u}) \in \mathcal{V}$. Condition 3 of Theorem 7.6 is therefore fulfilled.

It holds that $D_{\boldsymbol{q}}^{\mathrm{nc}} = 0$ and using Proposition 7.8 it follows that $D_{\boldsymbol{q}}^{\lambda_T^{-1}}(0) = \ker \boldsymbol{W}_T^{\mathrm{T}}(\boldsymbol{q})$. Furthermore, for $\boldsymbol{q} \in \mathcal{C} = \{\boldsymbol{q} \in \mathbb{R}^n \mid g_{N1} = g_{N2} = 0\}$ follows the implication $\boldsymbol{W}_N^{\mathrm{T}}(\boldsymbol{q})\boldsymbol{u} = \boldsymbol{0} \Longrightarrow \dot{y} = 0 \wedge \dot{\varphi} = 0$ and similarly $\boldsymbol{W}_T^{\mathrm{T}}(\boldsymbol{q})\boldsymbol{u} = \boldsymbol{0} \Longrightarrow$ $\dot{x} = 0$. We conclude that there is always dissipation when both contacts are closed and $\boldsymbol{u} \neq \boldsymbol{0}$ because

$$\ker \boldsymbol{W}_T^{\mathrm{T}}(\boldsymbol{q}) \cap \ker \boldsymbol{W}_N^{\mathrm{T}}(\boldsymbol{q}) = \{\boldsymbol{0}\} \qquad \forall \boldsymbol{q} \in \mathcal{C}, \tag{7.101}$$

and condition 4 of Theorem 7.6 is therefore fulfilled. The largest level set of $V = T(\boldsymbol{q}, \boldsymbol{u}) + Q(\boldsymbol{q})$ which lies entirely in $\mathcal{Q} = \{(\boldsymbol{q}, \boldsymbol{u}) \in \mathbb{R}^n \times \mathbb{R}^n \mid \boldsymbol{q} \in \mathcal{U}\}$ is given by $V(\boldsymbol{q}, \boldsymbol{u}) < mga$. The largest level set of V which lies entirely in $\mathcal{V}$ is determined by $V(\boldsymbol{q}, \boldsymbol{u}) < \frac{1}{2}J_S \frac{g}{a}$ and $V(\boldsymbol{q}, \boldsymbol{u}) < \frac{mga}{\sqrt{1+\mu^2}}$. We therefore choose the set Ω_{ρ^*} as

$$\Omega_{\rho^*} = \{(\boldsymbol{q}, \boldsymbol{u}) \in \mathbb{R}^n \times \mathbb{R}^n \mid V(\boldsymbol{q}, \boldsymbol{u}) < \rho^*\}, \text{ with } \rho^* = \min \left(\frac{1}{2} J_S \frac{g}{a}, \frac{mga}{\sqrt{1+\mu^2}} \right). \tag{7.102}$$

If additionally

$$\frac{1}{2} \frac{(\mu mg)^2}{k} < \rho^*, \tag{7.103}$$

then it holds that $\mathcal{E} \subset \Omega_{\rho^*}$, which is condition 6 of Theorem 7.6. Conditions 5 and 7 are fulfilled because all normal restitution coefficients are equal and strictly smaller than one. Condition 8 is fulfilled because of (7.102). The only limit set in Ω_{ρ^*} is the equilibrium set $\mathcal{E}$ because $e_N < 1$. The equilibrium set is positively invariant and condition 9 is therefore fulfilled. We conclude that Theorem 7.6 proves conditionally the local attractivity of the equilibrium set $\mathcal{E}$ and that Ω_{ρ^*} is a conservative estimate of the region of attraction. Naturally, the attractivity is only local, because the system has also other attractive equilibrium sets for $\varphi = n\pi$ with $n \in \mathbb{Z}$ and unstable equilibrium sets around $\varphi = \frac{\pi}{2} + n\pi$. As in the previous example of the falling block, the equilibrium set is symptotically attractive.

In a similar way we can study the stability properties of equilibrium sets of a rocking block (Figure 7.4). Although being a slight modification of the rocking bar, the verification of condition 3 of Theorem 7.6 becomes very elaborate for the rocking block example [103].

7.5.3 Constrained Bar

We now study an example with bilateral constraints. Consider a bar with mass m, length $2l$ and moment of inertia J_S around its centre of mass S, see Figure 7.5. The gravitational acceleration is denoted by g. The bar is subject to two holonomic constraints: Point 1 of the bar is constrained to the vertical slider and Point 2 of the bar is constrained to the horizontal slider. Coulomb friction is present in the contact between these endpoints of the bar and the grooves (friction coefficient μ_1 in the vertical slider and friction coefficient μ_2 in the horizontal slider). It should be noted that the realised friction forces

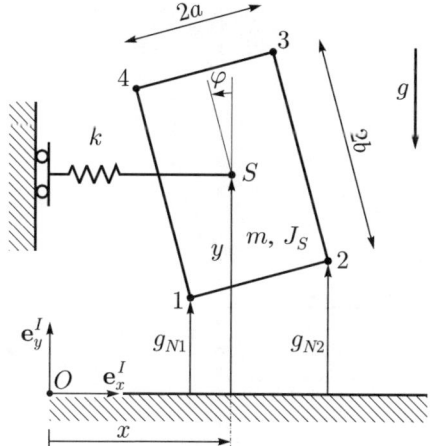

Fig. 7.4. Rocking block.

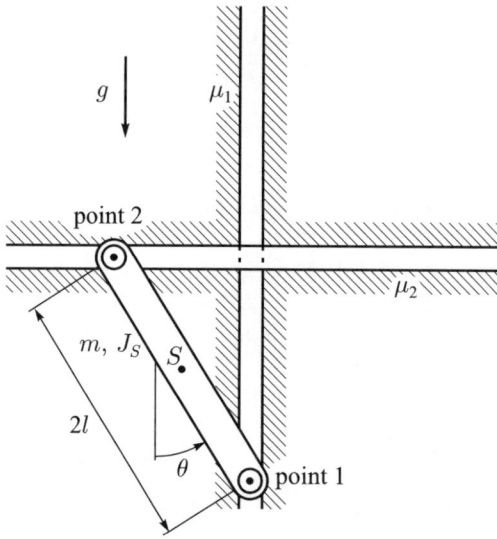

Fig. 7.5. Constrained bar.

depend on the constraint forces in the grooves (i.e. the friction is described by the non-associated Coulomb's law (5.49)). The dynamics of the system will be described in terms of the (independent) coordinate θ, see Figure 7.5. The corresponding equation of motion is given by

$$\left(ml^2 + J_S\right)\ddot{\theta} + mgl\sin\theta = 2l\sin\theta\lambda_{T_1} - 2l\cos\theta\lambda_{T_2}, \qquad (7.104)$$

where λ_{T_1} and λ_{T_2} are the friction forces in the vertical and horizontal sliders, respectively. Equation (7.104) can be written in the form (7.76), with

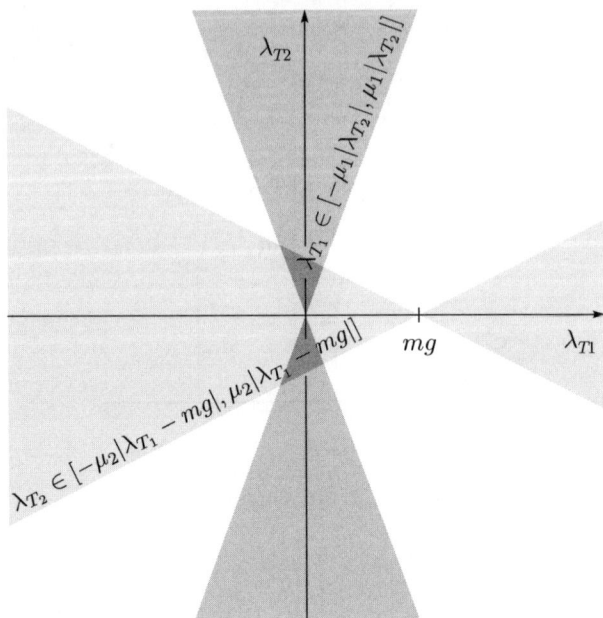

Fig. 7.6. Attainable friction forces in equilibrium.

$$\boldsymbol{M}(\boldsymbol{q}) = ml^2 + J_S, \quad \boldsymbol{h}(\boldsymbol{q}, \boldsymbol{u}) = -mgl\sin\theta, \quad \boldsymbol{W}_T(\boldsymbol{q}) = \begin{bmatrix} 2l\sin\theta & -2l\cos\theta \end{bmatrix}.$$
(7.105)

The equilibrium set of (7.104) is given by (7.77), with $C_{T_i} = \{-\lambda_{T_i} \mid -\mu_i|\lambda_{N_i}| \le \lambda_{T_i} \le +\mu_i|\lambda_{N_i}|\}$, $i = 1, 2$. Note that C_{T_i} depends on the normal force λ_{N_i}, which in turn may depend on the friction forces. The static equilibrium equations of the bar yield:

$$\lambda_{N_1} + \lambda_{T_2} = 0,$$
$$\lambda_{N_2} + \lambda_{T_1} - mg = 0, \qquad (7.106)$$
$$l\cos\theta\lambda_{N_1} - l\sin\theta\lambda_{N_2} + l\sin\theta\lambda_{T_1} - l\cos\theta\lambda_{T_2} = 0.$$

Based on the first two equations in (7.106) and the non-associated Coulomb's law (5.49), the following algebraic inclusions for the friction forces in equilibrium can be derived:

$$\lambda_{T_1} \in [-\mu_1|\lambda_{T_2}|, \mu_1|\lambda_{T_2}|],$$
$$\lambda_{T_2} \in [-\mu_2|\lambda_{T_1} - mg|, \mu_2|\lambda_{T_1} - mg|]. \qquad (7.107)$$

The resulting set of friction forces in equilibrium is depicted schematically in Figure 7.6. The equilibrium set $\mathcal{E}$ in terms of the independent generalised coordinate θ now follows from the equation of motion (7.104):

$$mgl\sin\theta = 2l\sin\theta\lambda_{T_1} - 2l\cos\theta\lambda_{T_2}. \qquad (7.108)$$

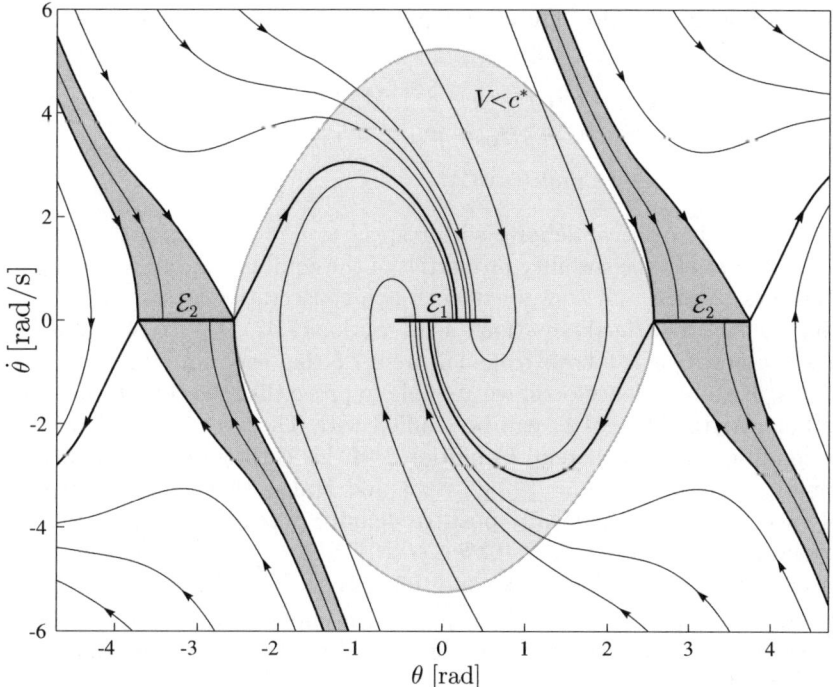

Fig. 7.7. Phase plane and the set in which $V = T + U < c^*$.

For values of θ such that $\cos\theta \neq 0$ (we assume that, for given values for m, g and l, the friction coefficients μ_1 and μ_2 are small enough to guarantee that this assumption is satisfied) we obtain:

$$\theta = \arctan\left(\frac{\lambda_{T_2}}{-\frac{mg}{2} + \lambda_{T_1}}\right) + (k-1)\pi, \qquad k = 1, 2, \tag{7.109}$$

for values of λ_{T_1} and λ_{T_2} taken from (7.107). Equation (7.109) describes the fact that there exist two isolated equilibrium sets (an equilibrium set $\mathcal{E}_1$ around $\theta = 0$ and $\mathcal{E}_2$ around $\theta = \pi$) for small values of the friction coefficients. The equilibrium sets are given by

$$\mathcal{E}_k = \left\{ (\theta, \dot{\theta}) \,\middle|\, \dot{\theta} = 0, \, |\theta - (k-1)\pi| \leq \arctan\left(\frac{2\mu_2}{1 - \mu_1\mu_2}\right) \right\}, \tag{7.110}$$

for $k = 1, 2$ and $\mu_1\mu_2 < 1$. Note that for $\mu_1\mu_2 \geq 1$ these isolated equilibrium sets merge into one large equilibrium set, such that any value of θ can be attained in this equilibrium set. We will consider the case of two isolated equilibrium sets here.

The potential energy of the constrained bar system

$$U(\boldsymbol{q}) = -mgl(\cos(\theta) - 1) \tag{7.111}$$

is locally positive definite. The altered potential energy function can be written as

$$U_a(\boldsymbol{q}) = U(\boldsymbol{q}) - (\boldsymbol{q} - \boldsymbol{q}^*)^{\mathrm{T}} \boldsymbol{W}_T(\boldsymbol{q}^*) \boldsymbol{\lambda}_T^* - U(\boldsymbol{q}^*)$$
$$= -mgl(\cos\theta + (\theta - \theta^*)\sin\theta^* - \cos\theta^*) \qquad (7.112)$$
$$\approx mgl(\theta - \theta^*)^2$$

which is locally positive definite with respect to $\boldsymbol{q}^* = \theta^* \in \mathcal{E}_q$.

We first study the stability properties of the equilibrium set $\mathcal{E}_1$ around $\theta = 0$. The system does not contain smooth non-conservative forces, i.e. $\boldsymbol{f}^{nc} = \boldsymbol{0}$. Using the fact that the dissipation due to friction (7.47), $D_{\boldsymbol{q}}^{\lambda_T}(\boldsymbol{u}) + \boldsymbol{\gamma}_T^{\mathrm{T}} \boldsymbol{\lambda}_T^*$, is always non-negative, it follows from Theorem 7.5 that each equilibrium position $\boldsymbol{q}^* \in \mathcal{E}_{1q}$ is stable. Henceforth, we are able to prove that the equilibrium set $\mathcal{E}_1$ is stable. Attractivity of $\mathcal{E}_1$ can be studied with Theorem 7.10 and we check the conditions stated therein. Condition 1 of this theorem is clearly satisfied. Namely, take the set $\mathcal{U} = \{\theta \mid |\theta| < \pi\}$ and realise that indeed the potential energy U (7.111) is locally positive definite in $\mathcal{U}$ and $\partial U/\partial\theta = mgl\sin\theta$ satisfies the demand $\partial U/\partial\theta \neq 0, \forall\theta \in \mathcal{U}\backslash\{0\}$. Since there are no smooth non-conservative forces $D_{\boldsymbol{q}}^{nc}(\boldsymbol{u}) = 0$, condition 2 is satisfied. Finally, we note that $D_{\boldsymbol{q}}^{\lambda_T}(\boldsymbol{u}) > 0$ for $(\boldsymbol{q}, \boldsymbol{u}) \in \mathcal{D} = \mathbb{R}^n \times \mathbb{R}^n$, which implies that condition 3 is satisfied. The set $\mathcal{U}$ contains the equilibrium set $\mathcal{E}_1$ and part of the equilibrium set $\mathcal{E}_2$ (see Figure 7.8). We now consider the largest level set $V < c^*$ for which the set $\mathcal{E}_1$ is the only equilibrium set within the level set of $V = T + U$. This level set is an open set and the value

$$c^* = mgl\left(1 - \frac{1 - \mu_1\mu_2}{\sqrt{4\mu_2^2 + (1 - \mu_1\mu_2)^2}}\right) \qquad (7.113)$$

is chosen such that its closure touches the equilibrium set $\mathcal{E}_2$. Consequently, we can conclude that the equilibrium set $\mathcal{E}_1$ is locally attractive. The phase plane of the constrained bar system is depicted in Figure 7.7 for the parameter values $m = 1$ kg, $J_S = \frac{1}{3}$ kg m^2, $l = 1$ m, $\varepsilon_N = \varepsilon_T = 0$, $\mu_1 = \mu_2 = 0.3$, $g = 10$ m/s^2. The trajectories in Figure 7.7 have been obtained numerically using the time-stepping method (see [101] and references therein). The equilibrium sets $\mathcal{E}_1$ and $\mathcal{E}_2$ are indicated by thick lines on the axis $\dot\theta = 0$. It can be seen in Figure 7.7 that the level set $V < c^*$ is a fairly good (though conservative) estimate for the region of attraction of the equilibrium set $\mathcal{E}_1$. Moreover, we see that the equilibrium set $\mathcal{E}_1$ is symptotically attractive.

Subsequently, we study the stability properties of the equilibrium set $\mathcal{E}_2$ around $\theta = \pi$, under the condition that $\mathcal{E}_1$ and $\mathcal{E}_2$ are two distinct equilibrium sets. We apply Theorem 7.12 and check the conditions stated therein. Consider the most left boundary point $\theta_m \in$ bdry $\mathcal{E}_{q2}$ for which $U(\theta)$ becomes minimal on the boundary, i.e. $\theta_m = \pi - \arctan\left(\frac{2\mu_2}{1-\mu_1\mu_2}\right)$. We can find a position $\theta_a < \theta_m$, arbitrarily close to θ_m, such that $U(\theta_a) < U(\theta_m)$, due to the fact that the potential energy is locally negative definite with respect to the position

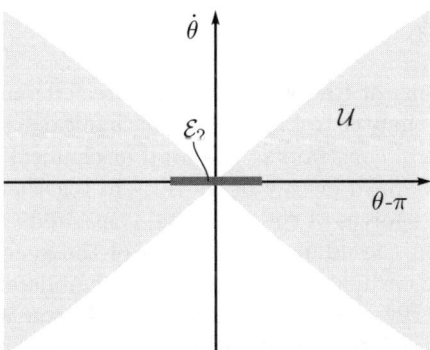

Fig. 7.8. Schematic representation of the set $\mathcal{U}$ in which $V \geq 0$ (V as in (7.114)).

$\theta = \pi$. The position θ_a is not an equilibrium position, because the equilibrium sets $\mathcal{E}_1$ and $\mathcal{E}_2$ are distinct. Using the fact that the dissipation in the system is nonnegative it follows from Theorem 7.12, that the equilibrium set $\mathcal{E}_2$ is unstable.

Theorem 7.12 follows from Theorem 6.33. If we would like to apply Theorem 6.33 directly, then the function V has to be chosen as follows:

$$V = -\frac{1}{2}\left(J_S + ml^2\right)\dot{\theta}^2 + mgl\left(1 + \cos\theta\right), \tag{7.114}$$

where $V \geq 0 \in \mathcal{U}$ with the set $\mathcal{U}$ depicted schematically in Figure 7.8. The time-derivative of V obeys

$$\dot{V} = -\dot{\theta}W_T(q)\lambda_T = -\gamma_T^T\lambda_T,$$

with the generalised force direction $W_T(q) = \begin{bmatrix} 2l\sin\theta & -2l\cos\theta \end{bmatrix}$, friction forces $\lambda_T^T = \begin{bmatrix} \lambda_{T_1} & \lambda_{T_2} \end{bmatrix}$ and sliding relative velocities $\gamma_T = W_T^T(q)\dot{\theta} = \begin{bmatrix} 2l\dot{\theta}\sin\theta & -2l\dot{\theta}\cos\theta \end{bmatrix}^T$ in the two frictional sliders. Note that $\dot{V} \geq 0$ for all $(\theta, \dot{\theta}) \in \mathcal{U}$ and $\dot{V} = 0$ if and only if $\dot{\theta} = 0$. We can easily show that solutions can not stay in $\mathcal{Z}\backslash\mathcal{E}_2$, with $\mathcal{Z} = \{(\theta, \dot{\theta}) \mid \dot{\theta} = 0\}$, using the equation of motion (7.104). The conditions of statement 3 of Theorem 6.33 are satisfied and it can be concluded that the equilibrium set $\mathcal{E}_2$ is unstable.

The equilibrium set $\mathcal{E}_2$ becomes a saddle point for $\mu_{1,2} = 0$. This saddle structure in the phase plane (see Figure 7.7) remains for $\mu_{1,2} > 0$, but $\mathcal{E}_2$ is a set instead of a point. Interestingly, the stable manifold of $\mathcal{E}_2$ is 'thick', i.e. there exists a bundle of solutions (depicted in dark grey) which are attracted to the unstable equilibrium set $\mathcal{E}_2$. Put differently: the equilibrium set $\mathcal{E}_2$ has a region of attraction, where the region is a set with a non-empty interior. The unstable half-manifolds of $\mathcal{E}_2$ originate at the tips of the set $\mathcal{E}_2$ and are heteroclinic orbits to the stable equilibrium set $\mathcal{E}_1$.

7.6 Summary

The stability theorems of Chapter 6 for measure differential inclusions have been applied in this chapter to Lagrangian mechanical systems with frictional unilateral and bilateral constraints. The total mechanical energy has been defined to be the mechanical energy of the system together with the potential of the normal contact forces of the unilateral constraints. This potential is an indicator function on the admissible domain of the generalised coordinates, which are minimal coordinates with respect to the bilateral constraints. The total mechanical energy is a natural choice for a Lyapunov (candidate) function and has been used to formulate Lyapunov-type theorems for stability, attractivity and instability of equilibrium points and sets. The stability and attractivity theorems rely on the energetic consistency, i.e. dissipativity of impacts. Sufficient conditions for energetic consistency have been given in terms of conditions on the system parameters. A number of examples have been presented which show the use of the aforementioned theorems, but also show their limitations. Notably, condition 3 of Theorem 7.6 might be extremely hard to prove and limits the use of this theorem considerably.

8

Convergence Properties of Monotone Measure Differential Inclusions

In this chapter, we present theorems which give sufficient conditions for the convergence of measure differential inclusions with certain maximal monotonicity properties. The framework of measure differential inclusions allows us to describe systems with state discontinuities, as has been shown in the previous chapters. The material presented in this chapter is based on the paper [104].

The chapter is organised as follows. First, we define the convergence property of dynamical systems in Section 8.1 and state the associated properties of convergent systems. Theorems are presented in Section 8.2 which give sufficient conditions for the convergence of measure differential inclusions with certain maximal monotonicity properties. Furthermore, we illustrate in Section 8.3 how these convergence results for measure differential inclusions can be exploited to solve tracking problems for certain classes of non-smooth mechanical systems with friction and one-way clutches. Illustrative examples of convergent mechanical systems are discussed in detail in Section 8.4. Finally, Section 8.5 presents concluding remarks.

8.1 Convergent Systems

In this section, we will briefly discuss the definition of convergence and certain properties of convergent systems. In the definition of convergence, the Lyapunov stability of solutions of (8.1) plays a central role. Definitions of (uniform) stability and attractivity of measure differential inclusions are given in the Section 6.4.

The definitions of convergence properties presented here are based on and extend the definition given in [44] (see also [132]). Consider a system described by the measure differential inclusion

$$\mathrm{d}\boldsymbol{x} \in \mathrm{d}\boldsymbol{\Gamma}(t, \boldsymbol{x}), \qquad (8.1)$$

where $\boldsymbol{x} \in \mathbb{R}^n$, $t \in \mathbb{R}$.

Let us formally define the property of convergence.

Definition 8.1. *System (8.1) is said to be*

- convergent *if there exists a solution $\bar{x}(t)$ satisfying the following conditions:*
  (i) $\bar{x}(t)$ *is defined for almost all $t \in \mathbb{R}$,*
  (ii) $\bar{x}(t)$ *is bounded for all $t \in \mathbb{R}$ for which $\bar{x}(t)$ exists,*
  (iii) $\bar{x}(t)$ *is globally attractively stable.*
- uniformly convergent *if it is convergent and $\bar{x}(t)$ is globally uniformly attractively stable.*
- exponentially convergent *if it is convergent and $\bar{x}(t)$ is globally exponentially stable.*

The wording 'attractively stable' has been used instead of the usual term 'asymptotically stable', because attractivity of solutions in (measure) differential inclusions can be asymptotic or symptotic (finite-time attractivity), see page 121.

The solution $\bar{x}(t)$ is called a *steady-state solution*. As follows from the definition of convergence, any solution of a convergent system "forgets" its initial condition and converges to some steady-state solution. In general, the steady-state solution $\bar{x}(t)$ may be non-unique. But for any two steady-state solutions $\bar{x}_1(t)$ and $\bar{x}_2(t)$ it holds that $\|\bar{x}_1(t) - \bar{x}_2(t)\| \to 0$ as $t \to +\infty$. At the same time, for *uniformly* convergent systems the steady-state solution is unique, as formulated below.

Property 8.2 ([131, 132]). If system (8.1) is uniformly convergent, then the steady-state solution $\bar{x}(t)$ is the only solution defined and bounded for almost all $t \in \mathbb{R}$.

In many engineering problems, dynamical systems excited by time-varying perturbations are encountered. Therefore, we will consider convergence properties for systems with time-varying inputs. So, instead of systems of the form (8.1), we consider systems of the form

$$\mathrm{d}x \in \mathrm{d}\boldsymbol{\Gamma}(x, w(t)), \tag{8.2}$$

with state $x \in \mathbb{R}^n$ and input $w \in \mathbb{R}^d$. The right-hand side of (8.2) is assumed to be continuous in w. In the following, we will consider the class $\overline{\mathbb{PC}}_d$ of piecewise continuous inputs $w(t) : \mathbb{R} \to \mathbb{R}^d$ which are bounded on $\mathbb{R}$. Below we define the convergence property for systems with inputs.

Definition 8.3. *System (8.2) is said to be* (uniformly, exponentially) *convergent if it is* (uniformly, exponentially) *convergent for every input $w \in \overline{\mathbb{PC}}_d$. In order to emphasize the dependency on the input $w(t)$, the steady-state solution is denoted by $\bar{x}_w(t)$.*

Uniformly convergent systems excited by periodic or constant inputs exhibit the following property, that is particularly useful in, for example, bifurcation analyses of periodically perturbed systems.

Property 8.4 ([44, 132]). Suppose system (8.2) with a given input $w(t)$ is uniformly convergent. If the input $w(t)$ is constant, the corresponding steady-state solution $\bar{x}_w(t)$ is also constant; if the input $w(t)$ is periodic with period T, then the corresponding steady-state solution $\bar{x}_w(t)$ is also periodic with the same period T.

8.2 Convergence of Maximal Monotone Systems

In this section we will consider the dynamics of measure differential inclusions (8.2) with certain maximal monotonicity conditions on $\boldsymbol{\Gamma}(\boldsymbol{x}, \boldsymbol{w}(t))$. In particular, we study systems for which $\boldsymbol{\Gamma}(\boldsymbol{x}, \boldsymbol{w}(t))$ can be split in a state-dependent part and an input-dependent part. The state-dependent part is, with the help of a maximal monotonicity requirement, assumed to be strictly passive with respect to the Lebesgue measure and passive with respect to the atomic measure. Such kind of systems will be simply referred to as 'maximal monotone systems' in the following.

We first formalise maximal monotone systems in Section 8.2.1, subsequently give sufficient conditions for the existence of a compact positively invariant set in Section 8.2.2 and finally give sufficient conditions for convergence in Section 8.2.3.

8.2.1 Maximal Monotone Systems

Let $\boldsymbol{x} \in \mathbb{R}^n$ be the state-vector of the system and $\boldsymbol{w} \in \mathbb{R}^m$ be the input vector. Consider the time-evolution of $\boldsymbol{x}$ to be governed by a measure differential equation of the form

$$\mathrm{d}\boldsymbol{x} = -\mathrm{d}\boldsymbol{a} - \boldsymbol{c}(\boldsymbol{x})\mathrm{d}t + \mathrm{d}\boldsymbol{b}(\boldsymbol{w}), \qquad (8.3)$$

where $\boldsymbol{c} : \mathbb{R}^n \to \mathbb{R}^n$ is a single-valued function and $\mathrm{d}\boldsymbol{a}$ and $\mathrm{d}\boldsymbol{b}(\boldsymbol{w})$ are differential measures with densities with respect to $\mathrm{d}t$ and $\mathrm{d}\eta$, i.e.

$$\mathrm{d}\boldsymbol{a} = \boldsymbol{a}_t'\mathrm{d}t + \boldsymbol{a}_\eta'\mathrm{d}\eta, \qquad (8.4)$$

and

$$\mathrm{d}\boldsymbol{b}(\boldsymbol{w}) = \boldsymbol{b}_t'(\boldsymbol{w})\mathrm{d}t + \boldsymbol{b}_\eta'(\boldsymbol{w})\mathrm{d}\eta. \qquad (8.5)$$

In the following, we will assume $\boldsymbol{x}^{\mathrm{T}}\boldsymbol{b}_\eta'(\boldsymbol{w})$ to be bounded from above by a constant β. Basically, this gives an upper-bound on the energy input of the impulsive inputs. Such an assumption makes sense from the physical point of view, see the example in Section 8.4.1. The quantities $\boldsymbol{a}_t'$ and $\boldsymbol{a}_\eta'$, which are functions of time, obey the set-valued laws

$$\boldsymbol{a}_t' \in \boldsymbol{\mathcal{A}}(\boldsymbol{x}), \qquad (8.6)$$

$$a'_\eta \in \boldsymbol{A}(\boldsymbol{x}^+), \tag{8.7}$$

where $\boldsymbol{A}$ is a set-valued mapping. The dynamics can be decomposed in a Lebesgue measurable part and an atomic part. The Lebesgue measurable part gives the differential equation

$$\dot{\boldsymbol{x}}(t) := \boldsymbol{x}'_t = -\boldsymbol{a}'_t(\boldsymbol{x}(t)) - \boldsymbol{c}(\boldsymbol{x}(t)) + \boldsymbol{b}'_t(\boldsymbol{w}(t)), \tag{8.8}$$

which forms with the set-valued law (8.6) a differential inclusion

$$\dot{\boldsymbol{x}} \in -\boldsymbol{A}(\boldsymbol{x}) - \boldsymbol{c}(\boldsymbol{x}) + \boldsymbol{b}'_t(\boldsymbol{w}) \quad \text{a.e.} \tag{8.9}$$

The atomic part gives the state-reset rule

$$\boldsymbol{x}^+ - \boldsymbol{x}^- := \boldsymbol{x}'_\eta = -\boldsymbol{a}'_\eta + \boldsymbol{b}'_\eta(\boldsymbol{w}). \tag{8.10}$$

In mechanics, the state-reset rule is called the impact equation. The above impact law (8.7), for which $\boldsymbol{A}$ is only a function of $\boldsymbol{x}^+$, corresponds to a completely inelastic impact equation. Because of the similarity between the laws (8.6) and (8.7), we can combine these laws into the measure law

$$\mathrm{d}\boldsymbol{a} \in \mathrm{d}\boldsymbol{A}(\boldsymbol{x}^+) = \boldsymbol{A}(\boldsymbol{x}^+)(\mathrm{d}t + \mathrm{d}\eta). \tag{8.11}$$

The equality of measures (8.3) together with the measure law (8.11) constitutes a measure differential inclusion

$$\mathrm{d}\boldsymbol{x} \in -\mathrm{d}\boldsymbol{A}(\boldsymbol{x}^+) - \boldsymbol{c}(\boldsymbol{x})\mathrm{d}t + \mathrm{d}\boldsymbol{b}(\boldsymbol{w}) := \mathrm{d}\boldsymbol{\Gamma}(\boldsymbol{x}, \boldsymbol{w}). \tag{8.12}$$

The set-valued operator $\boldsymbol{A}(\boldsymbol{x})$ models the non-smooth dissipative elements in the system. We assume that $\boldsymbol{A}(\boldsymbol{x})$ is a maximal monotone set-valued mapping, i.e. $\boldsymbol{A}(\boldsymbol{x})$ satisfies

$$(\boldsymbol{x}_2 - \boldsymbol{x}_1)^{\mathrm{T}} (\boldsymbol{A}(\boldsymbol{x}_2) - \boldsymbol{A}(\boldsymbol{x}_1)) \geq 0, \tag{8.13}$$

for any two states $\boldsymbol{x}_1, \boldsymbol{x}_2 \in \mathcal{X}$. Moreover, we assume that $\boldsymbol{0} \in \boldsymbol{A}(\boldsymbol{0})$. This last assumption together with the monotonicity assumption implies the condition

$$\boldsymbol{x}^{\mathrm{T}} \boldsymbol{A}(\boldsymbol{x}) \geq \boldsymbol{0} \tag{8.14}$$

for any $\boldsymbol{x} \in \mathcal{X}$, i.e. the action of $\boldsymbol{A}$ is passive. Furthermore, we assume that $\boldsymbol{A}(\boldsymbol{x}) + \boldsymbol{c}(\boldsymbol{x})$ is a strictly maximal monotone set-valued mapping, i.e. there exists an $\alpha > 0$ such that

$$(\boldsymbol{x}_2 - \boldsymbol{x}_1)^{\mathrm{T}} (\boldsymbol{A}(\boldsymbol{x}_2) + \boldsymbol{c}(\boldsymbol{x}_2) - \boldsymbol{A}(\boldsymbol{x}_1) - \boldsymbol{c}(\boldsymbol{x}_1)) \geq \alpha \|\boldsymbol{x}_2 - \boldsymbol{x}_1\|^2, \tag{8.15}$$

for any two states $\boldsymbol{x}_1, \boldsymbol{x}_2 \in \mathcal{X}$.

8.2.2 Existence of a Compact Positively Invariant Set

The existence of a compact positively invariant set is useful in the proof of convergence as will become clear in Section 8.2.3. Clearly, if the impulsive inputs are passive in the sense that $(\boldsymbol{x}^+)^{\mathrm{T}}\boldsymbol{b}'_\eta(\boldsymbol{w}(t)) \leq 0$ for all t, then the system is dissipative for large $\|\boldsymbol{x}\|$ and all solutions must be bounded. In the following theorem, we give a less stringent sufficient condition for the existence of a compact positively invariant set of (8.12) based on a dwell-time condition [75, 76].

Theorem 8.5. *A measure differential inclusions of the form (8.12) has a compact positively invariant set if*

1. $\boldsymbol{A}(\boldsymbol{x})$ *is a maximal monotone set-valued mapping with* $\boldsymbol{0} \in \boldsymbol{A}(\boldsymbol{0})$,
2. $\boldsymbol{A}(\boldsymbol{x})+\boldsymbol{c}(\boldsymbol{x})$ *is a strictly maximal monotone set-valued mapping, i.e. there exists an* $\alpha > 0$ *such that (8.15) is satisfied*,
3. *there exists a scalar* $\beta \in \mathbb{R}$ *such that* $(\boldsymbol{x}^+)^{\mathrm{T}}\boldsymbol{b}'_\eta(\boldsymbol{w}) \leq \beta$ *for all* $\boldsymbol{x} \in \mathcal{X}$, *i.e. the energy input of the impulsive inputs is bounded from above*,
4. *the time-instances* t_i *for which the input is impulsive are separated by the dwell-time* $\tau \leq t_{i+1} - t_i$, *with* $\tau = \frac{\delta}{2(\delta-1)\alpha}\ln(1+\frac{2\beta}{\delta^2\gamma^2})$ *and* $\gamma := \frac{1}{\alpha}\sup_{t\in\mathbb{R},\boldsymbol{a}'_t(\boldsymbol{0})\in\boldsymbol{A}(\boldsymbol{0})}\{-\boldsymbol{a}'_t(\boldsymbol{0}) - \boldsymbol{c}(\boldsymbol{0}) + \boldsymbol{b}'_t(\boldsymbol{w}(t))\}$ *for some* $\delta > 1$.

Proof: Consider the Lyapunov candidate function $W = \frac{1}{2}\boldsymbol{x}^{\mathrm{T}}\boldsymbol{x}$. The differential measure of W has a density $\dot{W}$ with respect to the Lebesgue measure $\mathrm{d}t$ and a density $W^+ - W^-$ with respect to the atomic measure $\mathrm{d}\eta$, i.e. $\mathrm{d}W = \dot{W}\mathrm{d}t + (W^+ - W^-)\mathrm{d}\eta$. We first evaluate the density $\dot{W}$:

$$
\begin{aligned}
\dot{W} &= \boldsymbol{x}^{\mathrm{T}}(-\boldsymbol{a}'_t - \boldsymbol{c}(\boldsymbol{x}) + \boldsymbol{b}'_t(\boldsymbol{w})) \\
&= \boldsymbol{x}^{\mathrm{T}}(-\boldsymbol{a}'_t - \boldsymbol{c}(\boldsymbol{x}) + \boldsymbol{a}'_t(\boldsymbol{0}) + \boldsymbol{c}(\boldsymbol{0})) + \boldsymbol{x}^{\mathrm{T}}(-\boldsymbol{a}'_t(\boldsymbol{0}) - \boldsymbol{c}(\boldsymbol{0}) + \boldsymbol{b}'_t(\boldsymbol{w})),
\end{aligned}
\tag{8.16}
$$

with $\boldsymbol{a}'_t \in \boldsymbol{A}(\boldsymbol{x})$ and $\boldsymbol{a}'_t(\boldsymbol{0}) \in \boldsymbol{A}(\boldsymbol{0})$. Due to strict monotonicity of $\boldsymbol{A}(\boldsymbol{x})+\boldsymbol{c}(\boldsymbol{x})$, there exists a constant $\alpha > 0$ such that

$$
\begin{aligned}
\dot{W} &\leq -\alpha\|\boldsymbol{x}\|^2 + \boldsymbol{x}^{\mathrm{T}}(-\boldsymbol{a}'_t(\boldsymbol{0}) - \boldsymbol{c}(\boldsymbol{0}) + \boldsymbol{b}'_t(\boldsymbol{w})), \\
&\leq -\|\boldsymbol{x}\|\left(\alpha\|\boldsymbol{x}\| - \sup_{t\in\mathbb{R},\boldsymbol{a}'_t(\boldsymbol{0})\in\boldsymbol{A}(\boldsymbol{0})}\{-\boldsymbol{a}'_t(\boldsymbol{0}) - \boldsymbol{c}(\boldsymbol{0}) + \boldsymbol{b}'_t(\boldsymbol{w}(t))\}\right).
\end{aligned}
\tag{8.17}
$$

Note that $\dot{W} < 0$ for $\boldsymbol{x}$ satisfying

$$
\|\boldsymbol{x}\| > \frac{1}{\alpha}\sup_{t\in\mathbb{R},\boldsymbol{a}'_t(\boldsymbol{0})\in\boldsymbol{A}(\boldsymbol{0})}\{-\boldsymbol{a}'_t(\boldsymbol{0}) - \boldsymbol{c}(\boldsymbol{0}) + \boldsymbol{b}'_t(\boldsymbol{w}(t))\}.
\tag{8.18}
$$

Let γ be

$$
\gamma = \max\left(0, \frac{1}{\alpha}\sup_{t\in\mathbb{R},\boldsymbol{a}'_t(\boldsymbol{0})\in\boldsymbol{A}(\boldsymbol{0})}\{-\boldsymbol{a}'_t(\boldsymbol{0}) - \boldsymbol{c}(\boldsymbol{0}) + \boldsymbol{b}'_t(\boldsymbol{w}(t))\}\right).
\tag{8.19}
$$

For $\|\boldsymbol{x}\| > \gamma$ we can prove an exponential decay of W (in between state jumps at $t = t_i$). The function $f(x) = -(1-\frac{1}{\delta})\alpha x^2$ is greater than $g(x) = -\alpha x^2 + \gamma\alpha x$

for $x > \delta\gamma$, where $\delta > 1$ is an arbitrary constant and $\gamma > 0$. It therefore holds that $\dot{W} \leq -(1 - \frac{1}{\delta})\alpha\|x\|^2$ for $\|x\| \geq \delta\gamma$, i.e.

$$\dot{W} \leq -2\left(1 - \frac{1}{\delta}\right)\alpha W \quad \text{for } \|x\| \geq \delta\gamma. \tag{8.20}$$

Subsequently, we consider the jump $W^+ - W^-$ of W:

$$W^+ - W^- = \frac{1}{2}(x^+ + x^-)^{\mathrm{T}}\left(x^+ - x^-\right) \tag{8.21}$$

with $x^+ - x^- = -a'_\eta + b'_\eta(w)$ and $a'_\eta \in \mathcal{A}(x^+)$. Elimination of x^- and exploiting the monotonicity of $\mathcal{A}(x)$ gives

$$\begin{aligned}
W^+ - W^- &= \frac{1}{2}(2x^+ + a'_\eta - b'_\eta(w))^{\mathrm{T}}\left(-a'_\eta + b'_\eta(w)\right) \\
&= (x^+)^{\mathrm{T}}\left(-a'_\eta + b'_\eta(w)\right) - \frac{1}{2}\left\|a'_\eta - b'_\eta(w)\right\|^2 \\
&\leq \beta,
\end{aligned} \tag{8.22}$$

in which we used the assumption that the energy input of the impulsive inputs $b'_\eta(w)$ is bounded from above by β (see condition 3 in the theorem) and the monotonicity and passivity of $\mathcal{A}$. Then, due to (8.17) and (8.18), for the non-impulsive part of the motion it holds that if $\|x(t_0)\| \leq \gamma$ then $\|x(t)\| \leq \gamma$ for all $t \in [t_0, t^*]$ (if no state resets occur in this time interval). Moreover, as far as the state resets are concerned, (8.22) shows that a state reset from a state $x^-(t_i) \in \mathcal{V}$ with $\mathcal{V} = \{x \in \mathcal{X} \mid \|x\| \leq \delta\gamma\}$ can only occur to $x^+(t_i)$ such that $W(x^+(t_i)) := \frac{1}{2}\|x^+(t_i)\|^2 \leq W(x^-(t_i)) + \beta \leq \frac{1}{2}\delta^2\gamma^2 + \beta$ (note hereto the specific form of $W = \frac{1}{2}x^{\mathrm{T}}x$). During the following open time-interval (t_i, t_{i+1}) for which $b'_\eta(w(t)) = 0$, the function W evolves as

$$W(x^-(t_{i+1})) = W(x^+(t_i)) + \int_{(t_i, t_{i+1})} \mathrm{d}W, \tag{8.23}$$

which may involve impulsive motion due to dissipative impulses a'_η. Let $t_\mathcal{V} \in (t_i, t_{i+1})$ be the time-instance for which $\|x^-(t_\mathcal{V})\| = \delta\gamma$. The function W will necessarily decrease during the time-interval $(t_i, t_\mathcal{V})$ due to (8.20) and $W^+ - W^- = -(x^+)^{\mathrm{T}}a'_\eta - \frac{1}{2}\|a'_\eta\|^2 \leq 0$ (the state-dependent impulses are passive). It therefore holds that

$$W(x^-(t_\mathcal{V})) \leq e^{-2(1-\frac{1}{\delta})\alpha(t_\mathcal{V}-t_i)}W(x^+(t_i)), \tag{8.24}$$

because $\mathrm{d}W \leq -2(1 - \frac{1}{\delta})\alpha W \mathrm{d}t + (W^+ - W^-)\mathrm{d}\eta \leq -2(1 - \frac{1}{\delta})\alpha W \mathrm{d}t$ for positive measures. Using $W(x^-(t_\mathcal{V})) = \frac{1}{2}\delta^2\gamma^2$ and $W(x^+(t_i)) \leq \frac{1}{2}\delta^2\gamma^2 + \beta$ in the exponential decrease (8.24) gives

$$\frac{1}{2}\delta^2\gamma^2 \leq e^{-2(1-\frac{1}{\delta})\alpha(t_\mathcal{V}-t_i)}\left(\frac{1}{2}\delta^2\gamma^2 + \beta\right) \tag{8.25}$$

or

$$t_{\mathcal{V}} - t_i \leq \frac{\delta}{2(\delta - 1)\alpha} \ln(1 + \frac{2\beta}{\delta^2\gamma^2}). \tag{8.26}$$

Consequently, if the next impulsive time-instance t_{i+1} of the input is after $t_{\mathcal{V}}$, then the solution $\boldsymbol{x}(t)$ has enough time to reach $\mathcal{V}$. Hence, if the impulsive time-instance of the input are separated by the dwell-time τ, i.e. $t_{i+1} - t_i \geq \tau$, with

$$\tau = \frac{\delta}{2(\delta - 1)\alpha} \ln(1 + \frac{2\beta}{\delta^2\gamma^2}), \tag{8.27}$$

then the set

$$\mathcal{W} = \left\{ \boldsymbol{x} \in \mathcal{X} \,\middle|\, \frac{1}{2}\|\boldsymbol{x}\|^2 \leq \frac{1}{2}\delta^2\gamma^2 + \beta \right\} \tag{8.28}$$

is a compact positively invariant set. □

Typically, we would like the invariant set $\mathcal{W}$ to be as small as possible, as it gives an upper-bound for the trajectories of the system. On the other hand, we also want the dwell-time to be as small as possible. The constant $\delta > 1$ plays in interesting role in the above theorem. By increasing δ, we allow the invariant set $\mathcal{W}$ to be larger, thereby decreasing the dwell-time τ. So, there is a kind of trade-off between the size of the invariant set and the dwell-time. Any *finite* value of δ is sufficient to prove the existence of a *compact* positively invariant set. We therefore can take the dwell-time τ to be an arbitrary small value, but not infinitely small. This brings us to the following corollary:

Corollary 8.6. *If the size of the compact positively invariant set is not of interest, then Condition 4 in Theorem 8.5 can be replaced by an arbitrary small dwell-time $\tau > 0$.*

Proof: Taking the limit of $\delta \to \infty$ gives the condition $\tau > 0$ for arbitrary γ and β. □

It therefore suffices to assume that the impulsive inputs are separated in time (which is not a strange assumption from a physical point of view) and simply put τ equal to the (unknown) minimal time-lapse between the impulsive inputs. Then, we calculate the corresponding value of δ and obtain the size of the compact positively invariant set.

In this section, we presented a sufficient condition for the existence of a compact positively invariant set, but the attractivity of solutions outside $\mathcal{W}$ to $\mathcal{W}$ is not guaranteed. If in addition the system is incrementally attractively stable, for which we will give a sufficient condition in Section 8.2.3, then it is also assured that all solutions outside $\mathcal{W}$ converge to $\mathcal{W}$.

8.2.3 Conditions for Convergence

In the following theorem, it is stated that strictly maximal monotone measure differential inclusions are uniformly convergent.

Theorem 8.7. *A measure differential inclusion of the form (8.12) is exponentially convergent if*

1. $\mathbf{A}(\mathbf{x})$ *is a maximal monotone set-valued mapping, with* $\mathbf{0} \in \mathbf{A}(\mathbf{0})$,
2. $\mathbf{A}(\mathbf{x}) + \mathbf{c}(\mathbf{x})$ *is a strictly maximal monotone set-valued mapping,*
3. *system (8.12) exhibits a compact positively invariant set.*

Proof:

Let us first show that system (8.12) is incrementally attractively stable, i.e. all solutions of (8.12) converge to each other for positive time. Consider hereto two solutions $\mathbf{x}_1(t)$ and $\mathbf{x}_2(t)$ of (8.12) and a Lyapunov candidate function $V = \frac{1}{2}\|\mathbf{x}_2 - \mathbf{x}_1\|^2$. Consequently, the differential measure of V satisfies:

$$\mathrm{d}V = \frac{1}{2}(\mathbf{x}_2^+ + \mathbf{x}_2^- - \mathbf{x}_1^+ - \mathbf{x}_1^-)^{\mathrm{T}} (\mathrm{d}\mathbf{x}_2 - \mathrm{d}\mathbf{x}_1), \qquad (8.29)$$

with

$$\mathrm{d}\mathbf{x}_1 = -\mathrm{d}\mathbf{a}_1 - \mathbf{c}(\mathbf{x}_1)\mathrm{d}t + \mathrm{d}\mathbf{b}(\mathbf{w}), \quad \mathrm{d}\mathbf{x}_2 = -\mathrm{d}\mathbf{a}_2 - \mathbf{c}(\mathbf{x}_2)\mathrm{d}t + \mathrm{d}\mathbf{b}(\mathbf{w}), \quad (8.30)$$

where $\mathrm{d}\mathbf{a}_1 \in \mathbf{A}(\mathbf{x}_1^+)$ and $\mathrm{d}\mathbf{a}_2 \in \mathbf{A}(\mathbf{x}_2^+)$. The differential measure of V has a density $\dot{V}$ with respect to the Lebesgue measure $\mathrm{d}t$ and a density $V^+ - V^-$ with respect to the atomic measure $\mathrm{d}\eta$, i.e. $\mathrm{d}V = \dot{V}\mathrm{d}t + (V^+ - V^-)\mathrm{d}\eta$. We first evaluate the density $\dot{V}$:

$$\begin{aligned}\dot{V} &= -(\mathbf{x}_2 - \mathbf{x}_1)^{\mathrm{T}}(\mathbf{a}_t'(\mathbf{x}_2) + \mathbf{c}(\mathbf{x}_2) - \mathbf{b}_t'(\mathbf{w}) - \mathbf{a}_t'(\mathbf{x}_1) - \mathbf{c}(\mathbf{x}_1) + \mathbf{b}_t'(\mathbf{w})) \\ &= -(\mathbf{x}_2 - \mathbf{x}_1)^{\mathrm{T}}(\mathbf{a}_t'(\mathbf{x}_2) + \mathbf{c}(\mathbf{x}_2) - \mathbf{a}_t'(\mathbf{x}_1) - \mathbf{c}(\mathbf{x}_1)), \end{aligned}$$
$$(8.31)$$

where $\mathbf{a}_t'(\mathbf{x}_1) \in \mathbf{A}(\mathbf{x}_1)$ and $\mathbf{a}_t'(\mathbf{x}_2) \in \mathbf{A}(\mathbf{x}_2)$, since both solutions $\mathbf{x}_1$ and $\mathbf{x}_2$ correspond to the same perturbation $\mathbf{w}$. Due to strict monotonicity of $\mathbf{A}(\mathbf{x}) + \mathbf{c}(\mathbf{x})$, there exists a constant $\alpha > 0$ such that

$$\dot{V} \leq -\alpha\|\mathbf{x}_2 - \mathbf{x}_1\|^2. \qquad (8.32)$$

Subsequently, we consider the jump $V^+ - V^-$ of V:

$$V^+ - V^- = \frac{1}{2}(\mathbf{x}_2^+ + \mathbf{x}_2^- - \mathbf{x}_1^+ - \mathbf{x}_1^-)^{\mathrm{T}} (\mathbf{x}_2^+ - \mathbf{x}_2^- - \mathbf{x}_1^+ + \mathbf{x}_1^-), \qquad (8.33)$$

with

$$\begin{aligned} \mathbf{x}_1^+ - \mathbf{x}_1^- &= -\mathbf{a}_\eta'(\mathbf{x}_1) + \mathbf{b}_\eta'(\mathbf{w}), \quad \mathbf{a}_\eta'(\mathbf{x}_1) \in \mathbf{A}(\mathbf{x}_1^+), \\ \mathbf{x}_2^+ - \mathbf{x}_2^- &= -\mathbf{a}_\eta'(\mathbf{x}_2) + \mathbf{b}_\eta'(\mathbf{w}), \quad \mathbf{a}_\eta'(\mathbf{x}_2) \in \mathbf{A}(\mathbf{x}_2^+). \end{aligned} \qquad (8.34)$$

Elimination of $\mathbf{x}_1^-$ and $\mathbf{x}_2^-$ and exploiting the monotonicity of $\mathbf{A}(\mathbf{x})$ gives

$$V^+ - V^- = \frac{1}{2}(2\boldsymbol{x}_2^+ + \boldsymbol{a}'_\eta(\boldsymbol{x}_2) - 2\boldsymbol{x}_1^+ - \boldsymbol{a}'_\eta(\boldsymbol{x}_1))^{\mathrm{T}} \left(-\boldsymbol{a}'_\eta(\boldsymbol{x}_2) + \boldsymbol{a}'_\eta(\boldsymbol{x}_1)\right)$$

$$= -(\boldsymbol{x}_2^+ - \boldsymbol{x}_1^+)^{\mathrm{T}} \left(\boldsymbol{a}'_\eta(\boldsymbol{x}_2) - \boldsymbol{a}'_\eta(\boldsymbol{x}_1)\right) - \frac{1}{2} \left\|\boldsymbol{a}'_\eta(\boldsymbol{x}_2) - \boldsymbol{a}'_\eta(\boldsymbol{x}_1)\right\|^2$$

$$\leq 0.$$

$$(8.35)$$

It therefore holds that V strictly decreases over every non-empty compact time-interval as long as $\boldsymbol{x}_2 \neq \boldsymbol{x}_1$. In turn, this implies that all solutions of (8.12) converge to each other exponentially (and therefore uniformly).

Finally we use Lemma 2 in [178], which formulates that if a system exhibits a compact positively invariant set, then the existence of a solution that is bounded for $t \in \mathbb{R}$ is guaranteed. We will denote this 'steady-state' solution by $\bar{\boldsymbol{x}}_w(t)$. The original lemma is formulated for differential equations (possibly with discontinuities, therewith including differential inclusions, with bounded right-hand sides). Here, we use this lemma for measure differential inclusions and would like to note that the proof of the lemma allows for such extensions if we only require continuous dependence on initial conditions. The latter is guaranteed for monotone measure differential inclusions, because incremental stability implies a continuous dependence on initial conditions.

Since all solutions of (8.12) are globally exponentially stable, also $\bar{\boldsymbol{x}}_w(t)$ is a globally exponentially stable solution. This concludes the proof that the measure differential inclusions (8.12) is exponentially convergent.   □

8.3 Tracking Control for Lur'e Type Systems

An important application of convergence theory is the tracking control of dynamical systems, i.e. the design of a controller, such that a desired trajectory $\boldsymbol{x}_d(t)$ of the system exists and is globally attractively stable. Tracking control of measure differential inclusions has received very little attention in literature [21, 27, 113].

In this section, we consider the tracking control problem of a nonlinear measure differential inclusion, which can be decomposed into a linear measure differential inclusion with a nonlinear maximal monotone operator in the feedback path. We allow the desired trajectory $\boldsymbol{x}_d(t)$ to have discontinuities in time (but assume it to be of locally bounded variation). The open-loop dynamics is described by an equality of measures:

$$\mathrm{d}\boldsymbol{x} = \boldsymbol{A}\boldsymbol{x}\mathrm{d}t + \boldsymbol{B}\mathrm{d}\boldsymbol{p} + \boldsymbol{D}\mathrm{d}\boldsymbol{s},$$

$$\boldsymbol{y} = \boldsymbol{C}\boldsymbol{x}$$

$$-\mathrm{d}\boldsymbol{s} \in \mathcal{H}(\boldsymbol{y}),$$

$$(8.36)$$

with $\boldsymbol{A} \in \mathbb{R}^{n \times n}$, $\boldsymbol{B} \in \mathbb{R}^{n \times n_p}$, $\boldsymbol{C} \in \mathbb{R}^{m \times n}$ and $\boldsymbol{D} \in \mathbb{R}^{n \times m}$. Herein, $\boldsymbol{x} \in \mathbb{R}^n$ is the system state (of locally bounded variation), $\mathrm{d}\boldsymbol{p} = \boldsymbol{w}\mathrm{d}t + \boldsymbol{W}\mathrm{d}\eta$ is the differential measure of the control action and $ds = \lambda dt + \Lambda d\eta$ is the differential measure of the nonlinearity in the feedback loop that is characterised by the set-valued maximal monotone mapping $\mathcal{H}(y)$. The problem that we consider here is the design of a control law dp such that the tracking of the desired trajectory $x_d(t)$ is assured. We propose to tackle the tracking problem by means of a combination of Lebesgue measurable linear tracking error-feedback and a possibly impulsive feedforward control:

$$dp = w_{fb}dt + dp_{ff}(t) = K\left(x - x_d(t)\right)dt + w_{ff}(t)dt + W_{ff}(t)d\eta, \quad (8.37)$$

with

$$w_{fb} = K\left(x - x_d(t)\right), \quad dp_{ff}(t) = w_{ff}(t)dt + W_{ff}(t)d\eta, \quad (8.38)$$

where $K \in \mathbb{R}^{n_p \times n}$ is the feedback gain matrix and $x_d(t)$ the desired state trajectory. We restrict the energy input of the impulsive control action $W_{ff}(t)$ to be bounded from above

$$(x^+)^{\mathrm{T}}BW_{ff} \leq \beta. \quad (8.39)$$

Note that this condition puts a bound on the jumps in the desired trajectory $x_d(t)$ which can be realised. Combining the control law (8.37) with the system dynamics (8.36) yields the closed-loop dynamics:

$$\begin{aligned} dx &= A_{cl}xdt + Dds + B(-Kx_d(t)dt + dp_{ff}(t)), \\ y &= Cx \\ -ds &\in \mathcal{H}(y), \end{aligned} \quad (8.40)$$

with

$$A_{cl} = A + BK. \quad (8.41)$$

We now propose a convergence-based control design. The main idea of this convergence-based control design is to find a controller of the form (8.37) that guarantees two properties:

a. the closed-loop system has a trajectory which is bounded for all t and along which the tracking error $x - x_d(t)$ is identically zero. In other words, the feedforward $w_{ff}(t)$ and $W_{ff}(t)$ has to be designed such that it induces the desired solution $x_d(t)$;

b. the closed-loop system is uniformly convergent. Hereto, the control gain matrix K should be designed appropriately.

Condition b) guarantees that the closed-loop system has a unique bounded UGAS steady-state solution, while condition a) guarantees that, by Property 8.2, this steady-state solution equals the bounded solution of the closed-loop system with zero tracking error. For other types of systems, the convergence property has been exploited to solve the output regulation problem, tracking problems and the synchronisation problem, see e.g. [132,134,169,172].

For the design of the feedback gain (to ensure that condition b is met), we employ the following strategy. First, we design $\boldsymbol{K}$ such that the linear part of system (8.40), (8.41) is strictly passive. Subsequently, using the fact that $\mathcal{H}(\boldsymbol{y})$ is maximal monotone, we show that this implies that the measure differential inclusion (8.40), (8.41) is (after a coordinate transformation) maximal monotone. Hence, exponential convergence for measure differential inclusions of the form (8.40) can be proven using Theorem 8.7. A similar result was found for a class of differential inclusions by Yakubovich [178]. In [178] it is shown that strict passivity of the linear part of the system is sufficient for exponential convergence for Lur'e-type systems with monotone set-valued nonlinearities and absolutely continuous state (i.e. for a class of differential inclusions).

Here, we will show that for *measure* differential inclusions (8.40), (8.41) that, if the triple $(\boldsymbol{A}_{cl}, \boldsymbol{D}, \boldsymbol{C})$ is strictly positive real (i.e. the linear part of the system (8.40) is strictly passive) and the nonlinearity $\mathcal{H}(\boldsymbol{y})$ is a monotone nonlinearity, then the system is uniformly convergent. Therefore, the feedback gain matrix $\boldsymbol{K}$ should be designed such that the triple $(\boldsymbol{A}_{cl}, \boldsymbol{D}, \boldsymbol{C})$ is strictly positive real.

Note that the triple $(\boldsymbol{A}_{cl}, \boldsymbol{D}, \boldsymbol{C})$ is rendered strictly passive by means of the feedback design. In other words we design $\boldsymbol{K}$ such that there exists a positive definite matrix $\boldsymbol{P} = \boldsymbol{P}^{\mathrm{T}} > 0$ for which the following conditions are satisfied:

$$\boldsymbol{A}_{cl}^{\mathrm{T}}\boldsymbol{P} + \boldsymbol{P}\boldsymbol{A}_{cl} < 0,$$
$$\boldsymbol{D}^{\mathrm{T}}\boldsymbol{P} = \boldsymbol{C}. \tag{8.42}$$

Let us now introduce a linear coordinate transformation $\tilde{\boldsymbol{x}} = \boldsymbol{S}\boldsymbol{x}$, where $\boldsymbol{P} = \boldsymbol{S}^{\mathrm{T}}\boldsymbol{S}$, i.e. $\boldsymbol{S}$ is the square root of $\boldsymbol{P}$. Using these transformed coordinates, the closed-loop dynamics can then be formulated in the form (8.12):

$$\mathrm{d}\tilde{\boldsymbol{x}} \in -\mathrm{d}\boldsymbol{\mathcal{A}}(\tilde{\boldsymbol{x}}^+) - \boldsymbol{c}(\tilde{\boldsymbol{x}})\mathrm{d}t + \mathrm{d}\boldsymbol{b}(\boldsymbol{w}) \tag{8.43}$$

with

$$\mathrm{d}\boldsymbol{\mathcal{A}}(\tilde{\boldsymbol{x}}^+) = \boldsymbol{S}\boldsymbol{D}\mathcal{H}(\boldsymbol{C}\boldsymbol{S}^{-1}\tilde{\boldsymbol{x}}^+)(\mathrm{d}t + \mathrm{d}\eta), \tag{8.44}$$

$$\boldsymbol{c}(\tilde{\boldsymbol{x}}) = -\boldsymbol{S}\boldsymbol{A}_{cl}\boldsymbol{S}^{-1}\tilde{\boldsymbol{x}}, \tag{8.45}$$

$$\mathrm{d}\boldsymbol{b}(\boldsymbol{w}) = \boldsymbol{S}\boldsymbol{B}(-\boldsymbol{K}\boldsymbol{x}_d(t)\mathrm{d}t + \mathrm{d}\boldsymbol{p}_{ff}(t)). \tag{8.46}$$

We will now show that condition (8.42) together with the monotonicity of the set-valued mapping $\mathcal{H}(\boldsymbol{y})$ implies strict monotonicity of the differential inclusion (8.43). Hereto, we prove the strict monotonicity of the set-valued operator $-\boldsymbol{S}\boldsymbol{A}_{cl}\boldsymbol{S}^{-1}\tilde{\boldsymbol{x}} + \boldsymbol{S}\boldsymbol{D}\mathcal{H}(\boldsymbol{C}\boldsymbol{S}^{-1}\tilde{\boldsymbol{x}})$. Using $\boldsymbol{\lambda}_i \in -\mathcal{H}(\boldsymbol{C}\boldsymbol{S}^{-1}\tilde{\boldsymbol{x}}_i)$, $i = 1, 2$, we can verify that it holds that

$$(\tilde{x}_1 - \tilde{x}_2)^{\mathrm{T}} \left(-SA_{cl}S^{-1}\tilde{x}_1 - SD\lambda_1 + SA_{cl}S^{-1}\tilde{x}_2 + SD\lambda_2 \right)$$
$$= (\tilde{x}_1 - \tilde{x}_2)^{\mathrm{T}} \left(-SA_{cl}S^{-1} \right)(\tilde{x}_1 - \tilde{x}_2) + (\tilde{x}_1 - \tilde{x}_2)^{\mathrm{T}} SD (\lambda_2 - \lambda_1)$$
$$= -(x_1 - x_2)^{\mathrm{T}} \left(S^{\mathrm{T}} SA_{cl} \right)(x_1 - x_2) + (x_1 - x_2)^{\mathrm{T}} S^{\mathrm{T}} SD (\lambda_2 - \lambda_1)$$
$$= -\frac{1}{2}(x_1 - x_2)^{\mathrm{T}} \left(PA_{cl} + A_{cl}^{\mathrm{T}}P \right)(x_1 - x_2) + (x_1 - x_2)^{\mathrm{T}} PD (\lambda_2 - \lambda_1).$$

$$(8.47)$$

Using the conditions (8.42), we can write $-x^{\mathrm{T}}(A_{cl}^{\mathrm{T}}P + PA_{cl})x \geq \alpha\|x\|^2$ for some $\alpha > 0$ and $x^{\mathrm{T}}PD = x^{\mathrm{T}}C^{\mathrm{T}} = y^{\mathrm{T}}$ and Eq. (8.47) becomes

$$(\tilde{x}_1 - \tilde{x}_2)^{\mathrm{T}} \left(-SA_{cl}S^{-1}\tilde{x}_1 - SD\lambda_1 + SA_{cl}S^{-1}\tilde{x}_2 + SD\lambda_2 \right)$$
$$\geq \frac{\alpha}{2}\|x_1 - x_2\|^2 + (y_1 - y_2)^{\mathrm{T}}(\lambda_2 - \lambda_1).$$

$$(8.48)$$

Finally, we use the fact that $\lambda_i \in -\mathcal{H}(y_i)$, $i = 1, 2$, and the monotonicity of the set-valued nonlinearity $\mathcal{H}(y)$ to conclude that

$$(\tilde{x}_1 - \tilde{x}_2)^{\mathrm{T}} \left(-SA_{cl}S^{-1}\tilde{x}_1 - SD\lambda_1 + SA_{cl}S^{-1}\tilde{x}_2 + SD\lambda_2 \right) \geq \frac{\alpha}{2}\|x_1 - x_2\|^2.$$

$$(8.49)$$

In other words, strict monotonicity of the $\tilde{x}$-dynamics is guaranteed. Earlier in the chapter we have shown that strict monotonicity implies uniform convergence. Moreover, the convergence property is conserved under smooth coordinate transformations (see [132]). Consequently, if the $\tilde{x}$-dynamics is uniformly convergent, then also the x-dynamics is uniformly convergent.

8.4 Illustrative Examples

In the next sections, examples concerning models for the control of mechanical systems with set-valued friction and one-way clutches illustrate the power of the result in Theorem 8.7. Moreover, the results of Section 8.3 on tracking control are applied to mechanical systems with friction and a one-way clutch in Sections 8.4.2 and 8.4.3.

8.4.1 One-way Clutch

The time-evolution of the velocity of a mass m (Figure 8.1) subjected to a one-way clutch, a dashpot $b > 0$ and an external input (considering both bounded and impulsive contributions) can be described by the equality of measures

$$m\mathrm{d}u = \mathrm{d}p + \mathrm{d}s - bu\mathrm{d}t.$$

$$(8.50)$$

We can decompose the differential measure $\mathrm{d}s$ of the one-way clutch in

$$\mathrm{d}s = \lambda\mathrm{d}t + \Lambda\mathrm{d}\eta,$$

$$(8.51)$$

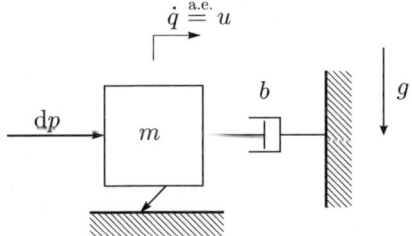

Fig. 8.1. Mass with one-way clutch and impulsive actuation.

where $\lambda := s'_t$ is the contact force and $\Lambda = s'_\eta$ is the contact impulse. The differential impulse measure ds of the one-way clutch obeys the set-valued force law

$$-ds \in \mathrm{Upr}(u^+), \qquad (8.52)$$

where $\mathrm{Upr}(x)$ is the unilateral primitive (2.24).

$$-y \in \mathrm{Upr}(x) \iff 0 \leq x \perp y \geq 0 \iff x \geq 0,\, y \geq 0,\, xy = 0, \qquad (8.53)$$

being a maximal monotone operator.

The input consists of a bounded force f and an impulse F

$$dp = f dt + F d\eta. \qquad (8.54)$$

We assume that an impulsive input $F > 0$ is transmitted by firing bullets with mass m_0 and constant speed $v \leq v_{\max}$ on the left side of the mass m. We assume a completely inelastic impact. If $u \geq v$, then the bullet is not able to hit the mass m and then the impulse F equals zero. If $u^+ < v$, then the impulse F equals the mass of the bullet multiplied with its velocity jump:

$$F = m_0(v - u^+). \qquad (8.55)$$

Similarly, we assume that an impulsive input $F < 0$ is transmitted by firing on the right side of the mass m with a speed $v < 0$, bounded by $|v| \leq v_{\max}$. The energy input $u^+ F = m_0 u^+ (v - u^+)$ of the impulse F is maximal when $u^+ = \frac{1}{2}v$ and is therefore bounded from above by $\beta := \frac{1}{4} m_0 v_{\max}^2 \geq |u^+ F|$.

We first prove the existence of a compact positively invariant set with Theorem 8.5. Theorem 8.5 uses the Lyapunov function $W(u) = \frac{1}{2}u^2$, which we recognise to be the kinetic energy divided by the mass m. The time-derivative $\dot{W}$ gives, using $u\lambda = 0$,

$$\dot{W} \leq -\frac{b}{m}u^2 + u \sup_{t \in \mathbb{R}}(f(t)), \qquad (8.56)$$

and it therefore holds that $\alpha = \frac{b}{m}$ and $\gamma = \frac{1}{b}\sup_{t \in \mathbb{R}}(f(t))$ with α and γ defined in Theorem 8.5. Theorem 8.5 states that if the time-instances t_i of the impulses F are separated by the dwell-time

$$\tau = \frac{\delta}{2(\delta - 1)\alpha} \ln(1 + \frac{2\beta}{\delta^2 \gamma^2}), \tag{8.57}$$

then the set

$$W = \left\{ u \in \mathbb{R}^+ \middle| \frac{1}{2} u^2 \le \frac{1}{2}\delta^2 \gamma^2 + \beta \right\} \tag{8.58}$$

is a compact positively invariant set for arbitrary $\delta > 1$. Following Corollory 8.6, we conclude that the dwell-time can be made arbitrary small by increasing δ. We therefore take τ to be smaller than the minimal time-lapse between two succeeding impulsive time-instances, which gives a lower bound for δ.

Just as in the proof of Theorem 8.7, we prove incremental stability using the Lyapunov function

$$V = \frac{1}{2}(u_2 - u_1)^2. \tag{8.59}$$

First, we consider the time-derivative $\dot{V}$:

$$\dot{V} = (u_2 - u_1)(\dot{u}_2 - \dot{u}_1)$$
$$= (u_2 - u_1)\frac{1}{m}(\lambda_2 - bu_2 - \lambda_1 + bu_1)$$
$$= (u_2 - u_1)\frac{1}{m}(\lambda_2 - \lambda_1) - \frac{b}{m}(u_2 - u_1)^2, \quad -\lambda_1 \in \mathrm{Upr}(u_1), \ -\lambda_2 \in \mathrm{Upr}(u_2)$$
$$\le -\frac{b}{m}(u_2 - u_1)^2. \tag{8.60}$$

Subsequently, we consider a jump in V:

$$V^+ - V^- = V(u_1^+, u_2^+) - V(u_1^-, u_2^-)$$
$$= \frac{1}{2}(u_2^+ - u_1^+)^2 - \frac{1}{2}(u_2^- - u_1^-)^2 \tag{8.61}$$
$$= \frac{1}{2}(u_2^+ + u_2^- - u_1^+ - u_1^-)(u_2^+ - u_2^- - u_1^+ + u_1^-).$$

Following the proof of Theorem 8.7, we eliminate u_1^- and u_2^- by substituting the impact equation $m(u_j^+ - u_j^-) = \Lambda_j + F$, $j = 1, 2$:

$$V^+ - V^- = \frac{1}{2}(2u_2^+ - \frac{1}{m}\Lambda_2 - 2u_1^+ + \frac{1}{m}\Lambda_1)\frac{1}{m}(\Lambda_2 - \Lambda_1)$$
$$= (u_2^+ - u_1^+)\frac{1}{m}(\Lambda_2 - \Lambda_1) - \frac{1}{2m^2}(\Lambda_2 - \Lambda_1)^2 \tag{8.62}$$
$$\le 0.$$

Hence, it holds for the differential measure dV that

$$dV = \dot{V}dt + (V^+ - V^-)d\eta \le -\alpha(u_2 - u_1)^2 dt, \quad \alpha = \frac{b}{m}. \tag{8.63}$$

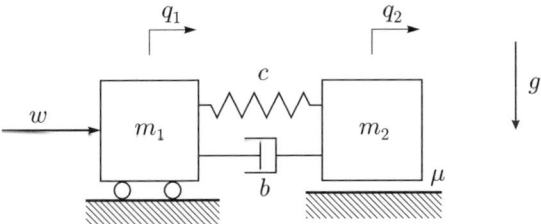

Fig. 8.2. Typical motor-load configuration with non-collocated friction and actuation.

Integration of dV over a non-empty time-interval therefore leads to a strict decrease of the function V as long as $u_2 \neq u_1$. This proves incremental stability. Consequently, the system is exponentially convergent (see Theorem 8.7).

8.4.2 Tracking Control with Set-valued Friction

In this section, we consider the tracking control problem for mechanical systems with set-valued friction. Hereto, we study a common motor-load configuration as depicted in Figure 8.2. The essential problem here is the fact that the friction and the actuation are non-collocated (i.e. the motor, mass m_1, is actuated and the load, mass m_2, is subject to friction). Note that the spring-damper combination, with stiffness c and viscous damping constant b, reflects a finite-stiffness coupling between the motor and load common in many motion systems. A common approach in tackling control problems for systems with friction is that of friction compensation. This angle of attack is clearly not feasible here since the actuation cannot compensate directly for the friction. Another common approach in compensating for nonlinearities can be recognised in the backstepping control schemes [85]. However, these generally require differentiability of the nonlinearity, which is not the case here due to the set-valued nature of the friction law.

In many applications, mainly the velocity of the load is of interest. In this context, one can think of controlling a printhead in a printer, where the printhead is to achieve a constant velocity when moving across the paper or drilling systems where the bottom-hole-assemble (including the drill bit) should achieve a constant cutting speed. From this perspective, the following third-order differential inclusion describes the dynamics of the system under study:

$$\dot{x} = \bar{A}x + Bw + D\bar{\lambda},$$
$$y = Cx \tag{8.64}$$
$$\bar{\lambda} \in -\bar{\mathcal{H}}(y),$$

with

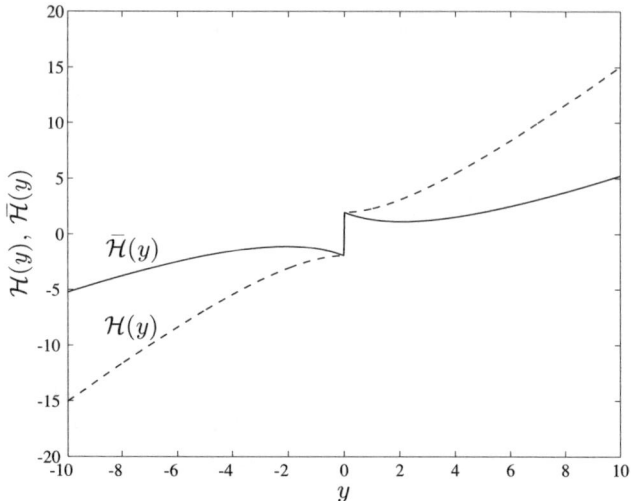

Fig. 8.3. Friction law $\mathcal{H}(y)$ and transformed friction law $\bar{\mathcal{H}}(y)$.

$$\bar{A} = \begin{bmatrix} 0 & -1 & 1 \\ \frac{c}{m_1} & -\frac{b}{m_1} & \frac{b}{m_1} \\ -\frac{c}{m_2} & \frac{b}{m_2} & -\frac{b}{m_2} \end{bmatrix}, \quad B = \begin{bmatrix} 0 \\ \frac{1}{m_1} \\ 0 \end{bmatrix}, \quad D = \begin{bmatrix} 0 \\ 0 \\ \frac{1}{m_2} \end{bmatrix}, \quad C^{\mathrm{T}} = \begin{bmatrix} 0 \\ 0 \\ 1 \end{bmatrix}. \quad (8.65)$$

Herein, $\boldsymbol{x} = \begin{bmatrix} q_2 - q_1 & \dot{q}_1 & \dot{q}_2 \end{bmatrix}^{\mathrm{T}}$ is the absolutely continuous system state, $w \in \mathbb{R}$ is the control action and $\bar{\lambda} \in \mathbb{R}$ is the friction force that is characterised by the set-valued mapping $\bar{\mathcal{H}}(\cdot) : \mathbb{R} \to \mathbb{R}$. The set-valued friction law adopted here includes a combination of Coulomb friction, viscous friction and the Stribeck effect:

$$\bar{\mathcal{H}}(y) = m_2 g \left(\mu_0 \operatorname{Sign}(y) + \mu_1 y - \frac{\mu_2 y}{1 + \mu_2 |y|} \right), \quad (8.66)$$

where g is the gravitational acceleration, $\mu_0 > 0$ is the Coulomb friction coefficient, $\mu_1 > 0$ is the viscous friction coefficient and μ_2 is an additional coefficient characterising the modelling of the Stribeck effect. It is well known that exactly such a Stribeck effect can induce instabilities, complicating the design of stabilising controllers, see e.g. [9]. In Figure 8.3, such type of set-valued static friction is depicted schematically. At this point, we will transform the friction law $\bar{\mathcal{H}}(y)$ to a strictly maximal monotone operator $\mathcal{H}(y)$

$$\mathcal{H}(y) = \bar{\mathcal{H}}(y) + \kappa y, \text{ with } \kappa = m_2 g(\mu_2 - \mu_1), \quad (8.67)$$

where the choice of κ ensures that the set-valued mapping $\mathcal{H}(y)$ is a maximal monotone mapping. System (8.64) can therefore be transformed into the form (8.36)

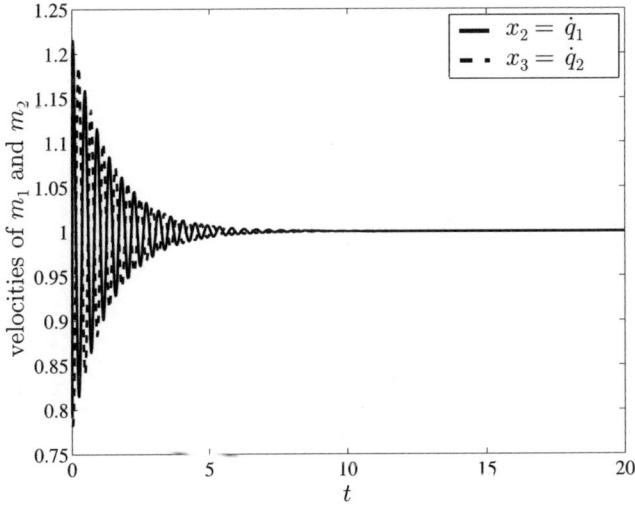

Fig. 8.4. Feedback and feedforward control.

$$\mathrm{d}x = \boldsymbol{A}x\mathrm{d}t + \boldsymbol{B}\mathrm{d}p + \boldsymbol{D}\mathrm{d}s,$$
$$y = \boldsymbol{C}x \qquad (8.68)$$
$$-\mathrm{d}s \in \mathcal{H}(y),$$

where $\boldsymbol{A} = \bar{\boldsymbol{A}} + \kappa\boldsymbol{D}\boldsymbol{C}$, $\mathrm{d}p = w\mathrm{d}t$, $\mathrm{d}s = \lambda\mathrm{d}t$ and $\mathcal{H}(y)$ is a maximal monotone mapping. We now use the convergence-based tracking control strategy proposed in Section 8.3 to solve the tracking problem of this mechanical system with friction. Hereto, we us a combination of linear error-feedback and feedforward control as in (8.37), where $\boldsymbol{K} = \begin{bmatrix} k_1 & k_2 & k_3 \end{bmatrix}$ is the feedback gain matrix which has to be chosen such that the triple $(\boldsymbol{A}_{cl}, \boldsymbol{D}, \boldsymbol{C})$ is strictly positive real.

We adopt the following system parameters: $g = 9.81$, $m_1 = 1$, $m_2 = 1$, $c = 100$, $b = 1$, $\mu_0 = 0.2$, $\mu_1 = 0.1$ and $\mu_2 = 0.2$. The resulting friction map and the transformed (monotone) friction map are shown in Figure 8.3. Moreover, we aim at tracking a constant velocity solution (with desired velocity v_d) for both the motor and the load; i.e. the desired state-trajectory is given by
$$x_d(t) = \left[\frac{1}{c} \left(-m_2 g \left(\mu_0 + \mu_1 v_d - \frac{\mu_2 v_d}{1+\mu_2|v_d|} \right) \right) \ v_d \ v_d \right]^{\mathrm{T}}, \text{ where } v_d = 1. \text{ Note that}$$
this velocity lies in the range in which the friction law exhibits a pronounced Stribeck effect. Let us design the controller in the form (8.37). Firstly, the feedforward which induces the desired solution is given by

$$w_{ff} = m_2 g \left(\mu_0 + \mu_1 v_d - \frac{\mu_2 v_d}{1 + \mu_2 |v_d|} \right). \qquad (8.69)$$

Secondly, by checking appropriate LMI conditions or frequency-domain conditions (see e.g. [85]) for the strictly positive realness of the triple $(\boldsymbol{A}_{cl}, \boldsymbol{D}, \boldsymbol{C})$,

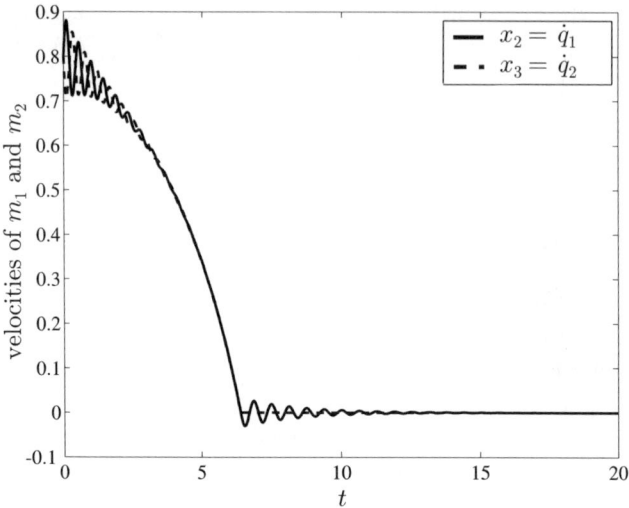

Fig. 8.5. Only feedforward control.

suitable controller gains can be selected: $k_1 = -30$, $k_2 = -150$ and $k_3 = -150$. The strict positive realness of the triple $(\boldsymbol{A}_{cl}, \boldsymbol{D}, \boldsymbol{C})$ can be proven using the following symmetric, positive definite matrix $\boldsymbol{P}$

$$\boldsymbol{P} = \begin{bmatrix} 50.98 & -0.33 & 0 \\ -0.33 & 0.005 & 0 \\ 0 & 0 & 1 \end{bmatrix} > 0. \tag{8.70}$$

satisfying the LMIs (8.42).

Next, we implement control law (8.37) on system (8.64), with these control gains and feedforward (8.69) and use numerical time-stepping schemes [1,101, 123] to numerically compute solution of the closed-loop system. In Figure 8.4, the velocities of both masses are depicted when the controller is active and asymptotic tracking of the constant velocity solution is achieved. Note that when only the feedforward is applied, the desired solution is still a solution of the system; however, no asymptotic tracking achieved, see Figure 8.5. In this figure, it is shown that both masses ultimately come to a standstill. Clearly, the system now exhibits at least two steady-state solutions; the desired solution and the solution on which $x_1 = -u_{ff}/c$, $x_2 = 0$ and $x_3 = 0$, as depicted in Figure 8.5. Consequently, the system without feedback is not convergent. For both cases the initial condition $\boldsymbol{x}(0) = \begin{bmatrix} 0 & 0.8 & 0.8 \end{bmatrix}^{\mathrm{T}}$ was used.

Note that we solve a stabilisation problem in this example. However, using the strategy discussed here, we can make any bounded feasible time-varying desired solution $\boldsymbol{x}_d(t)$ attractively stable using the same feedback gain matrix $\boldsymbol{K}$ and an appropriate feedforward.

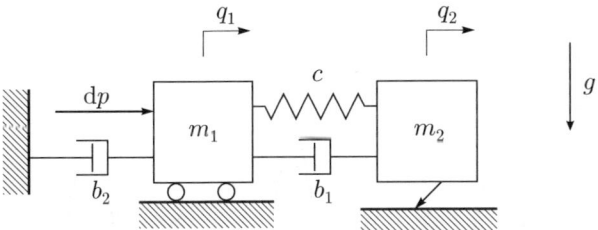

Fig. 8.6. Motor-load configuration with one-way clutch and impulsive actuation.

8.4.3 Tracking Control Using an Impulsive Input

In the example of Section 8.4.2 we solved the tracking problem mechanical motion system with set-valued friction. In the current example, we consider a similar system; however, the set-valued friction is replaced by a one-way clutch and impulsive control action is needed to achieve tracking of a periodic trajectory.

More specifically, we study a variant of the previous problem and replace the friction element by a one-way clutch and add an additional damper b_2. Moreover, we allow for impulsive inputs on the first mass. The open-loop dynamics is now described by an equality of measures:

$$\mathrm{d}\boldsymbol{x} = \boldsymbol{A}\boldsymbol{x}\mathrm{d}t + \boldsymbol{B}\mathrm{d}p + \boldsymbol{D}\mathrm{d}s,$$
$$y = \boldsymbol{C}\boldsymbol{x} \tag{8.71}$$
$$-\mathrm{d}s \in \mathcal{H}(y),$$

with

$$\boldsymbol{A} = \begin{bmatrix} 0 & -1 & 1 \\ \frac{c}{m_1} & -\frac{b_1+b_2}{m_1} & \frac{b_1}{m_1} \\ -\frac{c}{m_2} & \frac{b_1}{m_2} & -\frac{b_1}{m_2} \end{bmatrix}, \quad \boldsymbol{B} = \begin{bmatrix} 0 \\ \frac{1}{m_1} \\ 0 \end{bmatrix}, \quad \boldsymbol{D} = \begin{bmatrix} 0 \\ 0 \\ \frac{1}{m_2} \end{bmatrix}, \quad \boldsymbol{C}^{\mathrm{T}} = \begin{bmatrix} 0 \\ 0 \\ 1 \end{bmatrix}. \tag{8.72}$$

The evolution $\boldsymbol{x}(t)$ of the state vector $\boldsymbol{x} = \begin{bmatrix} q_2 - q_1 & u_1 & u_2 \end{bmatrix}^{\mathrm{T}}$ is of locally bounded variation. The differential measure of the control action $\mathrm{d}p = w\mathrm{d}t + W\mathrm{d}\eta$ now also contains an impulsive part W. The differential measure $\mathrm{d}s$ of the force in the one-way clutch is characterised by the scalar set-valued maximal monotone mapping $\mathcal{H}(x) = \mathrm{Upr}(x)$.

In this example, we try to let the velocity $x_3(t) = \dot{q}_2(t)$ approach the desired trajectory $x_{d3}(t)$. Hereto, we design trajectories $x_{d1}(t)$ and $x_{d2}(t)$ which generate the desired $x_{d3}(t)$. Subsequently, we aim at tracking of the desired state trajectory $\boldsymbol{x}_d(t)$. The state-tracking problem is solved by making the system uniformly convergent with a feedback $\boldsymbol{K}(\boldsymbol{x} - \boldsymbol{x}_d(t))$. As in Section 8.4.2, we can design $\boldsymbol{K}$ such that the triple $(\boldsymbol{A}_{cl}, \boldsymbol{D}, \boldsymbol{C})$ is rendered strictly passive, which, given the monotonicity of $\mathcal{H}(y)$, makes the system uniformly convergent.

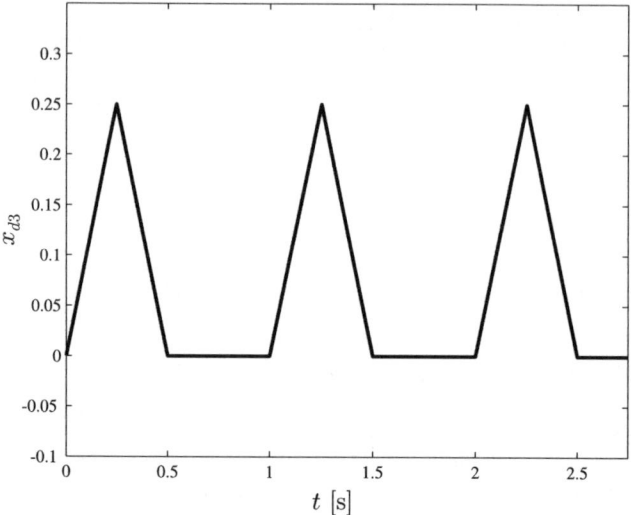

Fig. 8.7. Desired trajectory $x_{d3}(t)$.

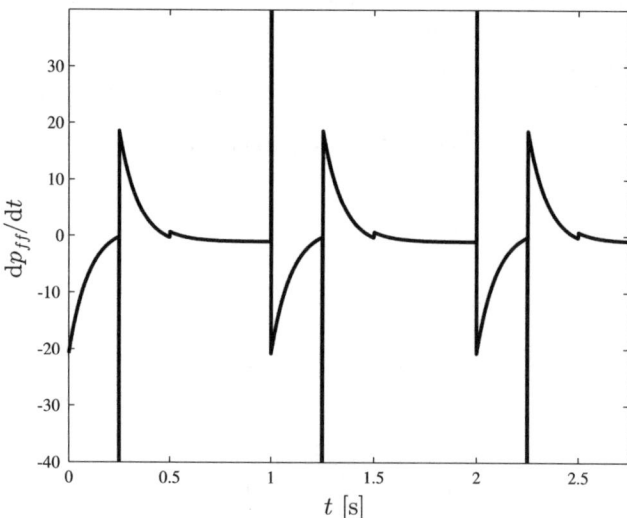

Fig. 8.8. Feedforward $\mathrm{d}p_{ff}/\mathrm{d}t$.

We adopt the following system parameters: $g > 0$, $m_1 = 1$, $m_2 = 1$, $c = 10$, $b_1 = 1$ and $b_2 = -1.4$. The negative damping $b_2 < 0$ causes the system matrix A to have a positive real eigenvalue. The desired velocity of the second mass is characterised by a periodic sawtooth wave with period time T:

$$x_{d3}(t) = \begin{cases} \mathrm{mod}(t,T) & \text{for } 0 \le \mathrm{mod}(t,T) \le \frac{T}{4} \text{ (ramp-up)} \\ -\mathrm{mod}(t,T) + \frac{T}{2} & \text{for } \frac{T}{4} \le \mathrm{mod}(t,T) \le \frac{T}{2} \text{ (ramp-down)} \\ 0 & \text{for } \frac{T}{2} \le \mathrm{mod}(t,T) \le T \text{ (deadband)} \end{cases} \quad (8.73)$$

The signal $x_{d3}(t)$ for $T = 1$ s is shown in Figure 8.7. The desired trajectory $x_{d3}(t)$ is a periodic signal which is time-continuous but has three kinks in each period. Kinks in $x_{d3}(t)$ can be achieved by applying an impulsive force on the first mass which causes an instantaneous change in the velocity $x_2 = \dot{q}_1$ and therefore a discontinuous force in the damper b_1. The one-way clutch on the second mass prevents negative values of x_{d3} and no impulsive force on the first mass is therefore necessary for the change from ramp-down to deadband. In a first step, the signals $x_{d1}(t)$, $x_{d2}(t)$ and $ds(t)$ are designed such that

$$\begin{aligned} \dot{x}_{d1}(t) &= -x_{d2}(t) + x_{d3}(t) \\ \mathrm{d}x_{d3}(t) &= (-\tfrac{c}{m_2}x_{d1}(t) - \tfrac{b_1}{m_2}(-x_{d2}(t) + x_{d3}(t)))\mathrm{d}t + \tfrac{1}{m_2}ds(t) \quad (8.74) \\ -ds(t) &\in \mathrm{Upr}(x_{d3}(t)), \end{aligned}$$

for the given periodic trajectory $x_{d3}(t)$. The solution of this problem is not unique as we are free to chose $ds(t) \ge 0$ for $x_{d3}(t) = 0$. By fixing $ds(t) = \dot{s}_0 dt$ to a constant value for $x_{d3}(t) = 0$ (i.e. $\dot{s}_0$ is a constant), we obtain the following discontinuous differential equation for $x_{d1}(t)$:

$$\dot{x}_{d1} = \begin{cases} \frac{m_2}{b_1}(-\dot{x}_{d3}(t) - \frac{c}{m_2}x_{1d}) & x_{d3}(t) > 0, \\ \frac{m_2}{b_1}(-\dot{x}_{d3}(t) - \frac{c}{m_2}x_{1d} + \frac{1}{m_2}\dot{s}_0) & x_{d3}(t) = 0. \end{cases} \quad (8.75)$$

The numerical solution of $x_{d1}(t)$ gives (after a transient) a periodic signal $x_{d1}(t)$ and $x_{d2}(t) = -\dot{x}_{d1}(t) + x_{d3}(t)$ (see the dotted lines in Figures 8.11 and 8.12 which are mostly below the solid lines). We have taken $\dot{s}_0 = 1$. Subsequently, the feedforward input $dp_{ff} = w_{ff}dt + W_{ff}d\eta$ is designed such that

$$dp_{ff} = m_1 dx_{d2} - (cx_{d1} + b_1(-x_{d2} + x_{d3}) - b_2 x_{d2})\, dt \quad (8.76)$$

and it therefore holds that $x(t) = x_d(t)$ for $t \ge 0$ if $x(0) = x_d(0)$, where $x(t)$ is a solution of (8.71), (8.72), with $dp = dp_{ff}$. The feedforward input dp_{ff}/dt is shown in Figure 8.8 and is equal to $w_{ff}(t)$ almost everywhere. Two impulsive inputs $W_{ff}(t)$ per period can be seen at the time-instances for which there is a 'change ramp-up to ramp-down' and 'ramp-down to deadband'. Next, we implement the control law (8.37) on system (8.71) with the feedforward dp_{ff} as in (8.76). We choose $K = \begin{bmatrix} 0 & -4 & 0 \end{bmatrix}$ which renders the

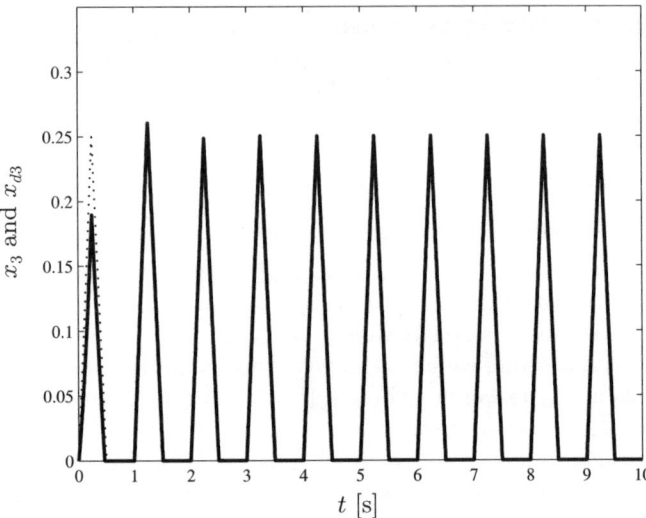

Fig. 8.9. Trajectories $x_3(t)$ (solid) and $x_{d3}(t)$ (dotted) for the case of feedback and feedforward control.

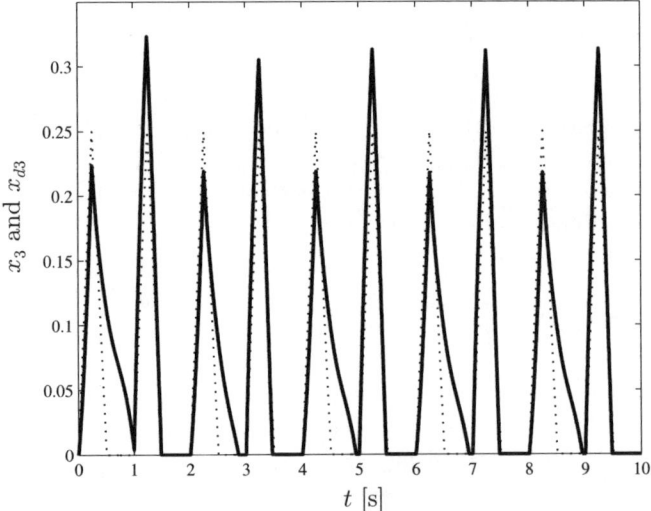

Fig. 8.10. Trajectories $x_3(t)$ (solid) and $x_{d3}(t)$ (dotted) for the case of only feedforward control.

triple $(\boldsymbol{A}_{cl}, \boldsymbol{D}, \boldsymbol{C})$ strictly positive real and, therefore, the closed-loop system (8.71), (8.72), (8.37), (8.76) has convergent dynamics. The strict positive realness of the triple $(\boldsymbol{A}_{cl}, \boldsymbol{D}, \boldsymbol{C})$ can be proven using the following matrix $\boldsymbol{P}$

$$\boldsymbol{P} = \begin{bmatrix} \frac{1}{5} & \frac{3}{5} & 0 \\ \frac{3}{5} & 4 & 0 \\ 0 & 0 & 1 \end{bmatrix} > 0, \quad \boldsymbol{Q} = -(\boldsymbol{A}_{cl}\boldsymbol{P} + \boldsymbol{P}\boldsymbol{A}_{cl}^{\mathrm{T}}) > 0, \quad \boldsymbol{D}^{\mathrm{T}}\boldsymbol{P} = \boldsymbol{C}, \quad (8.77)$$

where $\boldsymbol{A}_{cl} = \boldsymbol{A} + \boldsymbol{BK}$. Figure 8.9 shows the closed-loop dynamics for which the desired periodic solution $\boldsymbol{x}_d(t)$ is globally attractively stable. The attraction to the periodic solution from an arbitrary initial condition occurs in finite time (symptotic attraction). Figure 8.10 shows the open-loop dynamics for which there is no state-feedback. The desired periodic solution $\boldsymbol{x}_d(t)$ is not globally attractive, not even locally, and the solution from the chosen initial condition is attracted to a stable period-2 solution. Clearly, the system without feedback is not convergent. For both cases the initial condition $\boldsymbol{x}(0) = \begin{bmatrix} 0.16 & 2.17 & 0 \end{bmatrix}^{\mathrm{T}}$ was used. Figures 8.11 and 8.12 show the time-histories of $x_1(t)$ and $x_{d1}(t)$, respectively $x_2(t)$ and $x_{d2}(t)$, in solid and dotted lines. Jumps in the state $x_2(t)$ and desired state $x_{d2}(t)$ can be seen on time-instances for which the input is impulsive.

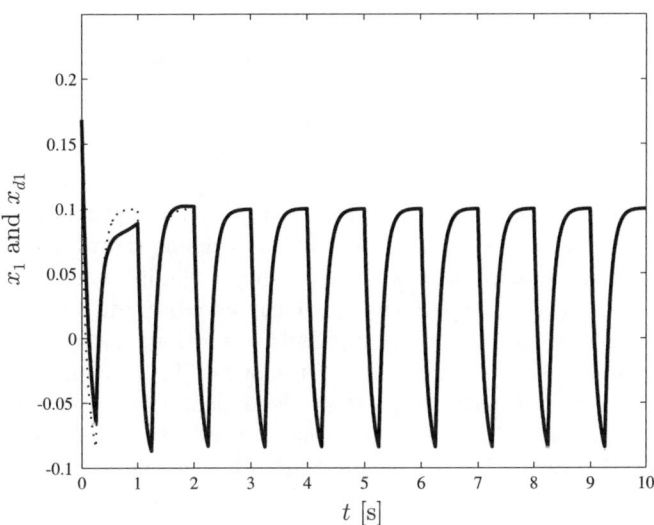

Fig. 8.11. Trajectories $x_1(t)$ (solid) and $x_{d1}(t)$ (dotted) for the case of feedback and feedforward control.

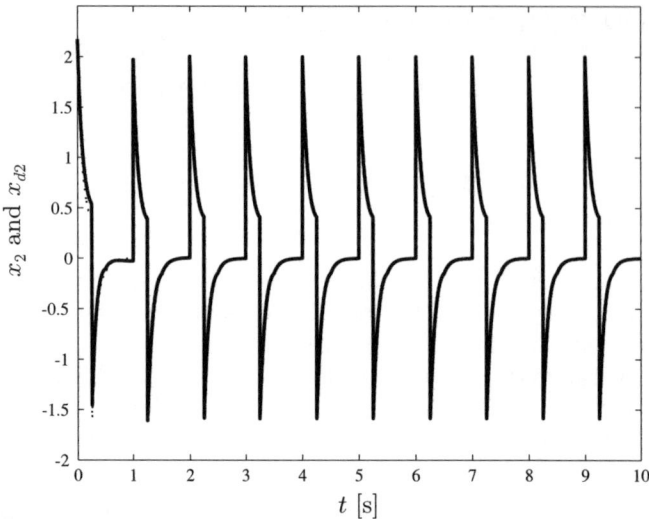

Fig. 8.12. Trajectories $x_2(t)$ (solid) and $x_{d2}(t)$ (dotted) for the case of feedback and feedforward control.

8.5 Summary

In the previous sections, sufficient conditions have been derived for the uniform convergence of a class of measure differential inclusions with certain maximal monotonicity properties. We will summarise the main ideas of the chapter.

First, sufficient conditions have been presented in Theorem 8.5 for the existence of a compact positively invariant set. Theorem 8.5 relies on a Lyapunov-based argument with the squared magnitude of the state as Lyapunov function, which acts as a kind of energy function. The assumption of strict monotonicity of the Lebesgue part of the state dependent right-hand side equals a strict passivity requirement with a quadratic dissipation. The quadratic dissipation can always outperform the linear energy input of bounded non-impulsive forces. Hence, during non-impulsive motion, the system dissipates energy for large enough magnitudes of the state. The assumption of monotonicity of the atomic part of the state dependent right-hand side equals a passivity requirement. Moreover, the energy input of the impulsive inputs is assumed to be bounded. This means that, for a given size of the compact positively invariant set of which we like to prove existence, we can find a dwell-time for the impulsive inputs. If the time-lapse between subsequent impulsive inputs is larger than this dwell-time, then the Lebesgue measurable dissipative forces have enough time to 'eat' the energy input of the impulsive input. This reasoning works also in the opposite direction. Given a certain dwell-time, there exists a certain compact positively invariant set of the system. This means that the dwell-time is not really a condition for the existence of a compact positively invariant set, but is merely a constant which relates to the size of such a set. The existence of such a set guarantees the existence of a solution that is bounded for all times.

Subsequently, sufficient conditions for incremental attractive stability have been derived in Theorem 8.7 using again a Lyapunov-based approach. The decrease of the Lyapunov function, which measures the distance between two arbitrary solutions, follows from a monotonicity condition. Incremental attractive stability implies that all solutions converge to one another. The aforementioned bounded (steady-state) solution must therefore be globally asymptotically stable for all bounded inputs, which rigourously proves uniform convergence of the system (Theorem 8.7).

The above theorems hinge on a few important assumptions, which we can give the following interpretations in the context of mechanical systems with impulsive right-hand sides:

1. *Separation of state-dependent forces and inputs.* In other words: no cross-talk between state-dependent forces and inputs. This excludes mixed terms in state and input, which for instance arise if the generalised force directions of the input forces are state-dependent.
2. *A strict monotonicity condition on the Lebesgue measurable right-hand side.* This implies that the state-dependent forces in the system are strictly passive.
3. *A monotonicity condition on the atomic (impulsive) right-hand side.* This implies that the state-dependent impulses in the system are passive.
4. *Bounded energy input of the impulsive inputs.* The physical meaning of this assumption has been elucidated in Section 8.4.1.
5. *A dwell-time condition.* The dwell-time can be chosen to be arbitrary small. In practice, there always exist a minimal time between two impulsive inputs which can be exerted on the system.

Condition 2 is the condition which may limit most of all the use of Theorem 8.7, simply because many systems are not dissipative. However, systems can be *made* dissipative using an appropriate control. In other words, the presented theorems give us the knowledge how to design controllers, such that the closed loop system is uniformly convergent. The uniform convergence can then be used for tracking control purposes, synchronisation etc. In Section 8.3 we presented such a convergence-based tracking control design for a class of measure differential inclusions in Lur'e form. Finally, we presented examples of mechanical systems with set-valued force laws. In these examples, it has been demonstrated that the tracking problem for a class of systems with non-collocated actuation and set-valued friction can be solved using the results presented in this chapter.

9

Concluding Remarks

An introductory part on non-smooth analysis and measure theory followed by a discussion of mechanical systems with frictional unilateral constraints within a setting of measure differential inclusions, led finally to a Lyapunov stability theory for non-smooth (Lagrangian) systems, being the very heart of this monograph. Which conclusions can we draw? The most remarkable and fundamental conclusion which we can give is, that we can indeed formulate a stability theory for non-smooth Lagrangian systems. It is sometimes believed that the mathematical apparatus is not yet available to rigorously deal with stability issues of non-smooth systems. The converse is true. Fundamental mathematical tools are non-smooth and convex analysis as well as the extension of the differential measure to include an atomic part, which leads to measure differential inclusions. Quintessential, however, is also the clear formulation of the mechanical modelling in terms of set-valued constitutive laws embedded in non-smooth potential theory. This last step requires in-depth knowledge in analytical mechanics as well as the aforementioned mathematical concepts. The generalisation of the Lyapunov stability framework to non-smooth Lagrangian systems follows in a natural way, but is certainly not trivial.

The previous chapters demonstrate that considerable progress has been made in the field of stability theory since the year 1644, in which Torricelli postulated his axiom which gave the onset to stability theory. In fact, with respect to the stability theory for smooth dynamical systems, this monograph only reveals the tip of the iceberg. Modern systems and control theory deals with stability-related topics such as input-output stability, input-to-state stability, absolute stability and convergent dynamics to name a few. Only a limited number of these stability-related issues have been studied in this monograph. The aim of this monograph is in a different direction. The previous chapters try to formulate a mathematical framework for the stability analysis of *non-smooth* systems within a setting of measure differential inclusions. The discussion has therefore been confined to the most fundamental issues: the definition of stability properties and stability analysis of equilibrium points and sets in autonomous measure differential inclusions using generalised Lyapunov-type theorems. The chosen Lyapunov approach results in constructive proofs of fundamental stability theorems, which paves the road towards the analysis of other stability properties.

The chosen focus of this work has on the one hand given the opportunity to discuss in-depth the basic philosophy behind non-smooth systems and stability theory, but opened on the other hand a Pandora's box of unresolved issues that may lead to new research directions. Clearly, further research has to deal with the stability analysis of other invariant sets such as periodic solutions in autonomous as well as non-autonomous systems.

The scientific merit of this monograph can be sought in the assistance to other theoretical or more practical fields of science. For instance, the theory of bifurcations might benefit from the stability framework for non-smooth systems, developed in the previous chapters. Although great interest is currently shown in the bifurcation analysis of non-smooth systems (mostly differential inclusions), no complete theory and classification of bifurcations in non-smooth systems exists to date. Bifurcations of equilibrium points and sets, as well as bifurcations of periodic solutions, are closely related to the gain or loss of stability under variation of a system parameter. A clear definition of stability, together with theorems to prove stability, is clearly of use for the further study on bifurcations in non-smooth systems.

Lagrangian mechanical systems, being an important class of physical systems that can be modelled using measure differential inclusions, have been given much attention in Chapters 5 and 7. Applications of the developed stability framework in mechanics may vary from the stability analysis of historical monuments to industrial applications such as brake systems, drilling systems and railway vehicles. The special structure of Lagrangian systems allows for a natural choice of the Lyapunov function and gives a physical interpretation of the theoretical results. Notably, the set-valued force laws in mechanics for unilateral frictional contact can be derived from non-smooth (pseudo-)potentials. Electrical circuits with diodes and similar switching elements, as well as hydraulic networks with one-sided valves, are other examples of engineering systems that can be modelled in the framework of non-smooth potentials and measure differential inclusions. The presented stability results therefore apply to a wide class of engineering systems. Naturally, specialised theorems for specific classes of engineering systems can be deducted from the results of Chapter 6 in the same way as has been done for Lagrangian mechanical systems in Chapter 7.

Clearly, stability theory is not only of importance from the point of view of dynamical systems analysis, but also plays a central role in control theory. The design of stabilising controllers for non-smooth systems, such as mechanical systems with unilateral contact, impact and friction (think of walking robots, drilling systems etc.), is currently receiving wide attention. We believe that the results in this monograph will prove to be useful in this respect. Moreover, in Chapter 8, on the convergence properties of the class of monotone measure differential inclusions, it has been shown how a symbiosis between non-smooth dynamics and recent developments from systems and control theory can be used to solve tracking control problems of mechanical systems with friction or impulsive inputs.

The above applications of the presented theory to other fields in science puts the work in a broader perspective and further justifies this monograph. However, of equal importance is the theory of stability and non-smooth dynamics in its own right. Both research fields, that of stability theory and non-smooth dynamics, put problems from different applied fields on an abstract mathematical level, but both fields also play their own role in science. Instead of considering quantitatively specific time-evolutions of reality, we are urged by stability theory to consider the qualitative behaviour of models of reality. Stability theory therefore leads science to a qualitative analysis of systems. Stability theory might be seen as a branch of nonlinear dynamics, which classically deals with the dynamics and bifurcations of differential and difference equations. The possibility of inequalities is blatantly and obstinately neglected in nonlinear dynamics. The relatively young field of non-smooth dynamics frees nonlinear dynamics from an established equality dogma and puts (variational) inequalities on the foreground. Stability theory and non-smooth dynamics therefore do not only sharpen the intellectual mind and catalyse mathematical development, but have a profound academic value. As a closing remark we would like to state that the preservation of academic values and continuous development of theoretical knowledge must be of prime importance to contemporary university politics when striving for further scientific success.

Sources and Translations

Various original texts have been studied while writing Section 1.2 of this book in which some historical notes are given on the theory of stability. We would like to thank the following libraries and institutions for their courtesy. Furthermore, the authors would like to thank S. Pleines (ETH Zurich) and Prof. em. F. Cerulus (K.U. Leuven) for their help with the translation of Latin texts.

p.3 E. Torricelli, *Opera Geometrica*: Max Planck Institute for the History of Science (digital library)

p.3 S. Stevin, *Byvough der Weeghconst*: Digital Library of the Royal Netherlands Academy of Arts and Sciences (KNAW).
Translation: The principal works of Simon Stevin, edited by E. Crone, E. J. Dijksterhuis, R. J. Forbes, M. G. J. Minnaert and A. Pannekoek

p.4 C. Huygens, *Œuvres Complètes de Christiaan Huygens*: La Bibliothèque Nationale de France/Gallica

p.4 D. Bernoulli, *Commentationes de statu aequilibrii corporum humido insidentium*: ETH-Bibliothek Zurich (Alte Drucke).
Translation: F. Cerulus [31]

p.5 L. Euler, *Scientia Navalis*: The Euler Archive (digital library). Translation: see [125]

p.5 J. L. Lagrange, *Méchanique Analytique*: ETH-Bibliothek Zurich (Alte Drucke)

References

1. ACARY, V., BROGLIATO, B., AND GOELEVEN, D. Higher order Moreau's sweeping process: mathematical formulation and numerical simulation. *Mathematical Programming, Series A* (2006), pp. 85. submitted.
2. ADLY, S. Attractivity theory for second order non-smooth dynamical systems with application to dry friction. *Journal of Mathematical Analysis and Applications 322* (2006), 1055–1070.
3. ADLY, S., AND GOELEVEN, D. A stability theory for second-order non-smooth dynamical systems with application to friction problems. *Journal de Mathématiques Pures et Appliquées 83* (2004), 17–51.
4. AEBERHARD, U., AND GLOCKER, CH. Energy considerations for excited perfect collisions. *Proceedings of the ENOC-2005 Conference, 7–12 August, Eindhoven* (2005), 422–431.
5. ALART, P., AND CURNIER, A. A mixed formulation for frictional contact problems prone to Newton like solution methods. *Computer Methods in Applied Mechanics and Engineering 92* (1991), 353–375.
6. AMBROSIO, L. Variational problems in SBV and image segmentation. *Acta Applicandae Mathematicae: An International Survey Journal on Applying Mathematics and Mathematical Applications 17*, 1 (1989), 1–40.
7. ANGELI, D. A Lyapunov approach to incremental stability properties. *IEEE Transactions on Automatic Control 47*, 3 (2002), 410–421.
8. APOSTOL, T. *Mathematical Analysis.* Addison Wessley, Reading MA, 1974.
9. ARMSTRONG-HÉLOUVRY, B., DUPONT, P., AND CANUDAS DE WIT, C. A survey of models, analysis tools and compensation methods for the control of machines with friction. *Automatica 30*, 7 (1994), 1083–1138.
10. AUBIN, J.-P., AND CELLINA, A. *Differential Inclusions*, vol. 264 of *Grundlehren der mathematischen Wissenschaften.* Springer-Verlag, Berlin, 1984.
11. AUBIN, J.-P., AND EKELAND, I. *Applied Nonlinear Analysis.* Pure and Applied Mathematics. Wiley, Berlin, 1984.
12. AUBIN, J.-P., AND FRANKOWSKA, H. *Set-valued Analysis*, vol. 2 of *Systems and Control: Foundations and Applications.* Birkhäuser, Boston, 1990.
13. BAINOV, D. D., AND SIMEONOV, P. S. *Systems with Impulse Effect: Stability, Theory and Applications.* Wiley, New York, 1989.
14. BALLARD, P. The dynamics of discrete mechanical systems with perfect unilateral constraints. *Archive for Rational Mechanics and Analysis 154* (2000), 199–274.

15. BARBASHIN, Y. A., AND KRASOVSKII, N. N. The stability of motion as a whole. *Dokl. Akad. Nauk SSSR 86*, 3 (1952), 453–456.
16. BASSEVILLE, S., LEGER, A., AND PRATT, E. Investigation of the equilibrium states and their stability for a simple model with unilateral contact and Coulomb friction. *Archive of Applied Mechanics 73* (2003), 409–420.
17. BERNARDIN, F., SCHATZMAN, M., AND LAMARQUE, C. H. Second-order multivalued stochastic differential equations on Riemannian manifolds. *Proceedings of the Royal Society of London A 460* (2004), 3095–3121.
18. BERNOULLI, D. Commentationes de statu aequilibrii corporum humido insidentium. *Comment. Acad. Scient. Imp. Petrop. X* (1738 (1747)), 147–163.
19. BHAT, S. J., AND BERNSTEIN, D. S. Finite-time stability of continuous autonomous systems. *SIAM Journal on Control and Optimization 38*, 3 (2000), 751–766.
20. BIRKHOFF, G. D. *Dynamical Systems.* A.M.S. Publications, Providence, 1927.
21. BOURGEOT, J. M., AND BROGLIATO, B. Tracking control of complementarity Lagrangian systems. *International Journal of Bifurcation and Chaos 15*, 6 (2005), 1839–1866.
22. BRÉZIS, H. *Operateurs Maximaux Monotones et Semi-groupes de Contractions dans les Espaces de Hilbert*, vol. 5 of *North-Holland Mathematics Studies.* North-Holland Publishing Company, Amsterdam, 1973.
23. BROGLIATO, B. *Nonsmooth Mechanics*, 2 ed. Springer, London, 1999.
24. BROGLIATO, B. Absolute stability and the Lagrange-Dirichlet theorem with monotone multivalued mappings. *Systems & Control Letters 51* (2004), 343–353.
25. BROGLIATO, B., DANIILIDIS, A., LEMARÉCHAL, C., AND ACARY, V. On the equivalence between complementarity systems, projected systems and unilateral differential inclusions. *Systems and Control Letters 55* (2006), 45–51.
26. BROGLIATO, B., AND GOELEVEN, D. The Krakovskii-LaSalle invariance principle for a class of unilateral dynamical systems. *Mathematics of Control, Signals, and Systems 17*, 1 (2005), 57–76.
27. BROGLIATO, B., NICULESCU, S.-I., AND ORHANT, P. On the control of finite-dimensional mechanical systems with unilateral constraints. *IEEE Transactions On Automatic Control 42*, 2 (1997), 200–215.
28. BROGLIATO, B., TEN DAM, A. A., PAOLI, L., GÉNOT, F., AND ABADIE, M. Numerical simulation of finite dimensional multibody nonsmooth mechanical systems. *ASME Applied Mechanics Reviews 55*, 2 (2002), 107–150.
29. CAI, C., TEEL, A. R., AND GOEBEL, R. Smooth Lyapunov functions for hybrid systems Part I: Existence is equivalent to robustness. *IEEE Transactions on Automatic Control* (2007). Accepted.
30. CAMLIBEL, M., PANG, J., AND SHEN, J. Lyapunov stability of complementarity and extended systems. *SIAM Journal on Optimization 17*, 4 (2006), 1056–1101.
31. CERULUS, F. private communication, to be published in vol. 6 of the 'Werke von Daniel Bernoulli', 2007.
32. CHAREYRON, S. *Stabilité des Systèmes Dynamiques Non-réguliers, Application aux Robot Marcheurs.* Ph.D. thesis, Institut National Polytechnique de Grenoble, Grenoble, France, 2005.
33. CHAREYRON, S., AND WIEBER, P. B. A LaSalle's invariance theorem for nonsmooth Lagrangian dynamical systems. *Proceedings of the ENOC-2005 Conference, 7-12 August, Eindhoven* (2005).

34. CHAREYRON, S., AND WIEBER, P. B. Stability and regulation of nonsmooth dynamical systems. *IEEE Transactions on Automatic Control* (2006). submitted.

35. CHETAEV, N. G. *The Stability of Motion.* Pergamon, New York, 1961.

36. CLARKE, F. H. *Optimization and Nonsmooth Analysis.* Wiley, New York, 1983.

37. CLARKE, F. H., LEDYAEV, Y. S., STERN, R. J., AND WOLENSKI, P. R. *Nonsmooth Analysis and Control Theory,* vol. 178 of *Graduate Texts in Mathematics.* Springer, New York, 1998.

38. CODDINGTON, E. A., AND LEVINSON, N. *Theory of Ordinary Differential Equations.* McGraw-Hill, New York, 1955.

39. CONTENSOU, P. Couplage entre frottement de glissement et frottement de pivotement dans la théorie de la toupie. In *Kreiselprobleme; Gyrodynamics* (Berlin, 1963), H. Ziegler, Ed., Springer-Verlag, pp. 201–216. IUTAM Symposium Celerina 1962.

40. CURNIER, A., Ed. *Proc. Contact Mechanics International Symposium, Lausanne, Switzerland, October 7-9, 1992* (Lausanne, 1992), Presses Polytechniques et Universitaires Romandes.

41. CURNIER, A. Unilateral contact. In Pfeiffer and Glocker [141], pp. 1–68.

42. DAVRAZOS, G. N., AND KOUSSOULAS, N. T. A review of stability results for switched and hybrid systems. *9th Mediterranean Conference on Control and Automation June 27-29, Dubrovnik, Croatia* (2001).

43. DEIMLING, K. *Multivalued Differential Equations.* De Gruyter, Berlin, 1992.

44. DEMIDOVICH, B. P. *Lectures on Stability Theory (in Russian).* Nauka, Moscow, 1967.

45. DIJKSTERHUIS, E. J. *De Mechanisering van het Wereldbeeld.* Meulenhoff, Amsterdam, 2000, originally printed in 1950.

46. EICH-SOELLNER, E., AND FÜHRER, C. *Numerical Methods in Multibody Dynamics.* Teubner, Stuttgart, 1998.

47. ELSTRODT, J. *Maß- und Integrationstheorie.* Springer, Berlin Heidelberg New York, 1996.

48. EULER, L. *Methodus inveniendi lineas curvas maximi minive proprietate gaudentes.* Bousquet, Lausanne & Geneva, 1744.

49. EULER, L. *Scientia navalis seu tractatus de construendis ac dirigendis,* vol. 1 and 2. Academiae Scientarum, St. Petersburg, 1749.

50. FILIPPOV, A. F. Differential equations with discontinuous right-hand side. *American Mathematical Society Translations, Series 2 42* (1964), 199–123.

51. FILIPPOV, A. F. *Differential Equations with Discontinuous Right-hand Sides.* Kluwer Academic, Dordrecht, 1988.

52. FROMION, V., MONACO, S., AND NORMAND-CYROT, D. Asymptotic properties of incrementally stable systems. *IEEE Transactions on Automatic Control 41* (1996), 721–723.

53. FROMION, V., SCORLETTI, G., AND FERRERES, G. Nonlinear performance of a PI controlled missile: an explanation. *International Journal of Robust and Nonlinear Control 9* (1999), 485–518.

54. FUSS, P. H. *Correspondance mathématique et physique de quelques célèbres géomètres du XVIIIème siècle, Tome II.* No. 35 in The Sources of Science. Johnson reprint corporation, New York London, 1968, originally printed in St.-Pétersbourg 1843.

55. GÉNOT, F., AND BROGLIATO, B. New results on Painlevé paradoxes. *European Journal of Mechanics – A/Solids 18* (1999), 653–677.
56. GAMKRELIDZE, R. V. *Analysis II, Convex Analysis and Approximation Theory*, vol. 14 of *Encyclopaedia of Mathematical Sciences*. Springer-Verlag, New York, 1989.
57. GLOCKER, CH. *Dynamik von Starrkörpersystemen mit Reibung und Stößen*, vol. 18, no. 182 of *Fortschr.-Ber. VDI*. VDI Verlag, Düsseldorf, 1995.
58. GLOCKER, CH. The principles of d'Alembert, Jourdain and Gauß in non-smooth mechanics, Part I: scleronomic multibody systems. *Zeitschrift für angewandte Mathematik und Mechanik 78*, 1 (1998), 21–37.
59. GLOCKER, CH. Discussion of d'Alembert's principle for non-smooth unilateral constraints. *Zeitschrift für angewandte Mathematik und Mechanik 79*, 1 (1999), S91–S94.
60. GLOCKER, CH. Formulation of spatial contact situations in rigid multibody systems. *Computer Methods in Applied Mechanics and Engineering 177* (1999), 199–214.
61. GLOCKER, CH. Scalar force potentials in rigid multibody systems. In Pfeiffer and Glocker [141], pp. 69–146.
62. GLOCKER, CH. On frictionless impact models in rigid-body systems. *Philosophical Transactions of the Royal Society of London A 359* (2001), 2385–2404.
63. GLOCKER, CH. *Set-Valued Force Laws, Dynamics of Non-Smooth Systems*, vol. 1 of *Lecture Notes in Applied Mechanics*. Springer-Verlag, Berlin, 2001.
64. GLOCKER, CH. Models of non-smooth switches in electrical systems. *International Journal of Circuit Theory and Applications 33*, 3 (2005), 205–234.
65. GOELEVEN, D., AND BROGLIATO, B. Stability and instability matrices for linear evolution variational inequalities. *IEEE Transactions on Automatic Control 49* (2004), 521–534.
66. GOELEVEN, D., AND BROGLIATO, B. Necessary conditions of asymptotic stability for unilateral dynamical systems. *Nonlinear Analysis 61*, 6 (2005), 961–1004.
67. GOELEVEN, D., MOTREANU, D., DUMONT, Y., AND ROCHDI, M. *Variational and Hemivariational Inequalities: Theory, Methods and Applications, Volume I: Unilateral Analysis and Unilateral Mechanics*, vol. 69 of *Nonconvex Optimization and its Applications*. Kluwer Academic Publishers, Dordrecht, 2003.
68. GOELEVEN, D., MOTREANU, D., DUMONT, Y., AND ROCHDI, M. *Variational and Hemivariational Inequalities: Theory, Methods and Applications, Volume II: Unilateral Problems*, vol. 70 of *Nonconvex Optimization and its Applications*. Kluwer Academic Publishers, Dordrecht, 2003.
69. GOLDSTINE, H. H. *A History of the Calculus of Variations from the 17th through the 19th Century*, vol. 5 of *Studies in the History of Mathematics and Physical Sciences*. Springer, New York, 1980.
70. GUCKENHEIMER, J., AND HOLMES, P. *Nonlinear Oscillations, Dynamical Systems, and Bifurcations of Vector Fields*, vol. 42 of *Applied Mathematical Sciences*. Springer-Verlag, New York, 1983.
71. HADDAD, W., CHELLABOINA, V., AND NERSESOV, S. *Impulsive and Hybrid Dynamical Systems: Stability, Dissipativity, and Control*. Princeton Series in Applied Mathematics. Princeton University Press, Princeton, 2006.
72. HEEMELS, M. *Linear Complementarity Systems: a Study in Hybrid Dynamics*. Ph.D. thesis, Eindhoven University of Technology, The Netherlands, 1999.

73. HEERTJES, M. F., PASTINK, H. A., VAN DE WOUW, N., AND NIJMEIJER, H. Experimental frequency-domain analysis of nonlinear controlled optical storage drives. *IEEE Transactions on Control Systems Technology 14*, 3 (2006), 389–397.

74. HESPANHA, J. P. Uniform stability of switched linear systems: Extensions of LaSalle's invariance principle. *IEEE Transactions on Automatic Control 49*, 4 (2004), 470–482.

75. HESPANHA, J. P., LIBERZON, D., AND TEEL, A. R. On input-to-state stability of impulsive systems. In *Proc. of the 44th IEEE Conf. on Decision and Control and the European Control Conf. 2005* (Sevilla, 2005).

76. HESPANHA, J. P., AND MORSE, A. S. Stability of switched systems with average dwell-time. *Proc. of the 38th Conf. on Decision and Control* (1999), 2655–2660.

77. HIRIART-URRUTY, J.-P., AND LEMARÉCHAL, C. *Convex Analysis and Minimization Algorithms I*, vol. 305 of *Grundlehren der mathematischen Wissenschaften*. Springer-Verlag, Berlin, 1993.

78. HIRIART-URRUTY, J.-P., AND LEMARÉCHAL, C. *Convex Analysis and Minimization Algorithms II*, vol. 306 of *Grundlehren der mathematischen Wissenschaften*. Springer-Verlag, Berlin, 1993.

79. HUYGENS, C. *De iis quae liquido supernatant libri tres, 1650*, vol. 11 of *Œuvres Complètes de Christiaan Huygens*. Martinus Nijhoff, The Hague, 1908.

80. JACOBI, C. G. J. *Vorlesungen über Dynamik.* (edited by A. Clebsch, reprinted by Chelsea, New York, 1969), Berlin, 1866.

81. JEAN, M., AND MOREAU, J. J. Unilaterality and dry friction in the dynamics of rigid body collections. In Curnier [40], pp. 31–48.

82. JOHANSSON, M. *Piecewise Linear Control Systems - A Computational Approach*, vol. 284 of *Lecture Notes in Control and Information Sciences*. Springer-Verlag, Heidelberg, 2002.

83. JOHNSON, K. L. *Contact Mechanics.* Cambridge University Press, Cambridge, 1985.

84. KANE, T. R., AND LEVINSON, D. A. *Dynamics, theory and applications.* McGraw-Hill, New York, 1985.

85. KHALIL, H. K. *Nonlinear Systems.* Prentice-Hall, New Jersey, 1996.

86. KLARBRING, A. Mathematical Programming and Augmented Lagrangian Methods for Frictional Contact Problems. In Curnier [40], pp. 409–422.

87. KOLMOGOROV, A., AND FOMIN, S. V. *Introductory Real Analysis.* Dover Publications, New York, 1975.

88. KUNZE, M., AND KÜPPER, T. Qualitative bifurcation analysis of a non-smooth friction oscillator model. *Zeitschrift für angewandte Mathematik und Physik 48*, 1 (1997), 87–101.

89. LA SALLE, J., AND LEFSCHETZ, S. *Stability by Liapunov's Direct Method.* Academic Press, New York, 1961.

90. LAGRANGE, J. L. Sur le mouvement des nœuds des orbites planétaires. *Mémoires de l'Académie Royale des Sciences et Belles-Lettres, Année 1774* (1776), 276–307.

91. LAGRANGE, J. L. *Méchanique Analytique.* La Veuve Desaint, Paris, 1788.

92. LAKSHMIKANTHAM, V., BAINOV, D. D., AND SIMEONOV, P. S. *Theory of Impulsive Differential Equations*, vol. 6 of *Series in Modern Applied Mathematics*. World Scientific, Singapore, 1989.

93. LAPLACE, P. S. Sur l'équation séculaire de la lune. *Mémoires de l'Académie Royale des Sciences de Paris, Année 1786* (1788), 243–271.
94. LAURSEN, T. A., AND SIMO, J. C. Algorithmic symmetrization of Coulomb frictional problems using augmented Lagrangians. *Computer Methods in Applied Mechanics and Engineering 108* (1993), 133–146.
95. LE SAUX, C., LEINE, R. I., AND GLOCKER, CH. Dynamics of a rolling disk in the presence of dry friction. *Journal of Nonlinear Science 15*, 1 (2005), 27–61.
96. LEINE, R. I. *Bifurcations in Discontinuous Mechanical Systems of Filippov-Type*. Ph.D. thesis, Eindhoven University of Technology, The Netherlands, 2000.
97. LEINE, R. I. Bifurcations of equilibria in non-smooth continuous systems. *Physica D 223* (2006), 121–137.
98. LEINE, R. I., BROGLIATO, B., AND NIJMEIJER, H. Periodic motion and bifurcations induced by the Painlevé paradox. *European Journal of Mechanics – A/Solids 21*, 5 (2002), 869–896.
99. LEINE, R. I., AND GLOCKER, CH. A set-valued force law for spatial Coulomb–Contensou friction. *European Journal of Mechanics – A/Solids 22* (2003), 193–216.
100. LEINE, R. I., GLOCKER, CH., AND VAN CAMPEN, D. H. Nonlinear dynamics and modeling of various wooden toys with impact and friction. *Journal of Vibration and Control 9*, 1 (2003), 25–78.
101. LEINE, R. I., AND NIJMEIJER, H. *Dynamics and Bifurcations of Non-Smooth Mechanical Systems*, vol. 18 of *Lecture Notes in Applied and Computational Mechanics*. Springer Verlag, Berlin, 2004.
102. LEINE, R. I., VAN CAMPEN, D. H., DE KRAKER, A., AND VAN DEN STEEN, L. Stick-slip vibrations induced by alternate friction models. *International Journal of Nonlinear Dynamics and Chaos in Engineering Systems 16*, 1 (1998), 41–54.
103. LEINE, R. I., AND VAN DE WOUW, N. Stability properties of equilibrium sets of nonlinear mechanical systems with dry friction and impact. *International Journal of Nonlinear Dynamics and Chaos in Engineering Systems* (2007).
104. LEINE, R. I., AND VAN DE WOUW, N. Uniform convergence of monotone measure differential inclusions: with application to the control of mechanical systems with unilateral constraints. *International Journal of Bifurcation and Chaos* (2007). accepted.
105. LEJEUNE-DIRICHLET, J. P. G. Über die Stabilität des Gleichgewichts. *CRELLE, Journal für die reine und angewandte Mathematik 32* (1846), 85–88.
106. LIAPOUNOFF, A. Problème général de la stabilité du mouvement. *Annales de la faculté des sciences de Toulouse, 2ᵉ série 9* (1907), 203–474.
107. LIBERZON, D. *Switching in Systems and Control*. Birkhäuser, Boston, 2003.
108. LOHMILLER, W., AND SLOTINE, J.-J. E. On contraction analysis for nonlinear systems. *Automatica 34* (1998), 683–696.
109. LYAPUNOV, A. M. *Problème général de la stabilité du mouvement*. No. 17 in Annals of Mathematics Studies. Princeton University Press, Princeton, 1947.
110. MAGNUS, K. *Kreisel; Theorie und Anwendungen*. Springer-Verlag, Berlin, 1971.
111. MATVEEV, A. S., AND SAVKIN, A. V. *Qualitative Theory of Hybrid Dynamical Systems*. Control Engineering Series. Birkhäuser, Boston, 2000.
112. MAXWELL, J. C. On governors. *Proceedings of the Royal Society of London 16* (1868), 270–283.

113. MENINI, L., AND TORNAMBÈ, A. Asymptotic tracking of periodic trajectories for a simple mechanical system subject to nonsmooth impacts. *IEEE Transactions on Automatic Control 46*, 7 (2001), 1122–1126.

114. MIHAJLOVIĆ, N. *Torsional and Lateral Vibrations in Flexible Rotor Systems with Friction.* Ph.D. thesis, Eindhoven University of Technology, The Netherlands, 2005.

115. MIHAJLOVIĆ, N., VAN VEGGEL, A. A., VAN DE WOUW, N., AND NIJMEIJER, H. Analysis of friction-induced limit cycling in an experimental drill-string system. *ASME Journal of Dynamic Systems, Measurements and Control 126*, 4 (2004), 706–720.

116. MONTEIRO MARQUES, M. *Differential Inclusions in Nonsmooth Mechanical Systems.* Birkhäuser, Basel, 1993.

117. MOREAU, J. J. Rafle par un convexe variable (première partie). *Séminaire d'Analyse Convexe*, exposé no 15, University of Montpellier (1971), 43 pp.

118. MOREAU, J. J. Rafle par un convexe variable (deuxième partie). *Séminaire d'Analyse Convexe,* exposé no 3, University of Montpellier (1972), 36 pp.

119. MOREAU, J. J. Problème d'évolution associé à un convexe mobile d'un espace hilbertien. *Comptes Rendus de l'Academie des Sciences, Séries A-B 276* (1973), 791–794.

120. MOREAU, J. J. Evolution problem associated with a moving convex set in a Hilbert space. *Journal of Differential Equations 26* (1977), 347–374.

121. MOREAU, J. J. Dynamique de systèmes à liaisons unilatérales avec frottement secéventuel; essais numériques. Tech. Rep. 85-1, LMGC, Montpellier, France, 1986.

122. MOREAU, J. J. Bounded variation in time. In *Topics in Nonsmooth Mechanics*, J. J. Moreau, P. D. Panagiotopoulos, and G. Strang, Eds. Birkhäuser Verlag, Basel, Boston, Berlin, 1988, pp. 1–74.

123. MOREAU, J. J. Unilateral contact and dry friction in finite freedom dynamics. In *Non-Smooth Mechanics and Applications*, J. J. Moreau and P. D. Panagiotopoulos, Eds., vol. 302 of *CISM Courses and Lectures*. Springer, Wien, 1988, pp. 1–82.

124. MOREAU, J. J. Jump functions of a real interval to a Banach space. *Annales de la faculté des sciences de Toulouse 5ᵉ série, (suppl.) S10* (1989), 77–91.

125. NOWACKI, H. *Leonhard Euler and the Theory of Ships.* Max Planck Society - eDocument Server, http://edoc.mpg.de/ ID: 314833.0, 2007.

126. PANAGIOTOPOULOS, P. D. A nonlinear programming approach to the unilateral contact-, and friction-boundary value problem in the theory of elasticity. *Ingenieur Archiv 44* (1975), 421–432.

127. PANAGIOTOPOULOS, P. D. *Hemivariational Inequalities.* Springer, Berlin, 1993.

128. PANAGIOTOPOULOS, P. D., AND GLOCKER, CH. Inequality constraints with elastic impacts in deformable bodies. The convex case. *Archive of Applied Mechanics 70* (2000), 349–365.

129. PAPASTAVRIDIS, J. G. *Analytical Mechanics: A Comprehensive Treatise on the Dynamics of Constrained Systems: for Engineers, Physicists and Mathematicians.* Oxford University Press, New York, 2002.

130. PAVLOV, A., POGROMSKY, A., VAN DE WOUW, N., AND NIJMEIJER, H. On convergence properties of piecewise affine systems. *International Journal of Control* (2007). Accepted.

131. PAVLOV, A., VAN DE WOUW, N., AND NIJMEIJER, H. Convergent systems: Analysis and design. In *Control and Observer Design for Nonlinear Finite and Infinite Dimensional Systems* (Stuttgart, 2005), T. Meurer, K. Graichen, and D. Gilles, Eds., vol. 332 of *Lecture Notes in Control and Information Sciences*, pp. 131–146.

132. PAVLOV, A., VAN DE WOUW, N., AND NIJMEIJER, H. *Uniform Output Regulation of Nonlinear Systems: A Convergent Dynamics Approach.* Systems & Control: Foundations and Applications (SC) Series. Birkhäuser, Boston, 2005.

133. PAVLOV, A., VAN DE WOUW, N., AND NIJMEIJER, H. Frequency response functions for nonlinear convergent systems. *IEEE Transactions on Automatic Control 52*, 6 (2007), 1159–1165.

134. PAVLOV, A., VAN DE WOUW, N., AND NIJMEIJER, H. Global nonlinear output regulation: convergence-based controller design. *Automatica 43*, 3 (2007), 456–463.

135. PERCIVALE, D. Uniqueness in the elastic bounce problem I. *Journal of Differential Equations 56* (1985), 206–215.

136. PERCIVALE, D. Uniqueness in the elastic bounce problem II. *Journal of Differential Equations 90* (1991), 304–315.

137. PEREIRA, F. L., AND SILVA, G. N. Lyapunov stability of measure driven impulsive systems. *Differential Equations 40*, 8 (2004), 1122–1130.

138. PETTERSSON, S., AND LENNARTSON, B. Hybrid system stability and robustness verification using linear matrix inequalities. *International Journal of Control 75*, 16–17 (2002), 1335–1355.

139. PFEIFFER, F., AND GLOCKER, CH. *Multibody Dynamics with Unilateral Contacts.* Wiley, New York, 1996.

140. PFEIFFER, F., AND GLOCKER, CH., Eds. *Proc. IUTAM Symposium on Unilateral Multibody Contacts, Munich, Germany, August 3–7, 1998* (Dordrecht, 1998), Kluwer Academic Pulishers.

141. PFEIFFER, F., AND GLOCKER, CH., Eds. *Multibody Dynamics with Unilateral Contacts*, vol. 421 of *CISM Courses and Lectures.* Springer, Wien, 2000.

142. PFEIFFER, F., AND HAJEK, M. Stick-slip motion of turbine blade dampers. *Philosophical Transactions of the Royal Society A 338* (1992), 503–517.

143. PLISS, V. A. *Nonlocal Problems of the Theory of Oscillations.* Academic Press, London, 1966.

144. POINCARÉ, H. *Les Méthodes Nouvelles de la Mécanique Céleste, Tome I, II, III.* Gautier-Villars, Paris, 1892, 1893, 1899.

145. POISSON, S. D. Mémoire sur les inégalités séculaires des moyens mouvements des planètes. *Journal de l'École Polytechnique XV*, 1 (1808).

146. QUITTNER, P. An instability criterion for variational inequalities. *Nonlinear Analysis 15*, 12 (1990), 1167–1180.

147. ROCKAFELLAR, R. T. *Convex Analysis.* Princeton Landmarks in Mathematics. Princeton University Press, Princeton, New Jersey, 1970.

148. ROCKAFELLAR, R. T. *The Theory of Subgradients and its Applications to Problems of Optimization. Convex and Nonconvex Functions*, vol. 1 of *Research and Education in Mathematics.* Heldermann Verlag, Berlin, 1970.

149. ROCKAFELLAR, R. T., AND WETS, R.-B. *Variational Analysis.* Springer, Berlin, 1998.

150. RUDIN, W. *Real and Complex Analysis.* McGraw-Hill, New York, 1987.

151. SANFELICE, R. G., GOEBEL, R., AND TEEL, A. R. Invariance principles for hybrid systems with connections to detectability and asymptotic stability. *IEEE Transactions on Automatic Control* (2007). Accepted.

152. SASTRY, S. *Nonlinear Systems: Analysis, Stability and Control*, vol. 10 of *Interdisciplinary Applied Mathematics*. Springer, New York, 1999.

153. SCHATZMAN, M. A class of nonlinear differential equations of second order in time. *Nonlinear Analysis, Theory, Methods and Applications 2*, 3 (1978), 355–373.

154. SCHATZMAN, M. Uniqueness and continuous dependence on data for one dimensional impact problems. *Mathematical and Computational Modelling 28* (1998), 1–18.

155. SIMO, J. C., AND LAURSEN, T. A. An Augmented Lagrangian treatment of contact problems involving friction. *Computers & Structures 42*, 1 (1992), 97–116.

156. SMIRNOV, G. V. *Introduction to the Theory of Differential Inclusions*, vol. 41 of *Graduate Studies in Mathematics*. American Mathematical Society, Providence, 2002.

157. SONTAG, E. D. On the input-to-state stability property. *European Journal of Control 1* (1995), 24–36.

158. STEVIN, S. *Wisconstighe Ghedachtenissen*. Bouwensz, Leiden, 1608.

159. STEWART, D. E. Formulating measure differential inclusions in infinite dimensions. *Set-valued Analysis 8* (2000), 273–293.

160. STIEGELMEYR, A. Chimney dampers. In Pfeiffer and Glocker [140], pp. 299–308.

161. SZÁBO, I. *Geschichte der mechanischen Prinzipien und ihrer wichtigsten Anwendungen*, vol. 32 of *Wissenschaft und Kultur*. Birkhäuser, Basel, 1996.

162. TORNAMBÈ, A. Modeling and control of impact in mechanical systems: Theory and experimental results. *IEEE Transactions on Automatic Control 44*, 2 (1999), 294–309.

163. TORRICELLI, E. *Opera Geometrica*. Florentiæ typis Amatoris Masse & Laurentij de Landis, Firenze, 1644.

164. TRUESDELL, C. A. *Essays in the History of Mechanics*. Springer, Berlin, 1968.

165. UTKIN, V. I. Equations of slipping regimes in discontinuous systems I. *Automation and Remote Control 32* (1971), 1897–1907.

166. UTKIN, V. I. Equations of slipping regimes in discontinuous systems II. *Automation and Remote Control 33* (1972), 211–219.

167. UTKIN, V. I. *Sliding Modes in Control and Optimization*. Springer-Verlag, Berlin, 1992.

168. UTKIN, V. I., GULDNER, J., AND SHI, J. *Sliding Mode Control in Electromechanical Systems*. Taylor & Francis, London, 1999.

169. VAN DE WOUW, A., PAVLOV, A., AND NIJMEIJER, H. Controlled synchronisation of continuous PWA systems. In *Group Coordination and Cooperative Control* (Trondheim, Norway, 2006), K. Y. Pettersen, J. T. Gravdahl, and H. Nijmeijer, Eds., vol. 336 of *Lecture Notes in Control and Information Sciences*, pp. 271–289.

170. VAN DE WOUW, N., AND LEINE, R. I. Attractivity of equilibrium sets of systems with dry friction. *International Journal of Nonlinear Dynamics and Chaos in Engineering Systems 35*, 1 (2004), 19–39.

171. VAN DE WOUW, N., PASTINK, H. A., HEERTJES, M. F., PAVLOV, A. V., AND NIJMEIJER, H. Performance of convergence-based variable-gain control of optical storage drives. *Automatica* (2007). Accepted.

172. VAN DE WOUW, N., AND PAVLOV, A. Output tracking control of PWA systems. In *Proc. of the 44th IEEE Conf. on Decision and Control and the European Control Conf. 2006* (San Diego, 2006), pp. 2637–2642.

173. VAN DEN BERG, R. A., POGROMSKY, A. Y., LEONOV, G. A., AND ROODA, J. E. Design of convergent switched systems. In *Group Coordination and Cooperative Control*, K. Y. Pettersen, H. Nijmeijer, and T. Gravdahl, Eds., vol. 336 of *Lecture Notes in Control and Information Sciences*. Springer, 2006.

174. VAN DER SCHAFT, A. J., AND SCHUMACHER, J. M. *An Introduction to Hybrid Dynamical Systems*, vol. 251 of *Lecture Notes in Control and Information Sciences, vol. 251*. Springer, London, 2000.

175. WANG, Y. Dynamic modeling and stability analysis of mechanical systems with time-varying topologies. *ASME Journal of Mechanical Design 115* (1993), 808–816.

176. WILLEMS, J. L. *Stability Theory of Dynamical Systems*. Thomas Nelson and Sons Ltd., London, 1970.

177. WOLFSTEINER, P. *Dynamik von Vibrationsfördern*, vol. 2, no. 511 of *Fortschr.-Ber. VDI*. VDI Verlag, Düsseldorf, 1999.

178. YAKUBOVICH, V. Matrix inequalities method in stability theory for nonlinear control systems: I. absolute stability of forced vibrations. *Automation and Remote Control 7* (1964), 905–917.

179. YAKUBOVICH, V. A., LEONOV, G. A., AND GELIG, A. K. *Stability of Stationary Sets in Control Systems with Discontinuous Nonlinearities*, vol. 14 of *Series on Stability, Vibration and Control of Systems. Series A*. World Scientific, Singapore, 2004.

180. YE, H., MICHEL, A., AND HOU, L. Stability theory for hybrid dynamical systems. *IEEE Transactions on Automatic Control 43*, 4 (1998), 461–474.

Index

234   Index